Fuzzy Logic

Fuzzy Logic

Implementation and Applications

Edited by

M J Patyra
*University of Minnesota,
USA*

D M Mlynek
*Swiss Federal Institute of Lausanne
Switzerland*

WILEY TEUBNER
A Partnership between John Wiley & Sons and B. G. Teubner Publishers
Chichester · New York · Brisbane · Toronto · Singapore · Stuttgart · Leipzig

Copyright © 1996 jointly by John Wiley & Sons Ltd. and B.G. Teubner

John Wiley & Sons Ltd
Baffins Lane
Chichester
West Sussex
PO19 1UD
England

B.G. Teubner
Industriestraße 15
70565 Stuttgart (Vaihingen)
Postfach 80 10 69
70510 Stuttgart
Germany

National Chichester 01243 779777
International   (+44) 1243 779777

National Stuttgart (0711) 789010
International   +49 711 789010

Reprinted November 1996
All rights reserved.

No part of this book may be reproduced by any means,
or transmitted, or translated into a machine language
without the written permission of the publisher.

Other Wiley Editorial Offices

John Wiley & Sons, Inc., 605 Third Avenue
New York, NY 10158-0012, USA

Brisbane • Toronto • Singapore

Other Teubner Editorial Offices

B.G. Teubner, Verlagsgesellschaft mbH, Johannisgaße 16
D-04103 Leipzig, Germany

Die Deutsche Bibliotheck - CIP-Einheitsaufnahme
Fuzzy logic : implementation and applications / ed. by M. J.
Patyra : D. M. Mlynek. - Stuttgart ; Leipzig ; Teubner ;
Chichester ; New York ; Brisbane ; Toronto ; Singapore :
Wiley, 1996
    ISBN 3 519 06450 2 (Teubner)
    ISBN 0 471 95059 9 (Wiley)
NE: Patyra, Marek J. (Hrsg.)

WG: 37	DBN 94.719152.6	96.03.26
2790	nh V: Teubner	

Library of Congress Cataloging in Publication Data
Fuzzy logic : implementation and applications / edited by M. J. Patyra,
  D. M. Mlynek.
    p. cm.
    Includes bibliographical references and index.
    ISBN 0 471 95059 9
    1. Automatic control. 2. Fuzzy logic. I. Patyra, M. J. (Marek J.)
II. Mlynek, D. M.
TJ213.F88 1996
629.8 – dc20   95-45241
                                                                         CIP

British Library Cataloguing in Publication Data

A catalogue record for this book is available from the British Library

ISBN Wiley 0 471 95059 9
ISBN Teubner 3 519 06450 2

Typeset in 10/12pt Times by Thomson Press (India) Ltd, New Delhi
Printed and bound in Great Britain by Bookcraft (Bath) Ltd.
This book is printed on acid-free paper responsibly manufactured from sustainable forestation
for which at least two trees are planted for each one used for paper production.

Contents

Editor's Preface	xi
List of Contributors	xvii
Acknowledgments	xix

THEORY — 1

1 Fuzzy Sets in Approximate Reasoning: a Personal View — 3

1.1 Introduction	3
1.2 Graduality- and Similarity-based Approximate Reasoning	4
1.2.1 Comparison of Fuzzy Relations, Extension Principle and Similarity	4
1.2.2 Interpolative Reasoning	6
1.2.3 Qualitative Reasoning	9
1.3 Uncertainty Management	11
1.3.1 Background	12
1.3.2 Uncertain Fuzzy Rules	15
1.3.3 Approximate Reasoning with Fuzzy Rules	18
1.3.4 Possibilistic Logic	21
1.3.5 Default Reasoning	24
1.3.6 Abductive Reasoning	26
1.4 Concluding Remarks	32
References	32

FUZZY LOGIC CONTROL — 37

2 Fuzzy Logic Control: a Systematic Design and Performance Assessment Methodology — 39

2.1 Introduction	39
2.2 The Phase Portrait Assignment Algorithm	41
2.2.1 Fuzzy Logic control	41
2.2.2 The Automatic Rule-Generation Method	42
2.3 Performance Assessment	49
2.3.1 Stability Analysis	50
2.3.2 Robustness Analysis	51
2.4 Application Examples	53
2.4.1 The Engine Model	53
2.4.2 The Fuzzy Controller	55
2.4.3 Simulation Results	59

2.5	Stability and Robustness Results	59
2.6	Conclusions	61
Acknowledgment		61
References		62

3 On the Compatibility of Fuzzy Control and Conventional Control Techniques ... 63

3.1	Introduction	63
3.2	Sliding Mode Fuzzy Control	65
	3.2.1 The Principle of Sliding Mode Control	65
	3.2.2 The Similarity Between SMC and FC	67
	3.2.3 The Sliding Mode FC (SMFC) as a State-dependent Filter	68
	3.2.4 Normalization and Denormalization	69
	3.2.5 FC with Boundary Layer	70
	3.2.6 FC of Higher Order	71
	3.2.7 Numerical Example	73
3.3	Scaling of Fuzzy Controllers Using the Cross-correlation	78
	3.3.1 Input-Output Correlation for an FC	81
	3.3.2 Application to a Redundant Manipulator Arm	87
3.4	Fuzzy inputs	90
	3.4.1 Some Useful Operations on Fuzzy Sets	91
	3.4.2 The sgn-function	97
	3.4.3 Sliding Mode Control and Related Control Strategies	99
	3.4.4 Simulation Results	107
References		113

4 On the Crisp-type Fuzzy Controller: Behaviour Analysis and Improvement ... 117

4.1	Introduction	117
4.2	The Crisp-Type Fuzzy Logic Controller	118
4.3	The Dynamic Analysis of the Crisp-Type Fuzzy Controller	119
4.4	The Static Analysis of the Crisp-Type Fuzzy Control System	127
4.5	An Improvement: Pid-Type Fuzzy Controller Structure	130
4.6	Further Improvement: The Parameter Adaptive Fuzzy Controller	134
4.7	Conclusions	137
References		138

FUZZY LOGIC HARDWARE IMPLEMENTATIONS ... 141

5 Design Considerations of Digital Fuzzy Logic Controllers ... 143

5.1	Introduction	143
5.2	Digital-based Fuzzy Logic Hardware	144
	5.2.1 Digital Fuzzification	144
	5.2.2 Digital Fuzzy Inferencing	146
	5.2.3 Digital Defuzzification	149
5.3	Fuzzy Logic Based Controllers	151
	5.3.1 Digital FLC Characteristics	151
	5.3.2 Single-Input Single-Output Fuzzy Logic Controllers	153
	5.3.3 Double-Input Single-Output Fuzzy Logic Controller	155
	5.3.4 Multiple-Input Single-Output Fuzzy Logic Controller	157
	5.3.5 Multiple-Input Multiple-Output Fuzzy Logic Controller	159

CONTENTS

5.4	Hardware Implementation: Comparative Study	163
	5.4.1 Hardware Mapping of FLC Models	164
	5.4.2 Hardware Implementation Issues	171
	5.4.3 Summary	172
5.5	Final Remarks	172
	References	173

6 Parallel Algorithm for Fuzzy Logic Controller — 177

6.1	Introduction	177
6.2	Mathematical Models for Fuzzy Model Building and Inference Computations	177
	6.2.1 Single-Input Single-Output System	177
	6.2.2 Multiple-Input Single-Output System	179
	6.2.3 Multiple-Input Multiple-Output System	180
6.3	Parallel Algorithm	181
6.4	Conceptual Hardware Implementation	187
	6.4.1 SISO System	187
	6.4.2 Hardware Architectures for MISO and MIMO Systems	190
	6.4.3 Fuzzy Controller Hardware Accelerator	191
6.5	Performance Characteristics	192
	6.5.1 Maximum Sustainable Processing Rate	192
	6.5.2 Improvements	193
6.6	Conclusions	194
	References	194

7 Fuzzy Flip-flop — 197

7.1	Introduction	197
7.2	Outline of Binary Flip-flop and Fundamental Fuzzy Operations	198
	7.2.1 A Binary Logic J-K Flip-flop	198
	7.2.2 Definition of Fuzzy Negation, t-norm and s-norm	199
7.3	Definition of Fuzzy Flip-flop	201
7.4	Fuzzy Flip-flop using Complementation, Min and Max Operations	203
7.5	Fuzzy Flip-flop using Complementation, Algebraic Product and Algebraic Sum	207
7.6	Fundamentals of Implementation of the Min Max Fuzzy Flip-flop	207
7.7	Discrete and Voltage Mode Min Max Fuzzy Flip-flop Circuits	210
7.8	Fundamentals of Implementation of the Algebraic Fuzzy Flip-flop	217
7.9	Discrete and Voltage Mode Algebraic Fuzzy Flip-flop Circuits	219
7.10	Comparison of the Performance of Min Max Type Versus Algebraic Type Fuzzy Flip-flop circuit	221
7.11	Fuzzy register circuit	222
7.12	VLSI design of the Fuzzy Register Circuit	224
	7.12.1 VLSI Design of the Min Max Type Fuzzy Flip-flop Circuit	224
	7.12.2 VLSI Design of the Fuzzy Register	230
7.13	Conclusion	235
	References	235

8 Design Automation of Fuzzy Logic Circuits — 237

8.1	Introduction	237
8.2	Basic Fuzzy Operators	238
	8.2.1 Terminology and Resolution Principle	238
	8.2.2 Fuzzy Inclusion as the Natural Extension of Boolean Inclusion	239
	8.2.3 Symbolic Implementation of Fuzzy Operators	242

8.3 CMOS Implementation 243
   8.3.1 CMOS Implementation of Fuzzy Operators 243
   8.3.2 Current Mirror-based Approach 244
   8.3.3 Case Study: Implementation of the Min Unit 247
8.4 Fuzzy Development System 247
   8.4.1 Basic Framework of the Fuzzy Logic Development Environment 248
   8.4.2 Graphical Simulation Interface 249
   8.4.3 Design Automation System 250
   8.4.4 Netlist 252
   8.4.5 Placement 256
   8.4.6 Route 256
   8.4.7 Superphenix 256
8.5 CMOS Fuzzy Logic-based Controller 259
8.6 Conclusion 260
   8.6.1 Features of the VLSI technique taken for the Integration of Fuzzy Circuits 260
   8.6.2 Improvement of the Structure of the Fuzzy Logic Development Environment 261
Acknowledgments 262
References 262

HYBRID SYSTEMS AND APPLICATIONS 265

9 Neuro-fuzzy Systems: Hybrid Configurations 267

9.1 Preliminaries 267
9.2 Main Classes of Fuzzy Systems 269
9.3 Fuzzy Systems 270
   9.3.1 Linguistic Systems and Fuzzy Systems 270
   9.3.2 Fuzzy Systems and Memory Processes 271
   9.3.3 Classic Neurons 272
   9.3.4 Linear Combiners as Neurons 274
   9.3.5 Elementary Recurrent Systems 275
9.4 Fuzzy Neurons 276
   9.4.1 Elementary Fuzzy Neurons 276
   9.4.2 Neurons with Fuzzy Weights 277
   9.4.3 Inclusion of Fuzzy Weights into the Conventional Neuron Model 277
9.5 Neural Networks 283
9.6 Discrete Systems and Generalizations to Neuro-fuzzy Systems 285
9.7 Invariant Neuro-fuzzy Systems 287
   9.7.1 Main Configurations 287
   9.7.2 Series Neuro-fuzzy Systems and their Interpretation 290
9.8 Several Recurrent Neuro-fuzzy System Configurations 292
   9.8.1 Elementary Loops: Models of Memory Effects (Output Memory) 292
   9.8.2 Implementation of Complex Equations and Connections with Chaos in Classic Systems 294
9.9 Final Remarks 296
References 296

10 A Fuzzy Logic Approach to Handwriting Recognition 299

10.1 Introduction 299
10.2 Human reading 300
10.3 Handwriting Recognition: Current Approaches 302
10.4 A Fuzzy Processor for Handwriting Recognition 304
   10.4.1 Data Extraction and Preprocessing 304

	10.4.2 The Feature Measures	305
	10.4.3 The Fuzzy Recognition Process	307
10.5	Training	308
	10.5.1 Fuzziness and Statistics	308
10.6	Rulebase quality	309
	10.6.1 Discriminability	309
	10.6.2 Usefulness of Measures and Rulebase Reduction	310
	10.6.3 Completeness	310
	10.6.4 Overall Quality and Self-tuning	312
10.7	Results	312
10.8	Conclusions	313
Acknowledgments		313
References		313

Index   **315**

Editor's Preface

This edited volume contains ten papers on the subject of fuzzy technology. Fuzzy technology emerged as a combination of fuzzy sets theory, fuzzy logic and fuzzy-based reasoning. As a technology it gained a very practical meaning through thousands of applications in different theoretical as well as practical disciplines, covering mathematics, physics, chemistry, biology, life science, social science, economy, computer science, and (foremost) electrical, electronic, mechanical, nuclear, chemical, textile, aeronautic, ocean, and many other engineering disciplines.

The goal of this book is to create an interest in fuzzy technology among researchers, engineers, professionals and students involved in the research and development in the broad area of artificial intelligence.

This book is also intended to bring the reader up-to-date in the area of implementations and applications of fuzzy technology, as well as to generate and stimulate new research ideas in this area. It may inspire and motivate the researcher in new directions, as well as creating a force for new efforts to make a fuzzy technology commonly known and used in science and engineering.

This volume appears at a time of unprecedented research interest in the field of fuzzy technology. I intentionally wrote research due to the events that have occurred during the last couple of years. To be more specific, I should describe this interest geographically. Without any doubts, it means industrial and scientific interest in Asia and Europe, but it is still 'only' a scientific interest in America. This paradox has been discussed on many occasions and is a subject of unofficial talks at almost all conferences covering fuzzy sets and fuzzy logic topics. According to industrial sources, the US market for fuzzy logic-based products 'isn't there yet'. A similar source admitted that most of the developments for fuzzy logic are going on in Asia (Japan) and Europe (Germany), mainly because 'US companies only look at the short term return', whereas Japanese and Europeans 'tend to look farther down the road'. One positive aspect of this situation is that the top management in US companies recognize tremendous opportunities for fuzzy technology, and it predicts an 'enormous market in the US within five years'. The home and popular electronic goods will mostly contribute to the success to come. On the other hand, in the area of research fuzzy technology has gained great attention due to its ability to cope with many ill-defined and/or artificial intelligence problems. Fuzzy technology has been recognized as one of the tools of so-called 'soft computing'. Neural network methods and genetic algorithms are among other tools that help in efficient problem-solving. Theory, application and implementation of fuzzy control is an arena where fuzzy technology has been most successful. This phenomenon also motivated the creation of this volume.

Henceforth we expect that this volume will be of great benefit to researchers, scientists and professionals developing fuzzy logic applications and working on the enhancements of the theory; within the five years it may still be a source of information and inspiration to managers and engineers helping them define features for their new products.

Many authors from around the world contributed to this volume. They are currently doing research, development and implementation at the cutting edge of fuzzy technology. All the authors deserve special recognition for making this volume possible and for providing such high-quality contributions.

The material is organized in four thematic sections. The first is an introduction to the theory of fuzzy sets; the second is devoted to fuzzy logic control; the third section covers unique examples of fuzzy logic implementation; finally, the fourth presents examples of neuro-fuzzy hybrid systems and their applications.

The introductory section contains the paper by top world experts in the theory of fuzzy sets and approximate reasoning, Didier Dubois and Henri Prade from Universite Paul Sabatier, Toulouse, France. Their contribution, entitled Fuzzy Sets in Approximate Reasoning: a Personal View, is absolutely unique because it presents an extraordinarily peronsal view inside the applications of fuzzy sets in approximate reasoning.

Due to their inherent abilities, fuzzy sets are capable of modelling uncertain situations and can be instrumental in the formalization of interpolative reasoning. There are at least two major advantages that can immediately be indicated in such an approach. First, similarity-based reasoning can benefit from fuzzy sets, since similarity is usually a matter of degree. Second, fuzzy sets can represent incomplete information; hence, they can be viewed as possibility distributions and can be used to generate possibility and necessity measures to assess the degree of possibility in various statements. This paper provides a personal overview throughout the last decade resulting from the research performed jointly by both authors. In this research they have explored two interpretations of fuzziness as either the description of a gradual property or as a model of incomplete state of information. The paper is an excellent introduction and a thorough guide to possibility theory serving as a convenient framework for modelling uncertainty in a qualitative way. It provides an overview through the methods where fuzzy sets serve reasoning purposes as well as the background necessary to understand basic methodological issues.

The second section of this volume is devoted to various aspects of fuzzy logic control.

The first paper in this section, Fuzzy Logic Control: a Systematic Design and Performance Assessment Methodology, is written by a noble scientist and researcher G. Vachtsevanos, from Georgia Institute of Technology, Atlanta, Georgia, and is co-authored by S. Farinvata, from Ford Electronics Division, Melvindale, MI. This paper sets a milestone in the systematic analysis and design approach to fuzzy dynamic systems. The lack of mathematical rigour in the analysis and design of fuzzy logic controllers motivated the presented research. As a result, an analytical background is laid out to avoid intuitive and *ad hoc* implementations of fuzzy logic in control. In the design area, the proposed approach combines the approximate system modelling and heuristic approach to develop a fuzzy logic controller that is complete and robust. Three measures of performance assessment are proposed: fuzzy stability, robustness, and optimality. Furthermore, the main objective in these areas is to formalize the analysis and design tools and to demonstrate their effectiveness in dealing with real-world applications. Usually, when the plant dynamics are ill-defined, such a system is subject to large disturbances. The proposed methodology provides an alternative solution to the available ones.

Examples from the automotive industry (e.g. car engine model) are widely used to illustrate the analysis and design tools. For the system in hand, the fuzzy rule base was developed. This rule base accommodates the approximate state space representation of the nonlinear dynamic system. The simulation results indicate the effectiveness of the fuzzy logic control tools in terms of stability and robustness to external disturbances. This approach is, however, not free of some drawbacks: the complexity increases with the increase of the number of inputs. Also, the search through the data base becomes time-consuming for a large number of categories. The developed system can easily be integrated into existing expert systems to provide an efficient way to control fuzzy dynamic systems.

An industrial approach to solving fuzzy logic based control issues is delineated in the next paper, On the Compatibility of Fuzzy Control and Conventional Control Techniques. This paper was written by the experienced researcher and developer R. Palm from Siemens, Munich, Germany. This paper continues the discussion on the compatibility of the crisp and fuzzy control systems. With crisp-type fuzzy systems the main question is again related to the determination of the stability, performance and robustness of such mixed systems. As proven, the conventional linear and nonlinear control theory can contribute a great deal to the strategy defining the mixed systems. To address problems of stability, robustness and performance, the necessary translation from the 'fuzzy' world into a 'crisp' world is required. That can be used as a common basis to make a fair comparison. This paper addresses three main issues: the similarities between fuzzy control and variable structure systems with sliding modes; the representation of the fuzzy controller as a nonlinear control element and its interpretation as an equivalent gain; and noisy signals in the control loop and their interpretations as fuzzy signals. These problems are thoroughly discussed and illustrated with many simulated examples. Also it is proved that there is a substantial similarity between fuzzy and conventional nonlinear control.

In the paper entitled On the Crisp-type Fuzzy Controller: Behaviour Analysis and Improvement, written by W. Z. Qiao and co-authored by one of the Japanese pioneer in fuzzy logic research and application, M. Mizumoto from Osaka Electro- Communication University, Nayagawa, Osaka, Japan, one specific type of fuzzy logic controller, the crisp fuzzy controller, is analyzed. Due to its simplicity, the crisp fuzzy controller has been widely used in a variety of industrial applications. In this type of controller the antecedent part of fuzzy control rules is standard (i.e. defined by a fuzzy set) but the consequent parts of these rules are crisp numbers as opposed to the classic fuzzy controller. The authors focus on the dynamic behaviour of such controllers. They studied the input–output characteristics of crisp-type fuzzy controllers with min–max and product–sum inference methods used. As discovered, both kinds of crisp fuzzy controllers have very similar input–output characteristics and the differences between them are minor. Also, the crisp-type fuzzy controller can be regarded as a parameter time varying PD controller. As a result, the analysis and design of fuzzy control systems can be performed with conventional PID control methodology. As shown previously the PD type fuzzy controller yields a steady-state error for the 'zero' system. This error can be eliminated with PI type fuzzy controller. The authors propose a structure that combines features of both PD type and PI type fuzzy controllers. Such a PID type fuzzy controller allows the control system to have a fast rise time and a small overshoot, as well as a shorter settling time than the PD or PI controllers. To further improve the performance of the proposed PID controller, the authors designed a parameter adaptive fuzzy controller. This controller decreases the equivalent integral control component of the fuzzy controller gradually with the system response process time. This is to increase the dumping of the system when the system is about to settle down, while keeping the proportional control component unchanged to guarantee the fast reaction against the system's error. For the parameter adaptive fuzzy controller, the oscillations of the system are strongly rejected and the settling time is substantially reduced. The presented results of the analysis and simulation of parameter adaptive fuzzy controller can be used to guide the design of more sophisticated fuzzy logic-based control systems.

The third section of this volume, Fuzzy Logic Hardware Implementations, provides an overview of state-of-the-art techniques that are used for the hardware implementations of fuzzy logic-based circuits and systems.

The paper written by M. Patyra from the University of Minnesota, Duluth, MN, is entitled Design Considerations of Digital Fuzzy Logic Controllers, and presents an overview of the design issues related to the digital implementations of fuzzy logic-based controllers. Fuzzy logic controllers are chosen for their unquestionable success in the area of applications and implementations of fuzzy technology. Since 1985 when the first digital implementation of a fuzzy logic controller emerged, there have been many successful implementations reported in the technical literature. Although common by means of used digital techniques, many of these implementations are hard to compare in terms of characteristic features.

The purpose of this paper is twofold. First, this paper provides a unified framework for the comparison of digital fuzzy logic controllers. Second, it presents an analytical formulation of hardware cost and performance of various configurations used to implement these controllers. This paper also looks into the most commonly used controller configurations and shows the fuzzy algorithms mapping into hardware. As recent developments show, digital fuzzy logic controllers could become inherent parts of larger-hierarchical control systems incorporating various technologies including classic control, neural networks and genetic algorithms.

The next paper in this section, entitled Parallel Algorithm for Fuzzy Logic Controller, is written by J. L. Grantner from Western Michigan University, Kalamazoo, MI. It discusses in great detail some of the issues mentioned in the previous paper. In this paper a parallel algorithm to build a fuzzy model and execute the fuzzy inference is proposed. Based on this algorithm, the inference engine and a conceptual hardware implementations for a high-speed fuzzy logic controller are discussed. The characteristic features of the algorithm include a high degree of parallelism in performing fuzzy operations, constant low memory requirement to store the complete knowledge base (independent of the number of rules), error detection in case the linguistic model fails, flexibility in supporting SISO, MISO and MIMO systems, and a convenient natural way of mapping the algorithm into the hardware structure. The presented approach allows the creation and a storage of the compact rule base. By analyzing the mathematical models for SISO, MISO and MIMO systems it shows that a single algorithm can be developed to parametrize the inputs and outputs to the controller, according to the operation to be executed (model building or fuzzy inference). As a result, the parallel architecture is proposed for the controller hardware accelerator which features a maximum sustainable performance. This architecture is implemented with the pipeline technique to ensure a high degree of hardware usability and boost performance.

The next paper, entitled Fuzzy Flip-flop, is written by world class experts who have been in the field of fuzzy technology since its very beginning: K. Ozawa and K. Hirota (Hosei University, Tokyo, Japan) and L. T. Koczy (Technical University of Budapest, Hungary). This paper is dedicated to a special class of digital fuzzy logic circuits, fuzzy flip-flops. As pointed out by the authors, a great deal of research has been devoted to the realization of the idea of a fuzzy computer. However, a few types of fuzzy inference engines (which are also called fuzzy processors) were prepared and implemented by Yamakawa, Tigai and Watanabe. Although these fuzzy inference devices opened a new realm of opportunity for fuzzy logic, especially in control applications, they performed basically a single step inference. In order to realize multistep inference, fuzzy memory modules are required. This motivated the research of which the results are described in this paper. The authors propose and define a fuzzy flip-flop which is an extension of the classic binary flip-flop, specifically a J-K flip flop. They formulate a truth table for the fuzzy J-K flip flop where binary negation, union, and intersection operations are extended by means of fuzzy negation, t-norm, and s-norm, respectively. Reset type and set type of equations for fuzzy J-K flip flop are formulated, and the results are graphically illustrated with respect to different t-norm and s-norm representations. These representations include fuzzy complement, minimum, maximum, algebraic product, and algebraic sum operations. The hardware implementation of a fuzzy J-K flip-flop with standard logic devices is also presented. These circuits are functionally tested and the results are illustrated. The results from the circuit testing also suggest a possible and efficient VLSI implementation for the fuzzy J-K flip-flops, as well as for more sophisticated circuits built from them, such as fuzzy registers and fuzzy memories.

The paper entitled Design of Fuzzy Logic Circuits, by L. Lemaitre from the Swiss Federal Institute of Technology, Lausanne, describes the methodology of generating structures for complex fuzzy logic circuits based on specified fuzzy functions. The methodology is based on the introduction of basic fuzzy operators. These operators have their immediate and simple representation in hardware, namely CMOS circuit representations. With such a background built up the design automation system takes the description of the fuzzy function and converts it into the form represented by basic operators. Such a form is ready to be used to generate a CMOS layout for the circuit performing the operations defined by a function in hand. The design automation system places the layout elements and solves the routing problems associated with the layout optimization. The optimization is performed on two levels, local and global. With such powerful tools in hand the current-mode analogue fuzzy logic controller was synthesized, developed and manufactured. The final chip reached a performance of 10MFLIPS while occupying only 0.4 mm^2 of silicon area.

In the fourth section Hybrid Systems and Applications the first paper, entitled. Neuro-fuzzy Systems: Hybrid Configurations, gives a thorough overview of neuro-fuzzy systems. Written by two experts currently specializing in neuro-fuzzy systems, H.-N. L. Teodorescu from the Technical University of Iasi, Romania, and T. Yamakawa from the Kyushu Institute of Technology, Iizuka, Fukuoka, Japan, this paper provides a systematic discussion of hybrid configurations of neuro-fuzzy systems. The merging of these two types of systems is possible due to the functional similarities that these systems feature. Five classes of neuro-fuzzy systems are selected based on their structure. The structure of neuro-fuzzy systems can be put into the following categories: neural network structure with fuzzy logic neurons, adaptive fuzzy systems that can be modified by the neural network during the learning process, classic neural network with the learning process determined by a fuzzy system, systems that are composed of independent neural networks and fuzzy systems, systems with neural network configurations that use fuzzy systems as neurons, and finally systems that are created by a mixture of two or more of the above systems. This paper discusses in detail three classes of systems; namely, neural network structures with fuzy logic neurons, systems that are composed of independent neural networks and fuzzy systems, and systems with neural network configurations that uses fuzzy systems as neurons. Such systems can be useful in various applications including the modelling of complex, empirical-driven processes, control of complex systems, pattern recognition and chaotic systems. Several examples of applications of such systems based upon the authors research results are provided.

A very interesting example of the direct application of fuzzy system is presented in the paper A Fuzzy Logic Approach to Handwriting Recognition. Written by D. J. Ostrowski and P. Y. K. Cheung from Imperial College, University of London, UK, this paper provides in-depth experience with fuzzy logic applied to the handwriting recognition problem. This paper describes the application of fuzzy logic to cursive script recognition. Handwriting recognition is crucial for many automated task problems. The fundamental problem of cursive script is segregation because the nature of cursive script makes reliable recognition of individual letters extremely difficult. Therefore, the use of prototypical models is not satisfactory. On the other hand, the cursive script recognition is a sophisticated process complexed by a variability of the handwriting of a particular individual. The approach described in this paper consists of recognizing letters in terms of the features of the individual letter and accepting that the feature separation is unreliable. The proposed solution to this issue uses the fuzzy rule-based approach. Such a rule base is capable of accepting the variability of data and its unreliable extraction from the original images. Moreover, it derives the characters features and ability to differentiate letters from training. This paper provides a brief description of current methods of machine recognition and fuzzy logic approach to such problems as well. The training of the rule base is discussed in detail. Finally, the examples are shown along with the statistical results obtained from the benchmark data set provided by the UNIPEN project.

Finally, the Editors wish the reader many educational, professional and fruitful experiences while studying this volume.

Marek J. Patyra
Duluth, Minnesota
1995

List of Contributors

Dominic Ostrowski and **Peter Cheung**
*Department of Electrical and Electronic Engineering
Imperial College
Exhibition Road
London SW7 2BT, UK*

Didier Dubois and **Henri Prade**
*IRIT
Universite Paul Sabatier
118 Rue de Narbonne
31062 Toulouse Cedex
France*

Shehu Farinwata
*Ford Research Laboratory
Dearborn, MI 48121-2053
USA*

Kaoru Hirota
*Department of Systems Science
Tokyo Institute of Technology
Tokyo
Japan*

Kazuhiro Ozawa
*Department of Instrument and Control Engineering
Hosei University
3-7-2 Kajino-cho
Koganei-city, Tokyo 184,
Japan*

Laszlo Koczy
*Department of Telecommunications and Telematics
Technical University of Budapest,
Stoczek u.2
H-1111 Budapest
Hungary*

Rainer Palm
*Corporate Research and Development
Siemens AG
Otto-Hahn-Ring 6
8000 Munich 83
Germany*

Marek Patyra
*Department of Electrical and
  Computer Engineering
University of Minnesota
Duluth, MN 55812-2496
USA*

George Vachtsevanos
*School of Electrical Engineering
Georgia Institute of Technology
Atlanta, GA 30332-0250
USA*

Wu Zhi Quiao and
Masaharu Mizumoto
*Department of Management Engineering
Osaka Electro-Communication University
Neyagawa, Osaka 572
Japan*

Laurent Lemaitre
Department of Computer Science
University of California
Berkeley, CA 94720-1776
USA

Janos Grantner
Department of Electrical Engineering
Western Michigan University
Kalamazoo, MI 49008-5066
USA

H.-N.L. Teodorescu
Technical University of Iasi
Iasi, Romania

T. Yamakawa
Kyushu Institute of Technology
Iizuka, Fukuoka
Japan

Acknowledgments

We are grateful to the staff of John Wiley and Sons, Europe for their advice, dedication and commitment to this project. Their high-quality assistance has made the production of this volume smooth and undisturbed. We would like to express our appreciation especially, to Anne-Marie Halligan and Peter Mitchell. Without their professional guidance and support, publication of this volume would have been almost impossible.

THEORY

1
Fuzzy Sets in Approximate Reasoning: a Personal View

D. Dubois and H. Prade
Université Paul Sabatier, Toulouse, France

1.1 INTRODUCTION

Fuzzy rule-based approximate reasoning is attracting more and more interest from researchers and practitioners nowadays. Zadeh (1973, 1975, 1979a) provided the basic machinery for fuzzy set-based approximate reasoning more than fifteen years ago. In his approach, each granule of knowledge is represented by a fuzzy set or a fuzzy relation on the appropriate universe. Then the fuzzy set representations of the different granules are combined and the result of this combination is projected onto the universe(s) of interest. A well-known particular case of this method is the pattern of inference, named 'generalized *modus ponens*', which enables us to deduce a fuzzy conclusion from a fuzzy rule and a fuzzy fact pertaining to the universe of discourse associated with the condition part of the rule. However, this framework is rather general, and a proper application of it requires a correct understanding of the intended meaning of the pieces of knowledge to be represented by fuzzy sets. In particular, fuzzy rules may have very different semantics, which lead to different choices concerning the multiple-valued connective to be used to model the rule. In this paper we distinguish between purely gradual rules and rules with uncertain conclusion parts. Purely gradual rules are of the form 'the more X is A, the more Y is B', which qualitatively describes a relation between the values of X and Y, but which is not pervaded with uncertainty. Gradual rules of the form 'the closer X is to..., the closer Y is to...' express pieces of knowledge which are of interest for interpolative reasoning purposes. Patterns of reasoning under uncertainty involve rules of the forms 'if p is true then q is somewhat certain', 'the more X is A, the more certain Y is B', or 'the more X is A, the more possible Y is B', which carry different kinds of uncertainty in their conclusion parts. We shall show below how these different kinds of rules can be represented in the fuzzy set framework.

In this paper we make a strong distinction between the use of fuzzy sets to model gradual properties, i.e. properties whose satisfaction is a matter of degree, or gradual notions like similarity, and the use of fuzzy sets to model incomplete, imprecise or uncertain information. In the latter case, the fuzzy set-based representation of the available information induces possibility and necessity measures for assessing the uncertainty of the conclusions which are computed. It is worth emphasizing that possibility theory offers an ordinal framework for the modelling of uncertainty due to the use in the theory of operations like max or min which are only sensitive to the ordering of the operands. This is also true for the order-reversing operation (the complementation to 1 when using the scale $[0, 1]$).

The paper is organized in two main parts. The first one is devoted to the use of the ideas of graduality expressed in terms of fuzzy sets for interpolative and qualitative reasoning purposes. The second main part deals with the management of uncertainty, and more particularly with the modelling of uncertain rules, the treatment of uncertainty and inconsistency in possibilistic logic, a possibility measure-based approach to default reasoning, and a possibilistic treatment of incomplete information in abductive reasoning for diagnosis.

1.2 GRADUALITY- AND SIMILARITY-BASED APPROXIMATE REASONING

One of the natural semantics for fuzzy sets is the expression of closeness, proximity, similarity. Then the elements in the core (or peak) of a fuzzy set (i.e. those with membership 1) are viewed as prototypical elements of the fuzzy set, while the other membership grades estimate the closeness of the elements to the prototypical ones. Proximity or similarity viewed as a graded relation between two elements of a referential can be represented in terms of a binary fuzzy relation, where the closer to 1 is the degree of membership of a pair of elements to the fuzzy relation, the greater their proximity or their similarity. In this section we focus on interpolative reasoning based on a set of fuzzy rules associated with similarity relations expressing closeness, and then we consider the use of approximate equality relations for qualitative reasoning on orders of magnitude. We first give the necessary background on fuzzy relations and related issues.

1.2.1 Composition of Fuzzy Relations, Extension Principle and Similarity

A binary fuzzy relation R is defined as a fuzzy subset on a Cartesian product $U \times V$. The composition of classical relations, namely $S \circ R = \{(u, w), \exists v, (u, v) \in R \text{ and } (v, w) \in S\}$ is extended to the fuzzy relations R and S defined on $U \times V$ and $V \times W$, respectively (we may have $U = V = W$) by

$$\forall u \in U, \forall w \in W, \mu_{S \circ R}(u, w) = \sup_{v \in V} \min(\mu_R(u, v), \mu_S(v, w)) \qquad (1.1)$$

More generally, min may be replaced by an operation $*$ and the $\circ*$ composition is then defined as

$$\mu_{S \circ * R}(u, w) = \sup_v \mu_R(u, v) * \mu_S(v, w) \qquad (1.2)$$

where $*$ is a binary operation defined on $[0, 1]$ expressing a conjunction such that $0*0 = 0*1 = 1*0 = 0$, $1*1 = 1$. The operation $*$ is usually chosen as being associative, monotonically increasing in each place and satisfying $a*1 = a$, $\forall a \in [0, 1]$. Such an operation is known as a triangular norm (Schweizer and Sklar, 1983) and can be shown to be commutative and such that $\forall (a, b) \in [0, 1]^2$, $a*b \leq \min(a, b)$. Noticeable triangular norms are min, product and the linear conjunction $\max(0, a + b - 1)$. When $*$ is associative, the composition $\circ *$ is also associative. We simply write $\circ$ if $* = \min$.

A particular case of (1.2) is when one of the fuzzy relations is replaced by a fuzzy set A (here on U), namely

$$\mu_{R \circ A}(v) = \sup_u \mu_R(u, v) * \mu_A(u) \qquad (1.3)$$

The composition $R \circ A$ extends to a fuzzy set A and to a fuzzy multiple-valued mapping Γ which associates with $u \in U$ the fuzzy set $\Gamma(u)$ defined by $\mu_{\Gamma(u)}(v) = \mu_R(u, v)$, the image of a classical subset by an ordinary multiple-valued mapping. A particular case of (1.3) is the extension principle (Zadeh, 1965, 1975) which enables the image $f(A)$ of a fuzzy set A via a function f from a set U to a set V to be computed as

$$\forall v \in V, \mu_{f(A)}(v) = \sup_{u:v=f(u)} \mu_A(u)$$
$$= 0 \text{ if } \not\exists u, v = f(u) \qquad (1.4)$$

When A is a crisp set, (1.1) reduces to the usual extension of a function to a set-valued argument, namely $f(A) = \{v, \exists u \in A, v = f(u)\}$. If f is bijective

$$\forall v \in V, \mu_{f(A)}(v) = \mu_A(f^{-1}(v)) \qquad (1.5)$$

When the function f has several arguments $u_1, u_2, \ldots, u_n$ an interesting situation is when the fuzzy set A is separable, so that A is a fuzzy Cartesian product $A_1 \times A_2 \times \cdots \times A_n$ with membership function $\mu_{A_1} * \mu_{A_2} * \cdots * \mu_{A_n}$, where $*$ is a triangular norm which models conjunctions in fuzzy set theory. Then the extension principle reads

$$\mu_{f(A)}(v) = \sup\{\mu_{A_1}(u_1) * \mu_{A_2}(u_2) * \cdots * \mu_{A_n}(u_n) | f(u_1, u_2, \ldots, u_n) = v\} \qquad (1.6)$$

which is an interactive form of the extension principle (Dubois and Prade, 1987a). In this paper we restrict ourselves to the non-interactive case, i.e. $* = \min$.

A fuzzy interval M is simply a fuzzy set of the real line such that $\mu_M(u) = 1$ for some u, and whose α-cuts $M_\alpha = \{u, \mu_M(u) \geq \alpha\}$ are closed intervals. The membership function of a fuzzy interval is thus upper semi-continuous (u.s.c.). The calculus of fuzzy intervals is based on the extension principle applied to real-valued real functions. Using u.s.c. fuzzy intervals M we are sure that any level cut $f(M)_\alpha$ of $f(M)$ is equal to the image of the corresponding level cut M_α of M through f when f is continuous, i.e. $f(M)_\alpha = f(M_\alpha)$. Let the interval $(M_i)_\alpha$ be denoted $[m_i^-(\alpha), m_i^+(\alpha)]$. We have the following result (Dubois and Prade, 1987a).

Let $M_1, \ldots, M_n$ be u.s.c. fuzzy intervals whose supports are bounded. Let $f: \mathbb{R}^n \to \mathbb{R}$ be continuous and non-decreasing with respect to each of its arguments. Then, in the non-interactive case ($* = \min$), the image of $M_1, \ldots, M_n$ by f is an upper semi-continuous fuzzy interval whose α-cuts are

$$[f(M_1, \ldots, M_n)]_\alpha = [f(m_1^-(\alpha), \ldots, m_n^-(\alpha)), f(m_1^+(\alpha), \ldots, m_n^+(\alpha))] \qquad (1.7)$$

for every α in $[0, 1]$. We shall see in Section 1.2.3 that the extension of arithmetic operations to fuzzy intervals can be useful to compute the composition of fuzzy relations expressing approximate equalities for instance. Taking $f(u_1, u_2) = u_1 + u_2$ in (1.7) yields for the extension of the addition to fuzzy intervals

$$(M_1 \oplus M_2)_\alpha = [m_1^-(\alpha) + m_2^-(\alpha), m_1^+(\alpha) + m_2^+(\alpha)] = (M_1)_\alpha \oplus (M_2)_\alpha$$

Thus if M_i is a trapezoid with support (a_i, d_i) (i.e. the set of elements with non-zero membership values) and core $[b_i, c_i]$, then $M_1 \oplus M_2$ is a trapezoid with support $(a_1 + a_2, d_1 + d_2)$ and core $[b_1 + b_2, c_1 + c_2]$.

A similarity relation S is a binary fuzzy relation defined on a set U whose membership function satisfies the three following properties (Zadeh, 1971)

(i) $\forall u \in U, \mu_S(u, u) = 1$ (reflexivity)
(ii) $\forall u \in U, \forall v \in U, \mu_S(u, v) = \mu_S(v, u)$ (symmetry)
(iii) $\forall u \in U, \forall v \in U, \forall w \in U, \mu_S(u, v) * \mu_S(v, w) \leq \mu_S(u, w)$ (max-$*$ transitivity)

Fuzzy relations satisfying (i) and (ii) are called proximity relations. See Ovchinnikov (1991) for a recent overview on similarity relations.

Similarity relations are closely related to the idea of distance. In particular if $*$ is the linear conjunction, max-$*$ transitivity is equivalent to the triangle inequality and $1 - \mu_S$ is a pseudometric; max-min transitivity, which is stronger, corresponds to ultrametrics. Max-$*$ transitivity can also be easily interpreted in terms of the composition of fuzzy relations, i.e. (iii) can be written $S \circ * S \subseteq S$, where fuzzy set inclusion is pointwisely defined by an inequality between membership functions. As we can see (iii) extends transitivity to degrees of similarity which are intermediate between 0 and 1 (transitivity of non-fuzzy relations is retrieved as a particular case). We shall see in Section 1.2.3 weaker forms of transitivity-like properties, where the similarity or approximate equality slightly deteriorates when applying transitivity.

1.2.2 Interpolative Reasoning

Many authors have pointed out that fuzzy control techniques perform an interpolation between the conclusions of the fuzzy control rules, on the basis of the degrees of matching of the situation under concern with the condition parts of the different rules. Interpolative reasoning naturally subsumes some idea of proximity, since intermediary values are computed from the known values corresponding to the situations which are closest to the current one. So, it is not surprising that the two main fuzzy control methods, those of Mamdani (1977) and Sugeno (1985), receive some justification, starting with a proper handling of proximity relations.

Let us consider a collection of n fuzzy rules if X_1 is $A_1^{(i)}$ and ... and X_p is $A_p^{(i)}$, then Y is $B^{(i)}$, $i = 1, n$. Then Klawonn and Kruse (1993a, 1993b) have shown that if the (fuzzy) relation R relating the X_j and Y satisfies some extensionality axioms, this is enough to derive the result given by Mamdani's method before defuzzification. Let S and T be two similarity relations equipping the domains U and V of X and Y, respectively. Then the extensionality of a relation R on $U \times V$ means

(i) $\forall u \in U, \forall s \in V, \forall v \in U, \mu_R(u,s) * \mu_S(u,v) \leq \mu_R(v,s)$;
(ii) $\forall u \in U, \forall s \in V, \forall t \in V, \mu_R(u,s) * \mu_T(s,t) \leq \mu_R(u,t)$.

This expresses that if u and s are in relation R, u (respectively, s) should be also in relation with the neighbours of s (respectively, u). Then let us assume that the $A_j^{(i)}$ and $B^{(i)}$ can be viewed as fuzzy sets of values close to peak values (here taken as unique for simplicity), i.e. $A_j^{(i)} = \{a_j^{(i)}\} \circ S_j, B^{(i)} = \{b^{(i)}\} \circ T$, where the $a_j^{(i)}$ and $b^{(i)}$ are peak values, S_j and T are similarity relations. Then, using the extensionality of the relation R relating the X_j and Y, as well as the fact that $\mu_R(a_1^{(i)}, \ldots, a_p^{(i)}, b^{(i)}) = 1$, $\forall i = 1, n$ (the peak values define points which completely belong to the graph of R), and defining the similarity relation on $U_1 \times \cdots U_p$ as $\mu_{S_1} * \cdots * \mu_{S_p}$, leads to

$$\mu_R(x_1, \ldots, x_p, v) \geq \max_{i=1,n} \mu_{A_1^{(i)}}(x_1) * \cdots * \mu_{A_p^{(i)}}(x_p) * \mu_{B^{(i)}}(v) \tag{1.8}$$

When $* = \min$, we recognize Mamdani's expression of the fuzzy set of V to be defuzzified for computing an output $Y = y$ when the observed input is $X_1 = x_1, \ldots, X_p = x_p$. However, at this point, no interpolation has been modelled in (1.8).

Sugeno's method starts from rules with non-fuzzy conclusion parts of the form 'if X is $A^{(i)}$ then $Y = b^{(i)}(x)$' (we do not deal here with compound conditions, for simplicity), and computes the output y as

$$y = \sum_i \mu_{A^{(i)}}(x) \cdot b^{(i)}(x) / \sum_i \mu_{A^{(i)}}(x) \tag{1.9}$$

which indeed looks like an interpolation. When $b^{(i)}(x)$ does not depend on x, this result can be obtained by applying Zadeh's approximate reasoning combination and projection approach, viewing the rules as gradual rules expressing that 'the closer X is to $a^{(i)}$, the closer Y is to $b^{(i)}$', i.e. modelling them as inequality constraints of the form $\mu_{A^{(i)}}(u) \leq \mu_{B^{(i)}}(v)$. Then the subset of V obtained by combining the results of the rules for the input $X = x_0$ is given by $\min_{i=1,n} \mu_{A^{(i)}}(x_0) \to \mu_{B^{(i)}}(v)$, where the implication defined by $a \to b = 1$ if $a \leq b$ and $a \to b = 0$ if $a > b$ encodes the above interpretation of the rules. When the $A^{(i)}$ and $B^{(i)}$ make suitable fuzzy partitions of U and V, respectively (it guarantees that $\forall u, \sum_i \mu_{A^{(i)}}(u) = 1$), which is in particular the case with the usual partitions made of triangular membership functions as in Figure 1.1, it can be shown that the fuzzy subset of V which is thus obtained in nothing but the singleton $\{y\}$ computed by Sugeno's method (1.9). Let us explain the situation in more detail.

Let us consider a collection of gradual rules of the form 'the closer X is to a_i, the closer Y is to b_i' where $(a_i, b_i), i = 1, n$ are pairs of scalar values. The first problem is to represent 'close to a_i', by means of a fuzzy set A_i. It seems natural to assume that $\mu_{A_i}(a_{i-1}) = \mu_{A_i}(a_{i+1}) = 0$, since there are special rules adapted to the cases $X = a_{i-1}$,

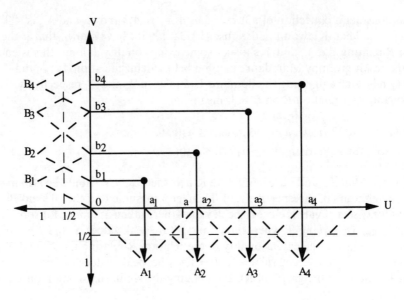

Figure 1.1 Triangular partitions for interpolation

$X = a_{i+1}$. Moreover if $u \neq a_i$, $\mu_{A_i}(u) < 1$ for $u \in (a_{i-1}, a_{i+1})$, since information is only available for $u = a_i$. Hence A_i should be a fuzzy interval with support (a_{i-1}, a_{i+1}) and core $\{a_i\}$. Besides, by symmetry, since the closer x is to a_{i-1}, the farther it is from a_i, $\mu_{A_{i-1}}$ should decrease when μ_{A_i} increases, and

$$\mu_{A_i}\left(\frac{a_i + a_{i+1}}{2}\right) = \mu_{A_{i-1}}\left(\frac{a_{i-1} + a_i}{2}\right) = 0.5$$

The simplest way of achieving this is to let

$$\forall u \in [a_{i-1}, a_i], \mu_{A_{i-1}}(u) + \mu_{A_i}(u) = 1$$

an example of which is triangular-shaped fuzzy sets, as in Figure 1.1. Clearly the conclusion parts of the rules should involve fuzzy sets B_i whose meaning is 'close to b_i', with a similar convention. In other words, each rule is understood as 'the more X is A_i, the more Y is B_i', i.e. here 'the closer X is to a_i, the closer Y is to b_i'. In that case the output associated with the precise input $X = a$, where $a_{i-1} < a < a_i$, is

$$B = (\alpha_{i-1} \to B_{i-1}) \cap (\alpha_i \to B_i) = (B_{i-1})_{\alpha_{i-1}} \cap (B_i)_{\alpha_i}$$

where $\alpha \to B$ corresponds to the level cut B_α, $\to$ is the implication defined above and $\alpha_{i-1} = \mu_{A_{i-1}}(a)$, $\alpha_i = \mu_{A_i}(a)$ with $\alpha_{i-1} + \alpha_i = 1$. Then it can easily be proved (without the assumption of a triangular shape) that there exists a unique value $y = b$ such that $\mu_B(b) = 1$, which exactly corresponds to the result of a linear interpolation, i.e. we have $b = \alpha_{i-1} \cdot b_{i-1} + \alpha_i \cdot b_i$. This is a theoretical justification for the inference method of Sugeno and Nishida (1985), where the conclusion parts of the rules are precise values

b_i and where a linear interpolation is performed on the basis of the degrees of matching $\alpha_i = \mu_{A_i}(a)$. Hence reasoning with gradual rules does model interpolation, linear interpolation being retrieved as a particular case. The more complicated case of gradual rules with compound conditions, i.e. rules of the form 'the more X_1 is $A_1, \ldots$, and the more X_n is A_n, the more Y is B' is studied in detail by Dubois, Grabisch and Prade (1994).

As we can see, these justifications of Mamdani's and Sugeno's methods, although they both rely on some closeness notion, are of different natures. The second one views interpolation as a conjunctive compromise between closeness constraints with respect to know points of the relation linking X and Y, and leads to a precise result. By contrast, Mamdani's approach is recovered through extensionality assumptions expressing that if two values are related, each one is also somewhat related to the neighbours of the other, which leads to an output calculated by a closeness-weighted disjunctive aggregation of the fuzzy sets appearing in the conclusion parts of the rules; the result is then a fuzzy range of possible values to be defuzzified, and interpolation is carried out only at the defuzzification level.

1.2.3 Qualitative Reasoning

Let us now turn towards another type of reasoning where proximity relations are explicitly handled. Let us for instance consider the two pieces of information 'X is approximately equal to Y' and 'Y is much larger than Z'. What can be said about Z with respect to X? This kind of situation can be for instance encountered in temporal reasoning when we know that two dates are close, and one of them takes place much later than a third one. An approximate equality between u and v can be modelled by a fuzzy relation E of the form $\mu_E(u, v) = \mu_L(u - v)$ which only depends on the values of the difference $u - v$, for instance

$$\forall u, \forall v, \mu_E(u,v) = \max\left(0, \min\left(1, \frac{\delta + \varepsilon - |u-v|}{\varepsilon}\right)\right) = \begin{cases} 1, & \text{if } |u-v| \leq \delta \\ 0, & \text{if } |u-v| > \delta + \varepsilon \\ (\delta + \varepsilon - |u-v|)/\varepsilon, & \text{otherwise.} \end{cases}$$

where δ and ε are, respectively, positive and strictly positive parameters which modulate the approximate equality. Following the same line, a more or less strong inequality can be modelled, for instance by a relation I of the form

$$\forall v, \forall w, \mu_I(v,w) = \mu_K(v-w) = \max\left(0, \min\left(1, \frac{v-w-\lambda}{\rho}\right)\right)$$

$$= \begin{cases} 1, & \text{if } v > w + \lambda + \rho; \ \lambda \geq 0, \ \rho > 0 \\ 0, & \text{if } v < w + \lambda; \\ (v - w - \lambda)/\rho, & \text{otherwise.} \end{cases}$$

The values of λ and ρ act on the meaning of the inequality and can express shades such as 'slightly greater', 'much greater', etc. When $\lambda = 0$ and $\rho \to 0$ the usual inequality relations

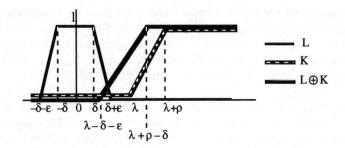

Figure 1.2 Combining 'appoximately equal' and 'much larger'

$>$ or $\geq$ are recovered (depending on how $\lim_{\rho \to 0}(v-w)/\rho$ is defined for small values of $v - w$). The composition of E and I yields

$$\forall u, \forall w, \mu_{E \circ I}(u, w) = \sup_{v \in \mathbb{R}} \min(\mu_L(u - v), \mu_K(v - w))$$

$$= \sup_{s, t : u - w = s + t} \min(\mu_L(s), \mu_K(t))$$

$$= \mu_{L \oplus K}(u - w)$$

where we recognize the expression of the extended sum of K and L, applying (1.6) to $f(s, t) = s + t$. Thus the possible values of the difference $x - z$ is restricted by $L \oplus K$. This result is represented in Figure 1.2 where the relations defined above are used. We see that it is certain that $x \geq z + \lambda - (\delta + \varepsilon)$, since the support of $L \oplus K$ is the interval $(\lambda - \delta - \varepsilon, +\infty)$ and that the value of the difference $x - z$ belongs to $L \oplus K$ at the degree 1 as soon as $x \geq z + \lambda + \rho - \delta$. Then, depending on the respective values of the parameters, x is still greater than z (but may be not as much as y with respect to z) (if $\lambda - \delta - \varepsilon > 0$), or we are only sure that x is not much smaller than z (if $\lambda + \rho - \delta > 0$).

As is suggested by this example, it is possible and even simple to derive inference rules in the case of fuzzy relations defined in terms of difference of values in order to perform these types of computation at the symbolic level, taking advantage of the fact that the composition of fuzzy relations then reduces to a simple arithmetic computation on the fuzzy numbers which express the semantics of the relations. This can be done as well if we use the quotient rather than the difference to compare numbers, as we are going to see.

A good example of reasoning with proximity relations is provided by qualitative reasoning on orders of magnitude, where the ideas of approximate equality and negligibility play an important role. This can be captured in the following way (Dubois and Prade, 1989). The relationship 'X is close to Y' (between real vaues) is understood as 'the quotient X/Y is close to 1', in the sense that the possible values of X/Y are restricted by a fuzzy set P such that $\mu_P(1) = 1$, modelling 'around 1'. Similarly, 'Z is negligible with respect to Y' will be modelled by '$1 + Z/Y$ is close to 1', where 'close' refers here to a fuzzy set P', possibly different from P, but still such that $\mu_{P'}(1) = 1$. In other words, we define a proximity relation $S(P)$ such that

$$\mu_{S(P)}(u, v) = \mu_P(u/v) \tag{1.10}$$

where P is a fuzzy number which satisfies $\mu_P(1) = 1$, $\mu_P(t) = \mu_P(1/t)$ (to ensure the symmetry of $S(P)$). Negligibility is expressed by

$$\mu_{N(P)}(u, v) = \mu_P(1 + u/v). \tag{1.11}$$

Then, given two relations $S(P)$ and $S(P')$, or $N(P)$ and $N(P')$, or $S(P)$ and $N(P')$, their composition results in a simple fuzzy arithmetic calculus on P and P'. Indeed

$$\mu_{S(P) \circ S(P')}(u, w) = \sup_v \min(\mu_P(u/v), \mu_{P'}(v/w))$$
$$= \sup_{(s,t): s \cdot t = u/w} \min(\mu_P(s), \mu_{P'}(t)) = \mu_{P \odot P'}(u/w)$$

where $\odot$ denotes the extension of product to fuzzy numbers. Thus

$$S(P) \circ S(P') = S(P \odot P') \tag{1.12}$$

This expresses that x is close to y in the sense of $S(P \odot P')$. It accounts for the degradation of the transitivity since $P \odot P \supset P$ usually, and thus $S(P \odot P) \supset S(P)$. Similarly, it can be established that

$$N(P) \circ N(P') = N((P \ominus 1) \odot (P' \ominus 1) \oplus 1) \tag{1.13}$$
$$S(P) \circ N(P') = N(P \odot (P' \ominus 1) \oplus 1) \tag{1.14}$$

where $\ominus$ and $\oplus$ denote extended subtraction and addition. It can be shown that there is, as expected, a strengthening of the negligibility relation obtained in (1.13) (i.e. for $P' = P$, $N(P) \supseteq N((P \ominus 1) \odot (P \ominus 1) \oplus 1)$ and a weakening obtained in (1.14) (i.e. $N(P) \subseteq N(P \odot (P \ominus 1) \oplus 1)$). It is thus possible to develop inference rules where the fuzzy sets which parametrize the closeness and negligibility relations are manipulated at a symbolic level. See Dubois and Prade (1989, 1991a) for details, and fuzzy set-based counterparts of approaches proposed in the qualitative reasoning literature.

1.3 UNCERTAINTY MANAGEMENT

Zadeh (1978a) was the first to propose the idea of representing an incomplete state of knowledge by means of a fuzzy set. If for instance we only know about quantity X that 'X is large', this means that the possible vaues for X are those compatible with the meaning of 'large' in the considered context, represented by the membership function of a fuzzy set, i.e. $\pi_X = \mu_{\text{large}}$, where π_X denotes the possibility distribution describing the more or less possible values for X according to what is known. As is recalled in the next section, two measures of uncertainty called possibility and necessity are associated with a possibility distribution. These measures turn out to be a convenient tool for the modelling of uncertainty which allows for the representation of imprecise pieces of information or partial states of ignorance. The lack of certainty in 'not A' does not entail the certainty of A in possibility theory, as it is the case with probability, where $\text{Prob}(\text{not } A) = 0 \Rightarrow \text{Prob}(A) = 1$.

1.3.1 Background

Let U be a set that represents the range of a variable X. Usually X stands for the unknown value taken by some single-valued attribute applied to an object under consideration. For instance X refers to the age of a man named Peter. A possibility distribution π_X on U is a mapping from U to the unit interval $[0, 1]$ attached to the single-valued variable X. The function π_X represents a flexible restriction which constrains the possible values of X according to the available information, with the following conventions:

$\pi_X(u) = 0$ means that $X = u$ is definitely impossible

$\pi_X(u) = 1$ means that absolutely nothing prevents that $X = u$.

Note that $\pi_X(u) = 1$ and $\pi_X(u') = 1$, with $u' \neq u$, are allowed at the same time. Intermediary levels of plausibility about the possible values of X are modelled by letting $\pi_X(u)$ be between 0 and 1 for some values u. The quantity $\pi_X(u)$ thus represents the degree of possibility of the assignment $X = u$. Thus we can acknowledge the fact that some values u are more possible than others, according to what is known. In other words, if we consider the logical propositions p_u whose semantic contents is '$X = u$', π_X encodes a preference relation among these propositions describing possible states of the world: if $\pi_X(u) < \pi_X(u')$, u' is strictly preferred to u as a possible candidate for the value of X according to the available information, among more or less possible (or allowed) values, i.e. among the u such that $\pi_X(u) > 0$. Clearly, if U exhaustively contains all the values which may be thought of for X, at least one of the elements of U should be fully possible as a value of X, so that $\exists u, \pi_X(u) = 1$ (normalization). To summarize, a possibility distribution π_X encodes a preference relation about the possible values of the variable X, according to the available knowledge.

Given a possibility distribution π_X, several set-valued functions of interest can be defined in order to estimate the tendency of X to belong to a given subset A of U. Two of these set-functions are well-known and widely used, namely: the so-called measure of possibility (Zadeh, 1978a)

$$\Pi(A) = \sup_{u \in A} \pi_X(u) \tag{1.15}$$

which estimates the consistency of the statement '$X \in A$' with what we know about the possible values of X; hence we look for the element(s) of A with the greatest possibility degree according to π_X (since A is all the more possible as there exists in A a value with a high degree of possibility). The second set-function is the dual measure of necessity (Zadeh, 1979b, Dubois and Prade, 1980)

$$N(A) = 1 - \Pi(\bar{A}) \tag{1.16}$$

$$= \inf_{u \notin A} 1 - \pi_X(u) \tag{1.17}$$

which estimates to what extent each value outside A, i.e. in the complement $\bar{A}$ of A, has a low degree of possibility; thus the values with the higher degrees of possibility should be among the elements of A, which makes us somewhat certain that indeed X belongs to A. The duality relation (1.16) between Π and N expresses that A is all the more certain as $\bar{A}$ is less consistent with the available knowledge. It assumes that π_X is normalized. The normalization of π_X entails $\max(\Pi(A), \Pi(\bar{A})) = 1$, and then $N(A) > 0$ entails $\Pi(A) = 1$, i.e. A should be completely consistent with what is known before being somewhat certain.

Note that, in agreement with the idea that only the ordering among the possibility degrees (and then among the possible values) is meaningful, the set functions Π and N only use the qualitative operations 'sup' and 'inf' (or, respectively, 'max' and 'min' when U is finite) and the order-reversing operation $1 - (\cdot)$.

Apart from Π and N, two other quantities can be defined with the same operations, namely (Dubois and Prade, 1992a): a measure of 'guaranteed possibility'

$$\Delta(A) = \inf_{u \in A} \pi_X(u) \qquad (1.18)$$

which estimates to what extent *all* the values in A are actually possible for X according to what is known, i.e. any value in A is at least possible for X at the degree $\Delta(A)$. Clearly Δ is a stronger measure than Π, i.e. $\Delta \leq \Pi$, since Π only estimates the existence of at least one value in A compatible with the available knowledge, whereas the evaluation provided by Δ concerns *all* the values in A. Note also that Δ and N are unrelated, since $N(A)$ estimates the certainty that the value of X is in A by checking the impossibility of all the values out of A, whereas $\Delta(A) > 0$ considers the possibility of all the values in A. The second quantity is a dual measure of 'potential certainty'

$$V(A) = 1 - \Delta(\bar{A}) \qquad (1.19)$$

$$= \sup_{u \notin A} 1 - \pi_X(u) \qquad (1.20)$$

which estimates to what extent there exists at least one value in the complement of A which has a low degree of possibility; this is a necessary condition for having '$X \in A$' somewhat certain (but in general far from being sufficient, except if $\bar{A}$ has only one element). Obviously we have $N \leq V$. The duality relation (1.19) expresses that the potential certainty of A corresponds to the absence of guaranteed possibility for $\bar{A}$. If we further have the information that there exists at least one value in U which is impossible for X (in other words, there is not a complete uncertainty about the value of X), i.e. $\exists u \in U, \pi_X(u) = 0$, or equivalently $1 - \pi_X(u) = 1$, we have the constraint $\min(\Delta(A), \Delta(\bar{A})) = 0$ and then $\Delta(A) > 0$ entails $V(A) = 1$. At the technical level, it is always possible to add an element to U if necessary in order to have $1 - \pi_X$ normalized. Δ and V are monotonically decreasing set functions (in the wide sense) with respect to set inclusion; this contrasts with Π and N, which are monotonically increasing.

Note that the four quantities $\Pi(A)$, $N(A)$, $\Delta(A)$ and $V(A)$ are only weakly related, since they are only constrained by

$$\max(\Pi(A), 1 - N(A)) = \sup_{u \in U} \pi_X(u) \quad (= 1 \text{ if } \pi_X \text{ is normalized})$$
$$\min(\Delta(A), 1 - V(A)) = \inf_{u \in U} \pi_X(u) \quad (= 0 \text{ if } 1 - \pi_X \text{ is normalized})$$

together with the duality relations (1.16) and (1.19). As can be seen, the four quantities $\Pi(A)$, $\Delta(A)$, $N(A)$ and $V(A)$ summarize all the information which can be extracted from π_X by taking the supremum or the infimum of π_X over A or over its complement.

A possibility distribution is not always specified as such, but often through the qualification of subsets of U. Let A be an ordinary subset of U. Two kinds of specification exist, as follows.

We know that it is (completely) certain that the value of X lies in A (for short, 'A is certain for X'). It means that any value outside A is (completely) impossible, i.e. $\forall u \notin A$, $\pi_X(u) = 0$ and π_X is unspecified over A, or if we prefer

$$\text{'}A \text{ is certain for } X\text{' is translated by } \forall u \in U, \pi_X(u) \leq \mu_A(u) \quad (1.21)$$

where μ_A is the $\{0, 1\}$-valued characteristic function of A. The second specification is that A is a (completely) possible range for X (for short 'A is possible for X'). This means that $\forall u \in A, \pi_X(u) = 1$ and π_X remains unspecified outside of A, or equivalently that

$$\text{'}A \text{ is possible for } X\text{' is translated by } \forall u \in U, \mu_A(u) \leq \pi_X(u) \quad (1.22)$$

In the first kind of specification we express our certainty, delimiting a subset A which certainly contains the value of X; the larger is A, the more imprecise we remain about the value of X; the smaller is A, the better the information. On the contrary, in the second kind of specification, we rather express our uncertainty, stating a range A of values which are indeed possible for X without rejecting any value outside of A; in some sense we are recognizing the extent of our ignorance with respect to the precise value of X. Then, the smaller is A, the less informative is the piece of knowledge; the larger is A, the more we know for sure the imprecision of our state of knowledge.

Let us now consider the cases of qualification where the possibility or the certainty is not complete but corresponds to an intermediary level α in the scale $[0, 1]$. This leads to the two following generalizations of (1.21) and (1.22).

The statement 'it is certain at least at the degree α that the value of X is in A', will be interpreted as 'any value outside A is at most possible at the complementary degree, namely $1 - \alpha$', i.e. $\forall u \notin A, \pi_X(u) \leq 1 - \alpha$, which leads to

$$\text{'}A \text{ is } \alpha\text{-certain for } X\text{' is translated by } \forall u \in U, \pi_X(u) \leq \max(\mu_A(u), 1 - \alpha). \quad (1.23)$$

It can be checked that this is equivalent to $N(A) \geq \alpha$. Note that for $\alpha = 1$, (1.21) is recovered. When α decreases from 1 to 0, our knowledge evolves from complete certainty in A to acknowledged ignorance about X.

The statement 'A is a possible range for X at least at the degree α' will be understood as $\forall u \in A, \pi_X(u) \geq \alpha$, which leads to

$$\text{'}A \text{ is } \alpha\text{-possible for } X\text{' is translated by } \forall u \in U, \min(\mu_A(u), \alpha) \leq \pi_X(u). \quad (1.24)$$

It can be checked that this is equivalent to $\Delta(A) \geq \alpha$. Note that for $\alpha = 1$, (1.22) is recovered. When α decreases from 1 to 0, our knowledge evolves from the certainty that A is the minimal range of our ignorance to a total lack of information whatsoever. This kind of possibility-qualification goes back to Zadeh (1978b) and Sanchez (1978).

The above repressentations (1.23) and (1.24) can be generalized to the case where A is a fuzzy set, namely we have Dubois and Prade (1992b)

$$\forall u \in U, \pi_X(u) \leq \max(\mu_A(u), 1 - \alpha) \quad (\alpha\text{-certainty qualification})$$
$$\forall u \in U, \pi_X(u) \geq \min(\mu_A(u), \alpha) \quad (\alpha\text{-possibility qualification})$$

This is still equivalent to $N(A) \geq \alpha$ and $\Delta(A) \geq \alpha$, provided that we use the following appropriate extensions of N and Δ when A is fuzzy

$$N(A) = \inf_{u \in U}(1 - \mu_A(u)) \to (1 - \pi_X(u)) \tag{1.25}$$

$$\Delta(A) = \inf_{u \in U} \mu_A(u) \to \pi_X(u) \tag{1.26}$$

where $\to$ is the so-called Gödel implication

$$r = s \to t = \begin{cases} 1, & \text{if } x \leq t \\ t, & s > t \end{cases}$$

Indeed we have the equivalences $\max(s, 1-r) \geq t \Leftrightarrow r \leq (1-s) \to (1-t)$ and $\min(s, r) \leq t \Leftrightarrow r \leq s \to t$, which are the basis of fuzzy relational equation solving (Di Nola et al., 1989). When A is a crisp subset (1.25) and (1.26) reduce to (1.17) and (1.18), respectively.

1.3.2 Uncertain Fuzzy Rules

Certainty- and possibility-qualification provide a tool to distinguish between different intended meanings that a fuzzy rule of the form 'if X is A then Y is B' may convey. Indeed such a rule can be interpreted in various ways according to how the if part qualifies the then-part. Since the satisfaction of the condition part by a precise state of fact $X = x_0$ is a matter of degree when A is fuzzy, this degree may act as a lower bound of the certainty or the possibility of the conclusion part of the rule. Thus, the expressions of the α-certainty and the α-possibility qualification give birth to two kinds of fuzzy rules when the coefficient α becomes dependent on the evaluation of some condition. First we consider rules which relate a variable X ranging on U to a variable Y ranging on V, and which express that 'the more X is A, the more certainly Y lies in B'. It may mean something like '$\forall u$, if $X = u$, it is at least $\mu_A(u)$-certain that B is a range for Y'. This can be represented by the following constraint on the conditional possibility distribution $\pi_{Y|X}$ relating the possible vaues of X and Y, due to (1.23):

$$\forall u \in U, \forall v \in V, \pi_{Y|X}(v, u) \leq \max(\mu_B(v), 1 - \mu_A(u)) \tag{1.27}$$

Such a rule will be called a certainty rule. A second kind of fuzzy rule called the possibility rule expresses that 'the more X is A, the more possible is B as a range for Y' and which can be understood as '$\forall u$, if $X = u$, it is at least $\mu_A(u)$-possible that Y lies in B'. This corresponds to the following constraint on the conditional possibility distribution $\pi_{Y|X}$ representing the rule

$$\forall u \in U, \forall v \in V, \min(\mu_A(u), \mu_B(v)) \leq \pi_{Y|X}(v, u) \tag{1.28}$$

due to (1.24). This model of fuzzy rule is close to Mamdani's original proposal in fuzzy logic control (Mamdani, 1977). Note that in the particular case where A is an ordinary subset and where we know that, if X is in A, B is both a possible and a certain range for Y, (1.27) and (1.28) yield

$$\begin{cases} \forall u \in A, \pi_{Y|X}(v,u) = \mu_B(v) \\ \forall u \notin A, \pi_{Y|X}(v,u) \text{ is completely unspecified.} \end{cases} \quad (1.29)$$

This corresponds to the usual modelling of a fuzzy rule with a non-fuzzy condition part.

Possibility rules also read 'the more X is A, the more it is possible that Y takes values v which have a high degree of membership in B'. Reversing the roles of the expressions 'high degree of membership' and 'possible values' in this rule leads to the new rule 'the more X is A, the higher the degree of membership in B of the possible values of Y', or if we prefer, more briefly 'the more X is A, the more Y is B'. The constraint which represents this new kind of rule, called the 'gradual rule', is thus obtained in exchanging μ_B and $\pi_{Y|X}$ in (1.28). This yields the inequality

$$\forall u \in U, \forall v \in V, \min(\mu_A(u), \pi_{Y|X}(v,u)) \leq \mu_B(v) \quad (1.30)$$

When $\pi_{Y|X}(v,u)$ takes only the values 0 or 1, we recognize the particular case of gradual rules already encountered in Section 1.2.2. More precisely, the intended meaning of a gradual rule can be easily read in (1.30): 'the greater the degree of membership of the value of X to the fuzzy set A and the more the value of Y is possibly related to X, the greater should be the degree of membership to B for this value of Y'. Using again the equivalence $\min(a,t) \leq b \Leftrightarrow t \leq a \rightarrow b$, (1.30) can be rewritten in the form

$$\forall u \in U, \forall v \in V, \pi_{Y|X}(v,u) \leq \mu_A(u) \rightarrow \mu_B(v) = \begin{cases} 1, & \text{if } \mu_A(u) \leq \mu_B(v) \\ \mu_B(v), & \text{if } \mu_A(u) > \mu_B(v). \end{cases} \quad (1.31)$$

Fuzzy rules based on (1.30) (with possible combinations different from the minimum) have been considered in the past by Trillas and Valverde (1985), Bouchon (1987) and the authors.

Exchanging similarly μ_B and $\pi_{Y|X}$ in (1.27), we obtain another kind of gradual rule represented by

$$\forall u \in U, \forall v \in V, \max(\pi_{Y|X}(v,u), 1 - \mu_A(u)) \geq \mu_B(v) \quad (1.32)$$

$$\Leftrightarrow 1 - \mu_B(v) \geq \min(\mu_A(u), 1 - \pi_{Y|X}(v,u))$$

$$\Leftrightarrow \pi_{Y|X}(v,u) \geq \begin{cases} 0, & \text{if } \mu_A(u) + \mu_B(v) \leq 1 \\ \mu_B(v), & \text{if } \mu_A(u) + \mu_B(v) > 1 \end{cases}$$

using the equivalence $\min(a, 1-t) \leq 1 - b \Leftrightarrow t \geq 1 - (a \rightarrow (1-b)) = a \wedge b$, where $\wedge$ denotes a non-commutative conjunction. As can be seen above in the second expression of (1.32), this gradual rule expresses that 'the more X is A and the less Y is related to X, the less Y is B'.

The particular case of (1.30) where $\pi_{Y|X}$ reduces to a function f such that $\pi_{Y|X}(v,u) = 1$ if $v = f(u)$ and $\pi_{Y|X}(v,u) = 0$ otherwise, i.e. $\forall u \in U$, $\mu_A(u) \leq \mu_B(f(u))$, has been considered for a long time in the fuzzy set literature under the name of 'fuzzy function' (Negoita and Ralescu, 1975). More generally, viewing $\pi_{Y|X}$ as encoding a fuzzy relation R, i.e. $\pi_{Y|X}(v,u) = \mu_R(u,v)$, and noting that (1.30) is equivalent to $\forall v \in V, \sup_u \min(\mu_A(u), \mu_R(u,v)) \leq$

$\mu_B(v)$, we recognize that this expresses the fuzzy set inclusion $A \circ R \subseteq B$, where $A \circ R$ is the image of the fuzzy set A by the fuzzy relation R. When A, B and R are non-fuzzy, we have the equivalence $A \circ R \subseteq B \Leftrightarrow R \subseteq \overline{A} + B$ ($+$ denotes the Cartesian co-product $A + B = \overline{\overline{A} \times \overline{B}}$); this equivalence no longer holds in the fuzzy case and its right part is the counterpart of (1.27). Similarly, in the non-fuzzy case we have the equivalence $A \circ \overline{R} \subseteq \overline{B} \Leftrightarrow R \supseteq A \times B$ which no longer holds in the fuzzy case; the left and right parts respectively correspond to (1.32) and (1.28), and clearly (1.32) expresses that any element of U which is not in relation R with an element of A should be outside B.

These four types of rule correspond to modifications of μ_B by a function $\mu_{\tau(u)}$ from $[0, 1]$ to $[0, 1]$ which may interpreted as a fuzzy truth-value $\tau(u)$ (Zadeh, 1978b), namely

for (1.27), (1.30) and (1.31): $\pi_{Y|X}(v, u) \leq \mu_{\tau(u)}(\mu_B(v))$

for (1.28) and (1.32): $\pi_{Y|X}(v, u) \geq \mu_{\tau(u)}(\mu_B(v))$

where $\mu_{\tau(u)}$ in each case depends on $\mu_A(u)$. This possibility of interpreting fuzzy inference by means of fuzzy truth-values acting as modifiers has been noted for a long time (Tsukamoto, 1979, Baldwin, 1979). The corresponding modifier functions $\mu_{\tau(u)}$ are given for the four types of rules in Figure 1.3 with $\alpha = \mu_A(u)$.

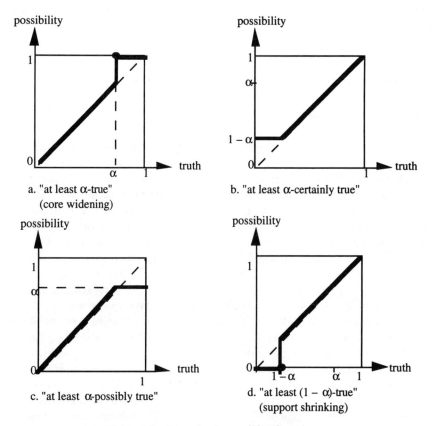

a. "at least α-true" (core widening)

b. "at least α-certainly true"

c. "at least α-possibly true"

d. "at least $(1 - \alpha)$-true" (support shrinking)

Figure 1.3 Four basic modifier functions

As is shown in Figure 1.3a, a gradual rule of the type of (1.30) and (1.31) can be interpreted in the following way: $\forall u \in U$, if $X = u$ then Y is B is at least $\mu_A(u)$-true. It is not a crisp 'at least α-true' (which would correspond to the ordinary subset $[\alpha, 1]$), but a fuzzy one in agreement with the fact that $\mu_t(t) = t, \forall t \in [0,1]$ is the identity modifier function accounting for the truth value 'true'. To summarize, in the cases of:

possibility rules, the possibility distribution $\pi_{Y|X}(\cdot, u)$ is bounded from below by $\varphi(\mu_B)$ with $\varphi(t) = \min(\mu_A(u), t)$, i.e. B is truncated up to the height $\alpha = \mu_A(u)$;

certainty rules, the possibility distribution $\pi_{Y|X}(\cdot, u)$ is bounded from above by $\varphi(\mu_B)$ with $\varphi(t) = \max(t, 1 - \mu_A(u))$, i.e. B is drowned in a level of indetermination $1 - \alpha$;

gradual rules (core widening), the possibility distribution $\pi_{Y|X}(\cdot, u)$ is bounded from above by $\varphi(\mu_B)$ with $\varphi(t) = \mu_A(u) \to t$ (where $\to$ denotes Gödel's implication), i.e. the core of B is enlarged;

gradual rule (support shrinking), the possibility distribution $\pi_{Y|X}(\cdot, u)$ is bounded from below by $\varphi(\mu_B)$ with $\varphi(t) = \mu_A(u) \wedge t = 0$ if $\mu_A(u) + t \leq 1$ and $\varphi(t) = t$ otherwise, i.e. the support of B is reduced, truncated.

Statements of rules involving 'the less' in place of 'the more' are easily obtained by changing A or B into their complements $\bar{A}$ and $\bar{B}$, due to the equivalence between 'the more X is A' and 'the less X is $\bar{A}$' (with $\mu_{\bar{A}} = 1 - \mu_A$).

1.3.3 Approximate Reasoning with Fuzzy Rules

1.3.3.1 Information principles

In the preceding section we have seen how the representation of fuzzy rules and more generally certainty and possibility qualifications lead to inequality constraints on possibility distributions. In the case of certainty qualification, the inequality is of the type $\pi_X \leq \mu_A$, where μ_A may be any $[0, 1]$-valued (normalized) membership function; it includes the case where μ_A is of the form $\mu_A = \max(\mu_F, 1 - \alpha)$, i.e. '$F$ is α-certain for X' is equivalent to 'A is certain for X' (where A is necessarily a fuzzy set). The principle of minimal specificity stipulates that each value u in the domain U of a variable X (in the general case X is a vector and U a Cartesian product) should receive the largest degree of possibility which is in agreement with the constraint(s); i.e. it leads to take $\forall u \in U, \pi_X(u) = \mu_A(u)$ in the above case. Indeed choosing a particular π such that $\pi < \mu_A$ in order to represent our knowledge about X would be arbitrarily too precise. The principle of minimal specificity has a role in possibility theory similar to the use of the maximal entropy principle in probability theory. The notion of specificity of a possibility distribution has been introduced by Yager (1983). The absence of constraint corresponds to the situation of complete ignorance modelled by $\forall u \in U, \pi_X(u) = 1$. On the contrary, complete knowledge corresponds to a maximal constraint, such that $\exists u_0, A = \{u_0\}$, i.e. $\pi_X(u_0) = 1$ and $\pi_X(u) = 0, \forall u \neq u_0$. In that sense any possibility distribution π_X is provisional in nature and likely to be improved by further information, when the available one is not complete. When several pieces of information constraining the possible values of X become available, i.e. $\pi_X \leq \mu_{A_i}, i = 1, n$, the principle of minimum specificity leads us to assume

$$\pi_X = \min_{i=1,\ldots,n} \mu_{A_i} \tag{1.33}$$

as a straightforward consequence of the inequalities. Note that the complete reliability of the sources entails that $\sup_u \min_{i=1,\ldots,n} \mu_{A_i}(u) = 1$. Otherwise, the sources conflict as soon as $\pi_X(u) < 1$, $\forall u$. Hence subnormalization corresponds to partial inconsistency. Clearly if we have two constraints $\pi_X \leq \mu_{A_1}$ and $\pi_X \leq \mu_{A_2}$, such that $\mu_{A_1} \leq \mu_{A_2}$, (i.e. we have the fuzzy set inclusion $A_1 \subseteq A_2$) the second piece of information is redundant and A_1 is said to be more specific than A_2.

Obviously a different information principle needs to be applied in case of possibility qualification where we deal with inequalities of the type $\pi_X \geq \mu_A$. Here μ_A does not need to be normalized, indeed μ_A may be of the form $\mu_A = \min(\alpha, \mu_F)$, but as already said it is expected that $\exists u \in U, \mu_A(u) = 0$. A principle of maximal specificity can be advocated in this case in order to limit the scope of the possible values for X to the values which are indeed known as (somewhat) possible. While the principle of minimal specificity embodies the commonsense claim that anything that is not impossible should not be ruled out, the converse principle expresses that anything which is not established as being possible can be neglected. The role of information principles in possibility and other non-standard theories of uncertainty is particularly stressed in the book by Klir and Folger (1988).

1.3.3.2 Combination and projection

Let us now assume that there are two variables X and Y ranging on U and V, respectively, with possibility distributions π_X and π_Y, respectively. The principle of minimal specificity leads us to define the joint possibility distribution $\pi_{X,Y}$ as the combination

$$\pi_{X,Y} = \min(\pi_X, \pi_Y) \tag{1.34}$$

as a consequence of the inequalities $\pi_{X,Y}(u,v) \leq \pi_X(u)$, $\forall v$ and $\pi_{X,Y}(u,v) \leq \pi_Y(v)$, $\forall u$, which express that the possibility that $X = u$ and $Y = v$ is necessarily upper-bounded by the possibility that $X = u$ (nothing being said about Y) on the one hand and the possibility that $Y = v$ on the other hand. Note that (1.34) is in agreement with (1.33). Indeed the possibility distribution π_X on U (respectively, π_Y on V) can be extended to $U \times V$ into a possibility distribution $\pi^1_{X,Y}$ (respectively, $\pi^2_{X,Y}$) by stating that $\forall u, \forall v, \pi^1_{X,Y}(u,v) = \pi_X(u)$ (respectively, $\pi^2_{X,Y}(u,v) = \pi_Y(v)$), since nothing is said about Y (respectively, X); this is called cylindrical extension. Then $\pi_{X,Y} = \min(\pi^1_{X,Y}, \pi^2_{X,Y})$ from (1.33). As can be seen it is equivalent to first apply the minimal specificity principle for each possibility distribution representing a piece of information and then to combine by (1.34), or to first combine the constraints corresponding to certainty-qualified statements and then to apply the minimal specificity principle to the result.

Conversely, given a possibility distribution $\pi_{X,Y}$ restricting the possible values of a pair of variables (X, Y) the induced restriction on the possible values of X (respectively, Y) can be obtained by the projection of $\pi_{X,Y}$ on U (respectively, V), i.e.

$$\pi_X(u) = \sup_{v \in V} \pi_{X,Y}(u,v); \quad \pi_Y(v) = \sup_{u \in U} \pi_{X,Y}(u,v). \tag{1.35}$$

Again, this is in agreement with the principle of minimal specificity since we compute the largest possibility degrees under the constraint $X = u$, $\forall u \in U$ (respectively, $Y = v$, $\forall v \in V$). The combination can be straightforwardly extended to n-tuples of variables and the projection can be performed on any subset of such tuples. Combination and projection as defined above are the basis of the theory of approximate reasoning introduced by Zadeh (1979a). The combination and projection operations in possibility theory play roles which are, respectively, similar to the construction of a joint distribution (under a stochastic independence assumption) and to the computation of a marginal distribution in probability theory. The simultaneous satisfaction of (1.34) and (1.35) supposes that X and Y are logically independent (i.e. that the degree of possibility that $X = u$ does not depend on the value of Y, $\forall u$, and similarly exchanging X and Y).

Note also that viewing fuzzy relations as possibility distributions, the composition of fuzzy relations (see Section 1.2.1) can be interpreted as a combination followed by a projection in the above sense.

1.3.3.3 Generalized modus ponens

An important particular case of the application of the combination/projection principle is the generalized *modus ponens* where a factual piece of information X is A' is combined with a fuzzy rule relating X and Y, in order to obtain a conclusion of the form Y is B', after projection. We only consider the cases of certainty and possibility rules in the following. Inference with gradual rules has been already presented in Section 1.2.2 in the particular case where $A' = \{x_0\}$ corresponds to a precise input. See Dubois, Martin-Clouaire and Prade (1988) for computational issues in the cases of a fuzzy input A' and of rules with compound conditions.

Certainty rules   From the two pieces of information 'X is A'' and 'the more X is A, the more certain Y is B', i.e. $\pi_X = \mu_{A'}$, $\pi_{Y|X}(v, u) = \max(\mu_B(v), 1 - \mu_A(u))$, by applying (1.27) we obtain by combination and projection

$$\pi_Y(v) = \sup_u \min(\mu_{A'}(u), \max(\mu_B(v), 1 - \mu_A(u)))$$
$$= \max(\mu_B(v), 1 - N_{A'}(A)) \qquad (1.36)$$

where $N_{A'}(A)$ is a necessity measure of the fuzzy event A, defined from the possibility distribution $\pi = \mu_{A'}$ (not to be confused with (1.25)), such that $N_{A'}(A) = 1$ if and only if the support of A' is included in the core of A, i.e. $\{u, \mu_{A'}(u) > 0\} \subseteq \{u, \mu_A(u) = 1\}$. Hence (1.36) expresses that the conclusion 'Y is B' is all the more certain as the fuzzy set A' of possible values of X corresponds to elements highly compatible with A. The use of certainty rules thus lead to the propagation of uncertainty coefficients.

Possibility rules   Let us consider the generic situation of two objects described in terms of attribute variables X and Y. The indices 1 and 2 will refer to the values of these attributes for each of these objects, respectively. Let us suppose we have the following information: $\pi_{X_1} = \mu_{A_1}$, $\pi_{Y_1} = \mu_{B_1}$, $\pi_{X_2} = \mu_{A_2}$ and the rule 'the more similar the values of X_1 and X_2, the more possible the approximate equality of the values of Y_1 and Y_2' (e.g. the more similar two second-hand cars, the more possibly do they have approximately equal prices). Let S and T be the fuzzy relations 'similar' and 'approximately equal' in the above rule. Then by combination/projection we obtain for Y_2, by applying (1.28)

$$\begin{aligned}\pi_{Y_2}(v_2) &= \sup_{u_2,u_2,v_1} \min(\mu_{A_1}(u_1), \mu_{A_2}(u_2), \mu_{B_1}(v_1), \mu_S(u_1,u_2), \mu_T(v_1,v_2)) \\ &= \min(\mu_{B \circ T}(v_1), \sup_{u_1,u_2} \min(\mu_S(u_1,u_2), \mu_{A_1}(u_1), \mu_{A_2}(u_2))) \\ &= \min(\mu_{B \circ T}(v_2), \Pi_{A_1 \times A_2}(S))\end{aligned} \quad (1.37)$$

where $\Pi_{A_1 \times A_2}(S)$ is the possibility measure of the fuzzy event 'X_1 and X_2 are similar', given that 'X_1 is A_1' and 'X_2 is A_2' and (1.34) is used to combine μ_{A_1} and μ_{A_2}. The conclusion obtained by (1.37) means that the values in B or approximately equal to a value in B are all the more possible for Y_2 as X_1 and X_2 are indeed possibly similar. This kind of weak conclusion is in the spirit of analogical reasoning which can lead only to tentative conclusions.

It is worth noting that the effect of parallel certainty rules leads to an increase in the precision of the response, whereas in the case of possibility rules, the response becomes wider. This is in conformity with the semantics of the rules. Namely in the case of certainty rules, given that X is totally compatible with both A_1 and A_2, we conclude that both B_1 and B_2 are certain, i.e. $X \in B_1 \cap B_2$. However, with possibility rules if X is totally compatible with both A_1 and A_2, then both B_1 and B_2 are possible responses, i.e $B_1 \cup B_2$ is a possible range for X. This can be easily checked by combining two parallel rules (either two certainty rules or two possibility rules) in the presence of a precise input.

1.3.4 Possibilistic Logic

In the preceding sections we have dealt with fuzzy predicates and with uncertainty measures together. In the remaining sections we focus on the use of possibility and/or necessity measures for the handling of uncertainty pertaining to classical, i.e. non-fuzzy, formulae $p, q, r...$ (which can be only true or false).

Several approaches have been proposed to deal with uncertainty and/or vagueness in automated reasoning; see Dubois and Prade (1991c, Part 2) for an overview. However, a large number of them are based on fuzzy logic, which completely departs from the possibilistic logic we present in this section. Fuzzy logic deals with propositions involving vague predicates (or properties whose satisfaction can be a matter of degree) and manipulates truth degrees which are truth-functional with respect to each connective, whereas possibilistic logic involves certainty and possibility degrees which are not compositional for all connectives and which are attached to classical formulae, i.e. containing only non-vague propositions or predicates (in the simplest case). The lack of complete certainty about the truth of a considered formula is to be understood as a consequence of a lack of complete information.

A possibilistic logic formula is a first-order logic formula with a numerical weight between 0 and 1 which is a lower bound on a possibility measure Π or on a necessity measure N. Thus this lower bound should obey the characteristic axioms governing these measures, i.e. $\forall p, \forall q, N(p \wedge q) = \min(N(p), N(q))$ and $\Pi(p \vee q) = \max(\Pi(p), \Pi(q))$, respectively, for necessity and possibility measures, with the duality relation $N(p) = 1 - \Pi(\neg p)$. We have the usual limit conditions $\Pi(\bot) = N(\bot) = 0$, $\Pi(T) = N(T) = 1$, where $\bot$ and T stand for the contradiction and the tautology, respectively. However, we only have $N(p \vee q) \geq \max(N(p), N(q))$ and $\Pi(p \wedge q) \leq \min(\Pi(p), \Pi(q))$ since for $q = \neg p$ we obtain $N(p \vee q) = 1$ and $\Pi(p \wedge q) = 0$, whereas $\max(N(p), N(\neg p))$ is less than 1 when the situation regarding the truth or falsity of p is uncertain, and $\min(\Pi(p), \Pi(\neg p))$ is strictly positive if p and $\neg p$ are both somewhat possible (state of partial ignorance). The weight attached to a formula represents to what extent it is possible or it is certain that the formula holds for true given the available information. Semantics have been proposed first when only lower bounds on a necessity measure are used (Dubois, Lang and Prade, 1992) and then extended to the general case where lower bounds of both possibility and necessity are allowed (Lang, Dubois and Prade, 1991). For the sake of brevity let us only indicate the semantics attached to necessity-valued formulae (p, α), where the intended meaning of α is $N(p) \geq \alpha$. Let $M(p)$ be the (classical) set of models of p, i.e. the set of interpretations which make p true. The semantics of (p, α) are represented by the fuzzy set of models $M(p, \alpha)$,

$$\mu_{M(p,\alpha)}(\omega) = 1 \text{ if } \omega \in M(p); \mu_{M(p,\alpha)}(\omega) = 1 - \alpha \text{ if } \omega \notin M(p) \qquad (1.38)$$

where ω denotes an interpretation. In other words, the interpretations compatible with (p, α) are restricted by the above possibility distribution. Those in $M(p)$ are considered as fully possible, whereas those outside are all the more possible as α is smaller, i.e. as the piece of knowledge is less certain. This is clearly in agreement with certainty qualification as discussed in Section 1.3.1.

In the case of several pieces of knowledge (p_i, α_i), $i = 1, n$, forming a knowledge base $\mathcal{K}$, in agreement with the minimal specificity principle, we associate the following possibility distribution, built by combining the membership functions $\mu_{M(p_i, \alpha_i)}$, namely

$$\pi_{\mathcal{K}}(\omega) = \min_{i=1,n} \mu_{M(p_i, \alpha_i)}(\omega) \qquad (1.39)$$

The lack of certainty in p_i, estimated by $1 - \alpha_i$, is committed to the interpretations which are not models of p_i. By performing the conjunction of the $M(p_i, \alpha_i)$, we associate each interpretation ω with a weight equal to $\pi_{\mathcal{K}}(\omega) = \min_{i=1,n} \mu_{M(p_i, \alpha_i)}(\omega)$. Thus the weights attached to formulae in the knowledge base $\mathcal{K} = \{(p_i, \alpha_i)\}$ induce an ordering among the interpretations (according to their level of possibility $\pi(\omega)$). This is very similar to Shoham's preferential model semantics (Shoham, 1988), see Dubois and Prade (1991b) on this point. It can be checked that $\forall i, N(p_i) = 1 - \Pi(\neg p_i) \geq \alpha_i$, with $\Pi(p_i) = \sup\{\pi_{\mathcal{K}}(\omega), \omega \in M(p_i)\}$.

The following deduction rules have been proved sound and complete for refutation with respect to the above semantics:

resolution rule
$$\frac{(\neg p \vee q, \alpha)(p \vee r, \beta)}{(q \vee r, \min(\alpha, \beta))}$$

particularization
$$\frac{(\forall x P(x), \alpha)}{(P(a), \alpha)}$$

(as well as more general substitutions). They are examples of local inferences expressed at the symbolic level.

If we want to compute the certainty degree which can be attached to a formula, we add to $\mathcal{K}$ the clause(s) obtained by refuting the proposition to evaluate with a necessity degree equal to 1. Then it can be shown that any lower bound obtained on the contradiction, by resolution, is a lower bound of the necessity of the proposition to evaluate. It can be proved that what is obtained is in complete agreement with the combination and projection approach taking into account the above semantics (Dubois and Prade, 1991c,Part 2).

The introduced semantics enable us to define the degree of partial inconsistency of a knowledge base $\mathcal{K}$ which is equal, in the case of necessity-weighted formulae, to $\text{Inc}(\mathcal{K}) = 1 - \sup_{\omega} \min_{i=1,n} \mu_{(p_i, \alpha_i)}(\omega)$. Then it can be shown that this degree estimates to what extent the lower bounds in the knowledge base violate the characteristic axiom of necessity measures and to what extent the fuzzy set of models of the knowledge base is empty. When $\text{Inc}(\mathcal{K}) = 0$, $\mathcal{K}$ is consistent and this is equivalent to the classical consistency of the set of formulae in $\mathcal{K}$ without taking into account the weights. It has been also shown that is it possible to reason with such partially inconsistent knowledge bases, still preserving the above-mentioned soundness and completeness results. Roughly speaking, the conclusions which can be obtained with a degree of uncertainty strictly higher than $\text{Inc}(\mathcal{K})$ are still meaningful since for sure only a consistent subpart of $\mathcal{K}$ is used for deducing them. Indeed in any inconsistent sub-base of $\mathcal{K}$ there is (at least) a clause with a weight less or equal to $\text{Inc}(\mathcal{K})$. In the case of partial inconsistency, it has been shown that possibilistic logic behaves in a non-monotonic way (Dubois and Prade, 1991b). More specifically, the possibility distribution $\pi_{\mathcal{K}}$ encodes a preference relation over interpretations, and inference in possibilistic logic corresponds to a preferential entailment notion as laid bare by Shoham (1988). This preferential entailment is different from Ruspini's approach to inference in fuzzy logic (Ruspini, 1991), which rather corresponds to an 'approximate' entailment based on a similarity relation between interpretations, and not to a preference relation.

Possibilistic logic implements a non-monotonic reasoning in the case of partial inconsistency. Indeed, it has been shown (Dubois and Prade, 1991b) that the preferential entailment (in the sense of Shoham (1988) $\vDash_{\pi}$, defined as the inclusion of the preferred models of p (which maximize $\Pi(p)$) in the set of models of q, by

$$p \vDash_{\pi} q \Rightarrow \{\omega | \omega \in M(p), \pi(\omega) = \Pi(p) > 0\} \subseteq M(q) \Leftrightarrow N(q|p) > 0$$

where

$$N(q|p) = 1 - \Pi(\neg q|p) \text{ and } \Pi(q|p) = \begin{cases} 1, & \text{if } \Pi(p) = \Pi(p \wedge q) \\ \Pi(p \wedge q), & \text{if } \Pi(p) > \Pi(p \wedge q) \end{cases}$$

and $\Pi(q|p)$ is the least specific solution of the equation $\Pi(p \wedge q) = \min(\Pi(q|p), \Pi(p))$ and where π is the possibility distribution associated with the semantics of the knowledge base $\mathcal{K}$, underlying Π. The entailment relation $\vDash_{\pi}$ is in complete agreement with non-monotonic consequence relations obeying the axiomatics of system P proposed by Kraus, Lehmann and Magidor (1990). For more details on possibilistic logic and its relation to symbolic approaches to nonmonotonic reasoning, see Benferhat et al. (1992) and Dubois, Lang and Prade (1994a).

As pointed out by Benferhat *et al.* (1993), the weighted clause $(\neg p \vee q, \alpha)$, understood as $N(\neg p \vee q) \geq \alpha$ is semantically equivalent to the weighted clause $(q, \min(\alpha, v(p)))$ where $v(p)$ is the truth value of p, i.e. $v(p) = 1$ if p is true and $v(p) = 0$ if p is false. This remark is very useful for hypothetical reasoning, since by 'transferring' a sub-formula from a clause to the weight part of the formula we are introducing explicit assumptions. Indeed changing $(\neg p \vee q, \alpha)$ into $(q, \min(v(p), \alpha))$ leads to state the piece of knowledge under the form 'q is certain at the degree α, provided that p is true'. More generally, the weight or label can be a function of logical (universally quantified) variables involved in the clause. The weight is no more just a degree but in fact a label which expresses the context in which the piece of knowledge is more or less certain. This is to be related to 'possibilistic assumption-based truth maintenance systems' (with weighted justifications and/or hypotheses, which have been defined and exemplified on a diagnosis problem (Benferhat et al., 1993). The approach contrasts with other uncertainty handling ATMSs in the sense that the symbolic processing and the calculus of uncertainty are no longer separated here.

Moreover the presence of logical variables in the weight also enables the expression of some graduality attached to vague predicates (as in the certainty rule 'the younger the person, the more certain he/she is single', where 'young' is a vague predicate) in a simple way, as $N(\text{single}(x)) \geq \mu_{\text{young}}(\text{age}(x))$ in our example. It would then allow for a flexible interface between the symbolic knowledge base and numerical inputs. Vague predicates can thus handled by introducing their characteristic functions in the weights; see Dubois, Lang and Prade (1994b).

1.3.5 Default Reasoning

The idea is to translate each default 'if p then q, generally' denoted by $p \sim > q$ into a constraint expressing that the situation where p and q is true has a greater plausibility than the one where p and $\neg q$ is true; if we prefer, we could express that in the context where p is true, q is more possible or plausible than $\neg q$. So we need a qualitative relation for comparing plausibility levels. We use a qualitative possibility relation $\geq_\Pi$ defined on a Boolean algebra of propositions (Dubois and Prade, 1991d, Yager, 1993). Such a relation is assumed to be: complete ($\forall p, \forall q, p \geq_\Pi q$ or $q \geq_\Pi p$); transitive; and to satisfy $\top >_\Pi \bot$, where $\top$ and $\bot$ represent tautology and contradiction, respectively, and $>_\Pi$ the strict part of the ordering $\geq_\Pi$; $\top \geq_\Pi p \geq_\Pi \bot, \forall p$; and the characteristic axiom $p \geq_\Pi q \Rightarrow r \vee p \geq_\Pi r \vee q$, $\forall p, \forall q, \forall r$. In the finite case, the only numerical counterparts to possibility relations are possibility measures, such that $\forall p, \forall q, \Pi(p \vee q) = \max(\Pi(p), \Pi(q))$; see (Dubois, 1986).

A default $p \sim > q$ is then understood formally as the constraint

$$\Pi(p \wedge q) > \Pi(p \wedge \neg q)$$

on a possibility measure Π describing the semantics of the available knowledge. This constraint can be shown to be equivalent to $N(q|p) > 0$.

Let us consider the following (usual) set of default rules d_1: 'birds fly', d_2: 'penguins do not fly', d_3: 'penguins are birds', symbolically written

$$d_1: b \sim > f$$
$$d_2: p \sim > \neg f$$
$$d_3: p \sim > b$$

The set of three defaults will be represented by the following set C of constraints:

$$b \wedge f >_\Pi b \wedge \neg f,$$
$$p \wedge \neg f >_\Pi p \wedge f,$$
$$p \wedge b >_\Pi p \wedge \neg b.$$

Let Ω be the finite set of interpretations of our propositional language. If this language is made of the literals $p_1, \ldots, p_n$, these interpretations correspond to the possible worlds where the conjunctions $*p_1 \wedge \cdots \wedge *p_n$ are true, where $*$ stands for the presence of the negation sign $\neg$ or its absence. In our example, $\Omega = \{\omega_0: \neg b \wedge \neg f \wedge \neg p, \omega_1: \neg b \wedge \neg f \wedge p, \omega_2: \neg b \wedge f \wedge \neg p, \omega_3: \neg b \wedge f \wedge p, \omega_4: b \wedge \neg f \wedge \neg p, \omega_5: b \wedge \neg f \wedge p, \omega_6: b \wedge f \wedge \neg p, \omega_7: b \wedge f \wedge p\}$. Then the set of constraints C' on models is

$$C'_1: \max(\pi(\omega_6), \pi(\omega_7)) > \max(\pi(\omega_4), \pi(\omega_5)),$$
$$C'_2: \max(\pi(\omega_5), \pi(\omega_1)) > \max(\pi(\omega_3), \pi(\omega_7)),$$
$$C'_3: \max(\pi(\omega_5), \pi(\omega_7)) > \max(\pi(\omega_1), \pi(\omega_3)).$$

Let $>_\pi$ be a ranking of Ω, such that $\omega >_\pi \omega'$ iff $\pi(\omega) > \pi(\omega')$ on Ω. Any finite consistent set of constraints: $p_i \wedge q_i >_\Pi p_i \wedge \neg q_i$ induces a partially defined ranking $>_\pi$ on Ω, that can be completed according to the principle of minimum specificity. The idea is to try to assign to each world ω the highest possibility level (in forming a well-ordered partition of Ω) without violating the constraints. The ordered partition of Ω associated with $>_\pi$ using the minimum specificity principle can easily be obtained by the following procedure (Benferhat, Dubois and Prade, 1992).

(a) $i = 0$

(b) While Ω is not empty repeat (b1)–(b4):

  (b1) $i \leftarrow i + 1$
  (b2) put in E_i every model which does not appear in the right side of any constraints of C'
  (b3) remove the elements of E_i from Ω
  (b4) remove from C' any constraint containing elements of E_i.

Let us apply now this algorithm. The models which do not appear on the right side of any constraint of C' are $\{\omega_0, \omega_2, \omega_6\}$, we call this the set E_1. We remove the elements of E_1 from Ω and we remove the constraint C'_1 from C' (since assigning to ω_6 the highest possibility level makes the constraint C'_1 always satisfied). We again start the procedure and we find successively the two following sets $\{\omega_4, \omega_5\}$ and $\{\omega_1, \omega_3, \omega_7\}$. Finally, the well-ordered partition of Ω is

$$\{\omega_0, \omega_2, \omega_6\} >_\pi \{\omega_4, \omega_5\} >_\pi \{\omega_1, \omega_3, \omega_7\}$$

Let $E_1, \ldots, E_m$ be the obtained partition. A numerical counterpart to $>_\pi$ can be defined by

$$\pi(\omega) = \frac{m + 1 - i}{m} \quad \text{if } \omega \in E_i, \quad i = 1, m$$

In our example we have $m = 3$ and $\pi(\omega_0) = \pi(\omega_2) = \pi(\omega_6) = 1; \pi(\omega_4) = \pi(\omega_5) = 2/3; \pi(\omega_1) = \pi(\omega_3) = \pi(\omega_7) = 1/3$. Note that this is purely a matter of convenience to use a numerical scale, and any other numerical counterpart such that $\pi(\omega) > \pi(\omega')$ iff $\omega >_\pi \omega'$ will work as well. Namely π is used as an ordinal scale. From this possibility distribution, we can compute for any proposition p its necessity degree $N(p)$. For instance, $N(\neg p \vee \neg f) = \min\{1 - \pi(\omega) | \omega \models p \wedge f\} = \min(1 - \pi(\omega_3), \ 1 - \pi(\omega_7)) = 2/3$, whereas $N(\neg b \vee f) = \min\{1 - \pi(\omega) | \omega \models b \wedge \neg f\} = \min(1 - \pi(\omega_4), \ 1 - \pi(\omega_5)) = 1/3$ and $N(\neg p \vee b) = \min\{1 - \pi(\omega_1), 1 - \pi(\omega_3)\} = 2/3$.

The method then consists in turning each default $p_i \to q_i$ into a possibilistic clause $(\neg p_i \vee q_i, N(\neg p_i \vee q_i))$ where N is computed from the possibility distribution π induced by the set of constraints corresponding to the default knowledge base. Then we apply the possibilistic inference machinery for reasoning with the defaults together with the available factual knowledge.

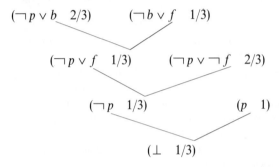

The possibilistic knowledge base equivalent to our set of defaults is $\Sigma = \{(\neg b \vee f \ 1/3), (\neg p \vee b \ 2/3), (\neg p \vee \neg f \ 2/3)\}$. Then the following derivation, knowing with the certainty that Tweety is a penguin (p 1) and a bird (b 1), gives the optimal degree of inconsistency of $\Sigma \cup \{(p \ 1), (b \ 1)\}$ which is equal to 1/3 by applying the resolution rule of possibilistic logic repeatedly:

By refutation, the following derivation shows that $\neg f$ is truly a logical consequence of $\Sigma \cup \{(p \ 1), (b \ 1)\}$, i.e. Tweety does not fly, since by adding the piece of information (f 1) we find the degree of inconsistency equal to 2/3 which is higher than 1/3:

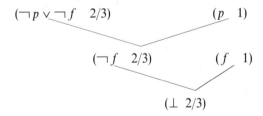

This simple example suggests how possibility theory can cope with default reasoning in a simple and rather elegant way.

1.3.6 Abductive Reasoning

This section points out the interest of possibility and necessity measures to describe plausible links between disorders and manifestations in a diagnosis problem. The presentation is made in terms of sets of manifestations and of relations between disorders and manifestations, rather than in logical terms, for simplicity.

Let $\mathscr{S}$ be a system whose current state is described by means of a n-tuple of binary attributes $(a_1, \ldots, a_i, \ldots, a_n)$. When $a_i = 1$ we shall say that the manifestation m_i is present; when $a_i = 0$, it means that m_i is absent. When there is no manifestation present, $\mathscr{S}$ is said to be in its normal state and its state is described by the n-tuple $(0, \ldots, 0, \ldots, 0)$. Let $\mathscr{M}$ denote the set of the n possible manifestations $\{m_1, \ldots, m_i, \ldots, m_n\}$. Let $\mathscr{D}$ be a set of possible disorders $\{d_1, \ldots, d_j, \ldots, d_k\}$. A disorder can be present or absent. We assume here that its presence is not a matter of degree. To each d_j we associate the set $M(d_j)$ of manifestations which are entailed, or if we prefer caused, produced, by the presence of d_j alone. We first consider the completely informed case where all the present manifestations are observed and where the set of manifestations which appear when a disorder is present is perfectly known. Thus if $m_i \notin M(d_j)$ it means that m_i cannot be caused by d_j. We thus define a relation R on $\mathscr{D} \times \mathscr{M}$, defined by $(d_j, m_i) \in R \Leftrightarrow m_i \in M(d_j)$, which associates manifestations and disorders. Such a relational approach has been particularly developed by Peng and Reggia (1990).

Given a set M^+ of present manifestations which are observed, the problem is to find what disorder(s) may have produced the manifestations in M^+. We suppose that the set $M^- = \mathscr{M} - M^+ = \overline{M^+}$ is the set of manifestations which are absent, i.e. all manifestations which are present are observed. Although deductive reasoning enables us to predict the presence of manifestation(s) from the presence of disorder(s), abductive reasoning looks for possible cause(s) of observed effects. In other words, we look for plausible explanations (in terms of disorders) of an observed situation. Clearly, while it is at least theoretically possible to find out all the possible causes which may have led to a given state of the system $\mathscr{S}$, the ordering of the possible solutions according to some levels of plausibility is out of the scope of logical reasoning, strictly speaking. However, one may consider that an explanation of a set of manifestations is more reasonable when it involves a small number of disorders being conjointly present: a single cause is more acceptable than several jointly present ones. This is called the principle of parsimony. For the case of a set of disorders which jointly causes the manifestations the reader is referred to Dubois and Prade (1993a). In the completely informed case described above, we have: $M^+ = \overline{M^-}$, i.e. all the present manifestations are observed, and equivalently all the manifestations which are not observed are indeed absent; and $\forall d$, $M(d) = M(d)^+ = \overline{M(d)^-}$, where $M(d)^+$ (respectively, $M(d)^-$) is the set of manifestations which are certainly present (respectively, certainly absent) when disorder d alone is present. In the general case, we only have $M^+ \cap M^- = \emptyset$ but $M^+ \cup M^- \neq \mathscr{M}$ and similarly $M(d)^+ \cap M(d)^- = \emptyset$, $M(d)^+ \cup M(d)^- \neq \mathscr{M}$, $\forall d$. Let $M^0 = M - M^+ - M^-$.

Let us consider the general case where both the information pertaining to the manifestations and the information relative to the association between disorders and manifestations is incomplete. Then d belongs to the set $\hat{D}$ of potential disorders, each of which can alone explain both M^+ and M^- if and only if d does not produce with certainty any manifestation which is certainly absent in the evidence, and no observed manifestation is ruled out by d. Formally we have

$$\hat{D} = \{d \in \mathscr{D}, M(d)^+ \subseteq \overline{M^-} \text{ and } M(d)^- \subseteq \overline{M^+}\} \qquad (1.40)$$

This also means that

$$\hat{D} = \{d \in \mathcal{D}, M(d)^+ \cap M^- = \emptyset \text{ and } M(d)^- \cap M^+ = \emptyset\} \quad (1.41)$$

$\hat{D}$ gathers all the disorders in $\mathcal{D}$ which cannot be ruled out from the observations. If $M^- = \emptyset$ and $M(d)^- = \emptyset$, i.e. no information is available on the manifestations certainly absent, it can be verified that $\hat{D} = \mathcal{D}$ and the whole set of possible disorders is obtained.

The unknown disorder d_0 always belongs to $\hat{D}$. The membership test (1.40) is thus very permissive, i.e. $\hat{D}$ can be very large, and contains what may look like irrelevant causes, when they cannot be ruled out. Namely, a disorder d may belong to $\hat{D}$ defined by (1.40) and (1.41) even if $M(d)^+ \cap M^+ = \emptyset$ and $M(d)^- \cap M^- = \emptyset$. Indeed, in this case, $M(d)^+ \cup M(d)^- \subseteq M^0$. This means that $\hat{D}$ includes disorders, no sure manifestations of which are observed and no forbidden manifestations are for sure absent. Such a disorder d may still be present when M^+ and M^- are observed, since all the available certain information we have about d pertains to manifestations in M^0 about which no observation exists (this is true for the unknown disorder d_0). Then, among the disorders in $\hat{D}$ we may prefer those, if any, which have at least a weak relevance to the observations, namely the subset $\hat{D}^*$ of $\hat{D}$, defined by

$$\hat{D}^* = \{d \in \hat{D}, M(d)^+ \cap M^+ \neq \emptyset \text{ or } M(d)^- \cap M^- \neq \emptyset\} \quad (1.42)$$

This eliminates the disorders such that $M(d)^+ \cup M(d)^- \subseteq M^0$, i.e. the disorders which are not suggested by the observations without being ruled out by them (and d_0 particularly).

In this section we present a graded extension of the above model. Namely, M^+ and M^- are now fuzzy sets of manifestations which are more or less certainly present, and more or less certainly absent, respectively. However we keep the requirement $M^+ \cap M^- = \emptyset$ (where the intersection is defined by the min operation), i.e. we cannot be somewhat certain both of the presence and of the absence of the same manifestation simultaneously. Similarly, $M(d)^+$ and $M(d)^-$ will denote the fuzzy sets of manifestations which are, respectively, more or less certainly present and more or less certainly absent when disorder d alone is present (more generally when the subset D of disorders is present). Obviously, we also assume $\forall d, M(d)^+ \cap M(d)^- = \emptyset$.

By complementation (defined by $\mu_{\bar{F}} = 1 - \mu_F$) of M^- and $M(d)^-$, we obtain the fuzzy sets $\overline{M^-}$ and $\overline{M(d)^-}$ of manifestations which are more or less possibly present, respectively in the considered situation and when d is present. This corresponds to the usual duality between what is (more or less) certain, i.e. necessarily true, and what is (more or less) possibly true. Indeed a pair of dual possibility and necessity measures Π and N are related by the relation $\Pi(A) = 1 - N(\bar{A})$, for any event A (here A represents the presence of a manifestation).

Note that $M^+ \subseteq \overline{M^-}$, $M(d)^+ \subseteq \overline{M(d)^-}$, in the sense of fuzzy set inclusion. An even stronger inclusion holds. Since $M^+ \cap M^- = \emptyset$, we have

$$\{m_i \in \mathcal{M}, \mu_{M^+}(m_i) > 0\} \subseteq \{m_i \in \mathcal{M}, \mu_{M^-}(m_i) = 0\} = \{m_i \in \mathcal{M}, \mu_{\overline{M^-}}(m_i) = 1\} \quad (1.43)$$

i.e. the support of M^+ (i.e. the elements with a positive membership grade) is included in the core of $\overline{M^-}$ (the elements which belong with degree 1); the same holds for $M(d)^+$ and

$\overline{M(d)^-}$. This is in agreement with the fact that for crisp events A, we have $N(A) > 0 \Leftrightarrow \Pi(\overline{A}) < 1 \Rightarrow \Pi(A) = 1$ since then one of A or $\overline{A}$, at least, should be completely possible. A pair of fuzzy sets (F, G) such that $F \cap \overline{G} = \emptyset$ is called a twofold fuzzy set (Dubois and Prade, 1987b). Twofold fuzzy sets (F, G) have been introduced to model incompletely known sets, i.e. sets for which we know elements gathered in F, which more or less certainly belong to it, as well as other elements, gathered in $\overline{G}$, which more or less certainly do not belong to it. However, $F \cup \overline{G}$ may not cover the whole referential. Similarly, the pair $(M(d)^+, \overline{M(d)^-})$ defines twofold fuzzy relations on $\mathscr{D} \times \mathscr{M}$.

The extension to fuzzy sets of equations (1.40) and (1.41) can be performed very simply. It requires that the extent to which two fuzzy sets F and G of $\mathscr{M}$ intersect be evaluated. The consistency between F and G is simply defined as (Zadeh, 1979a)

$$\text{cons}(F, G) = \sup_{m \in \mathscr{M}} \min(\mu_F(m), \mu_G(m)) \tag{1.44}$$

It computes the degree of existence of some common element for F and G. Equation (1.41) is based on checking the inconsistency level between fuzzy sets, that is $1 - \text{cons}(F, G)$. The fuzzy extension of (1.41) then leads to compute a fuzzy set $\hat{D}$ of potential unique disorders as

$$\forall d \in \mathscr{D}, \mu_{\hat{D}}(d) = \min(1 - \text{cons}(M(d)^+, M^-), 1 - \text{cons}(M(d)^-, M^+))$$
$$= 1 - \max(\text{cons}(M(d)^+, M^-), \text{cons}(M(d)^-, M^+)) \tag{1.45}$$

where the minimum operator expresses the conjunction of the conditions in (1.41). It is easy to check that (1.43) can be written in terms of inclusion indices, like (1.40), in the form

$$\mu_{\hat{D}}(d) = \min(\text{inc}(M(d)^+, \overline{M^-}), \text{inc}(M(d)^-, \overline{M^+})) \tag{1.46}$$

where the inclusion of fuzzy set F into fuzzy set G is defined by

$$\text{inc}(F, G) = 1 - \text{cons}(F, \overline{G}) = \inf_{m \in \mathscr{M}} \max(1 - \mu_F(m), \mu_G(m)) \tag{1.47}$$

Note that $\text{inc}(F, G) = 1 \Leftrightarrow \text{support}(F) \subseteq \text{core}(G) \Leftrightarrow F \cap \overline{G} = \emptyset$, i.e. the inclusion index extends the strong inclusion encountered in (1.43) and guarantees that d fully belongs to $\hat{D}$ (i.e. with degree 1) if and only if $M(d)^+$ and $\overline{M^-}$, $M(d)^-$ and $\overline{M^+}$ do not overlap at all. Equation (1.45) clearly expresses that a disorder d is all the less a candidate explanation as the fuzzy set of its more or less certain effects overlaps the fuzzy set of manifestations more or less certainly absent, or as the fuzzy set of effects which are more or less certainly absent when d is present overlaps the fuzzy set of manifestations which are more or less certainly present. This is intuitively satisfying.

In the particular case where $M^- = \emptyset$, i.e. we only have positive observations, $\hat{D}$ is defined by

$$\forall d \in \mathscr{D}, \mu_{\hat{D}}(d) = \min_{i=1,n} \max(1 - \mu_{M^+}(m_i), \mu_{\overline{M(d)^-}}(m_i)) \tag{1.48}$$

This case corresponds to the situation studied by Peng and Reggia (1990) and $\mu_D(d)$

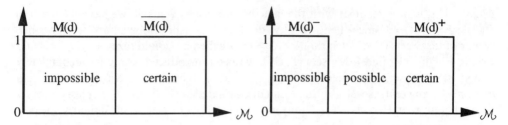

Figure 1.4 Completely informed situation   Figure 1.5 Partially informed situation

evaluates to what extent d covers M^+. The idea of relevance expressed by equation (1.42) in the crisp case can be easily extended using the consistency index. Namely

$$\mu_{\hat{D}^*}(d) = \min(\mu_{\hat{D}}(d), \max(\text{cons}(M(d)^+, M^+), \text{cons}(M(d)^-, M^-))) \qquad (1.49)$$

Note that the modelling of uncertainty remains qualitative in this approach, as in the preceding section. Indeed, we could use a finite completely ordered chain of levels of certainty ranging between 0 and 1, i.e. $l_1 = 0 < l_2 < \cdots < l_n = 1$ instead of $[0,1]$, with $\min(l_i, l_k) = l_i$ and $\max(l_i, l_k) = l_k$ if $i \leq k$, and $1 - l_i = l_{n+1-i}$.

Taking into account the incomplete nature of the information about the presence or absence of manifestations decreases the discrimination power when going from the completely informed case to the incomplete information case, since then the number of possible disorders in $\hat{D}$ increases. This is due to the fact that now there are manifestations which are neither certain nor impossible and consequences of the presence of a given disorder d which are only possible, as pictured in Figure 1.5, whereas $M(d)^+ = M(d)$ and $M(d)^- = \overline{M(d)}$ in Figure 1.4 (similar figures can be drawn for M^+ and M^-).

This suggests that, in order to improve the discrimination power of the model, we have to refine the non-fuzzy model in such a way that consequences (respectively, manifestations) previously expressed as certain (respectively, certainly present) and impossible (respectively, certainly absent) remain classified in the same way and where some possible consequences (respectively, possibly present manifestations) are now allowed to be either somewhat certain (respectively, somewhat certainly present) or somewhat impossible (respectively, somewhat certainly absent). See Figure 1.6. Then (1.46) enables us to rank-order the possible disorders which are compatible with the observations. This counterbalances the increase of candidates due to the incompleteness of the information.

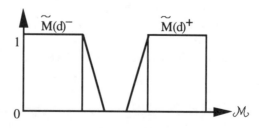

Figure 1.6 Partially informed situation with preferences

UNCERTAINTY MANAGEMENT   31

It can easily be shown that adding preference levels on top of incompleteness modelling can at the same time enable the discrimination power of the completely informed situation to be recovered, and also enable the extra disorders obtained on the partially informed situation to be ranked in terms of their plausibility levels. See Dubois and Prade (1993a).

Example Let us show the benefit of grading uncertainty on an example. Let $\mathscr{D} = \{d_1, d_2, d_3, d_4\}$, $\mathscr{M} = \{m_1, m_2, m_3\}$ and let us first consider the completely informed case

$$M(d_1)^+ = \{m_2\}, M(d_1)^- = \{m_1, m_3\}$$
$$M(d_2)^+ = \{m_3\}, M(d_2)^- = \{m_1, m_2\}$$
$$M(d_3)^+ = \{m_1, m_2, m_3\}, M(d_3)^- = \emptyset$$
$$M(d_4)^+ = \{m_2, m_3\}, M(d_4)^- = \{m_1\}$$
$$M^+ = \{m_2, m_3\}, M^- = \{m_1\}.$$

Then, we obtain $\hat{D} = \{d_4\}$.

Let us weaken our example into an incompletely informed case:

$$M(d_1)^+ = \emptyset, M(d_1)^- = \{m_1\}$$
$$M(d_2)^+ = \{m_3\}, M(d_2)^- = \emptyset$$
$$M(d_3)^+ = \{m_3\}, M(d_3)^- = \emptyset$$
$$M(d_4)^+ = \emptyset, M(d_4)^- = \{m_1\}.$$

Then $\hat{D} = \{d_1, d_2, d_3, d_4\}$, i.e. all causes are possible. Let us now consider the following table, where the previous incompletely informed case is completed with some gradual uncertainty information that is in accordance with the complete information case.

Thus we obtain that d_4 is much more plausible than d_2, itself more plausible than d_3, which is more plausible than d_1, i.e. ranking of the possible disorders is obtained. This is in accordance with the complete information case as well as the incomplete information case,

Table 1.1

		$M(d_1)^+$	$M(d_1)^-$	$M(d_1)^+$	$M(d_2)^-$	$M(d_3)^+$	$M(d_3)^-$	$M(d_4)^+$	$M(d_4)^-$	$\mathscr{M}$
membership	m_1	0	1	0	0.9	0.7	0	0	1	0	M^-
	m_2	0.8	0	0	0.6	0.5	0	0.7	0	1	M^+
grades	m_3	0	0.8	1	0	1	0	0.3	0	1	M^+
	m_3	0	0.8	1	0	1	0	0.3	0	1	M^+
cons$(M(d)^+, M^-)$		0		0		0.7		0			
cons$(M(d)^+, M^-)$			0.8		0.6		0		0		
form.(1.45): $1-\max$		0.2		0.4		0.3		1			

but significantly improves their findings. Note that all disorders are relevant in the sense of (1.49). The reader is referred to Cayrac et al. (1994) for the application of this approach to a satellite diagnosis problem.

1.4 CONCLUDING REMARKS

This overview paper has provided a description of some fuzzy set or possibility theory-based methods in approximate reasoning which correspond to two interpretations of the idea of a fuzzy set as either describing a gradual property (or if we prefer a fuzzy predicate) or modelling an incomplete state of information. This paper offers only a partial overview of the existing works in approximate reasoning, and is centred on the work done by the authors for about 10 years. There exists a third interpretation which has not been presented here for the sake of brevity, but which is also worth considering. This is when a fuzzy set models a flexible constraint that expresses what are the values which are allowed to be chosen for a given quantity or parameter. In such a situation the semantics of the possibility distribution is in terms of preference. This is not to be confused with a possibility distribution representing an incomplete state of knowledge. Indeed in that latter case the ill-known value restricted by the possibility distribution is not a matter of choice, but rather of randomness. The extension of constraint satisfaction problem methods to flexible constraints modelled in the fuzzy set framework has been thoroughly investigated; see Dubois, Fargier and Prade (1994).

Besides this, the paper has advocated possibility theory as a convenient framework to handle uncertainty in a rather qualitative way, but no comparison has been provided with probability-based methods. Readers interested in a comparative view and a discussion on the differences but also on the links between possibilistic and the probabilistic frameworks are referred to Bezdek (1994) and Dubois and Prade (1993b).

REFERENCES

Baldwin, J. F. (1979) A new approach to approximate reasoning using a fuzzy logic. *Fuzzy Sets and Systems*, **2**, 309–325.

Benferhat, S., Dubois, D., Lang, J., and Prade, H. (1993) Hypothetical reasoning in possibilistic logic: basic notions, applications and implementation issues. In Wang, P. Z., and Loe, K. F. (Eds) *Between Mind and Computer—Fuzzy Science and Engineering* (World Scientific: Singapore), pp. 1–29.

Benferhat, S., Dubois, D., and Prade, H. (1992) Representing default rules in possibilistic logic. In Nebel, B., Rich, C., Swartout, W. (Eds) *Proc. of the 3rd Int. Conf. on Principles of Knowledge Representation and Reasoning (KR'92)* Cambridge, MA, pp. 673–684.

Bezdek, J. C. (Ed.), (1994) Special issue: Fuzziness vs. probability—the nth round. *IEEE Transactions on Fuzzy Systems*, **2**, 1–42.

Bouchon, B. (1987) Fuzzy inferences and conditional possibility distributions. *Fuzzy Sets and Systems*, **23**, 33–41.

Cayrac, D., Dubois, D., Haziza, M., and Prade, H. (1994) Possibility theory in 'fault mode effect analyses'—a satellite fault diagnosis application. *Proc. of the 3rd IEEE Int. Conf. on Fuzzy Systems (FUZZ-IEEE'94)*, Orlando, pp. 1176–1181.

Di Nola, A., Sessa, S., Pedrycz, W., and Sanchez, E. (1989) *Fuzzy Relation Equations and Their Applications to Knowledge Engineering* (Dordrecht: Kluwer Academic).
Dubois, D. Belief structures, possibility theory and decomposable confidence measures on finite sets, *Comput. Artif. Intell.* (Bratislava), **5** (5) (1986), 403–416.
Dubois, D., Fargier, H., and Prade, H. (1994) Propagation and satisfaction of flexible constraints. In Yager, R. R., Zadeh, L. A. (Eds) *Fuzzy Sets, Neural Networks and Soft Computing* (Van Nostrand), pp. 166–187.
Dubois, D., Grabisch, M., Prade, H. (1994) Gradual rules and the approximation of control laws. Nguyen, H. T., et al. (Eds) *Theoretical Aspects of Fuzzy Control* (New York: Wiley).
Dubois, D., Lang, J., and Prade, H. (1992) Advances in automated reasoning using possibilistic logic. In Kandel, A. (Ed.) *Fuzzy Expert Systems* (Boca Raton, FL: CRC Press), pp. 12–134.
Dubois, D., Lang, J., and Prade, H. (1994a) Possibilistic logic. In Gabbay, D. M. et al. (Eds) *Handbook of Logic in Artificial Intelligence and Logic Programming* (Oxford: Clarendon Press), pp. 439–513.
Dubois, D., Lang, J., and Prade, H. (1994b) Automated reasoning using possibilistic logic: semantics, belief revision, and variable certainty weights. *IEEE Transactions on Knowledge and Data Engineering*, **6**, pp 64–71.
Dubois, D., Martin-Clouaire, R., and Prade, H. (1988) Practical computing in fuzzy logic. In Gupta, M. M., and Yamakawa, T. (Eds) *Fuzzy Computing—Theory, Hardware, and Applications* (Amsterdam: North-Holland), pp. 11–34.
Dubois, D., and Prade, H. (1980) *Fuzzy Sets and Systems: Theory and Applications* (New York: Academic Press).
Dubois, D., and Prade, H. (1987a) Fuzzy numbers: an overview. In Bezdek, J. C. (Ed.) *Analysis of Fuzzy Information—Vol. I: Mathematics and Logic* (Boca Raton: CRC Press), pp. 3–39.
Dubois, D., and Prade, H. (1987b) Twofold fuzzy sets and rough sets—some issues in knowledge representation. *Fuzzy Sets and Systems*, **23**, 3–18.
Dubois, D., and Prade, H. (with the collaboration of H. Farreny, R. Martin-Clouaire and C. Testemale), (1988) *Possibility Theory—An Approach to Computerized Processing of Uncertainty* (New York: Plenum Press).
Dubois, D., and Prade, H. (1989) Order-of-magnitude reasoning with fuzzy relations. *Revue d'Intelligence Artificielle*, **3**, 69–94.
Dubois, D., and Prade, H. (1991a) Semantic considerations on order of magnitude reasoning. In Singh, M. G., and Travé-Massuyès, L. (Eds), (1991) *Decision Support Systems and Qualitative Reasoning* (Amsterdam: North-Holland), pp. 223–228.
Dubois, D., and Prade, H. (1991b) Possibilistic logic, preference models, non-monotonicity and related issues. *Proc. of the 12th Int. Joint Conf. on Artificial Intelligence (IJCAI'91)*, Sydney, Australia, pp. 419–424.
Dubois, D., and Prade, H. (1991c) Fuzzy sets in approximate reasoning. *Fuzzy Sets and Systems*, Part 1, **40**, 143–202; Part 2 (with Lang, J.), **40**, 203–244.
Dubois, D., and Prade, H. (1991d) Epistemic entrenchment and possibilistic logic. *Artificial Intelligence*, **50**, 223–229.
Dubois, D., and Prade, H. (1992a) Possibility theory as a basis for preference propagation in automated reasoning. *Proc. of the 1st IEEE Int. Conf. on Fuzzy Systems (FUZZ-IEEE'92)* San Diego, CA, pp. 821–832. March 8–12, 1992.
Dubois, D., and Prade, H. (1992b) Fuzzy rules in knowledge-based systems—Modelling gradedness, uncertainty and preference. In Yager, R. R., and Zadeh, L. A. (Eds) *An Introduction to Fuzzy Logic Applications in Intelligent Systems* (Dordrecht: Kluwer Academic), pp 45–68.

Dubois, D., and Prade, H. (1993a) A fuzzy relation-based extension of Reggia's relational model for diagnosis handling uncertain and incomplete information. In Heckerman, D., and Mamdani, A. (Eds) *Proc. of the 9th Conf. on Uncertainty in Artificial Intelligence*, pages 106–113.

Dubois, D., and Prade, H. (1993b) Fuzzy sets and probability: misunderstandings, bridges and gaps. *Proc. of the 2nd IEEE Int. Conf. on Fuzzy Systems (FUZZ-IEEE'93)*, San Francisco, CA, pp. 1059–1068.

Klir, G. J., and Folger, T. A. (1988) *Fuzzy Sets, Uncertainty and Information* (Englewood Cliffs, NJ: Prentice Hall).

Klawonn, F., and Kruse, R. (1993a) Equality relations as a basis for fuzzy control. *Fuzzy Sets and Systems*, **54**, 147–156.

Klawonn, F., and Kruse, R. (1993b) Fuzzy control as interpolation on the basis of equality relations. *Proc. of the 2nd IEEE Int. Conf. on Fuzzy Systems (FUZZ-IEEE'93)*, San Francisco, CA, pp. 1125–1130.

Kraus, S., Lehmann, D., and Magidor, M. (1990) Nonmonotonic reasoning, preferential models and cumulative logics. *Artificial Intelligence*, **44**, 134–207.

Lang, J., Dubois, D., and Prade, H. (1991) A logic of graded possibility and certainty coping with partial inconsistency. *Proc. of the 7th Conf. on Uncertainty in Artificial Intelligence*, Los Angeles, CA, pp. 188–196.

Mamdani, E. H. (1977) Application of fuzzy logic to approximate reasoning using linguistic systems. *IEEE Transactions on Computers*, **26**, 1182–1191.

Negoita, C. V., and Ralescu, D. A. (1975) *Applications of Fuzzy Sets to Systems Analysis* (Basel: Birkhaeuser).

Ovchinnikov, S. V. (1991) Similarity relations, fuzzy partitions, and fuzzy orderings. *Fuzzy Sets and Systems*, **40**, 107–126.

Peng, Y., and Reggia, J. A. (1990) *Abductive Inference Models for Diagnostic Problem-Solving* (Berlin: Springer Verlag).

Ruspini, E. H. (1991) On the semantics of fuzzy logic. *International Journal of Approximate Reasoning*, **5**, 45–88.

Sanchez, E. (1978) On possibility-qualification in natural languages. *Information Sciences*, **15**, 45–76.

Schweizer, B., and Sklar, A. (1983) *Probabilistic Metric Spaces* (New York: North-Holland).

Shoham, Y. (1988) *Reasoning About Change—Time and Causation from the Standpoint of Artificial Intelligence* (Cambridge, MA: The MIT Press).

Sugeno, M., and Nishida, M. (1985) Fuzzy control of model car. *Fuzzy Sets and Systems*, **16**, 103–113.

Sugeno, M. (1985) An introductory survey of fuzzy control. *Information Sciences*, **36**, 59–83.

Tsukamoto, T. (1979) An approach to fuzzy reasoning method. In Gupta, M. M., Ragade, R. K., and Yager, R. R. (Eds) *Advances in Fuzzy Set Theory and Applications* (Amsterdam: North-Holland), pp. 137–149.

Trillas, E., and Valverde, L. (1985) On mode and implication in approximate reasoning. In Gupta, M. M., Kandel, A., Bandler, W., Kiszka, J. B. (Eds) *Approximate Reasoning in Expert Systems* (Amsterdam: North-Holland), pp. 157–166.

Yager, R. R. (1983) An introduction to applications of possibility theory. *Human Systems Management*, **3**, 246–269.

Yager, R. R. (1993) On the completion of qualitative possibility measures. *IEEE Transactions on Fuzzy Systems*, **1**, 184–194.

Zadeh, L. A. (1965) Fuzzy sets. *Information and Control*, **8**, 338–353.

Zadeh, L. A. (1971) Similarity relations and fuzzy orderings. *Information Sciences*, 177–200.

Zadeh, L. A. (1973) Outline of a new approach to the analysis of complex systems and decision processes. *IEEE Transactions on Systems, Man and Cybernetics*, **2**, 28–44.

Zadeh, L. A. (1975) The concept of a linguistic variable and its application to approximate reasoning. *Information Sciences*, Part 1, **8**, 199–249; Part 2, **2**, 301–357; Part 3, **9**, 43–80.

Zadeh, L. A. (1978a) Fuzzy sets as a basis for a theory of possibility. *Fuzzy Sets and Systems*, **1**, 3–28.

Zadeh, L. A. (1978b) PRUF—A meaning representation language for natural languages. *International Journal of Man–Machine Studies*, **10**, 395–460.

Zadeh, L. A. (1979a) A theory of approximate reasoning. In Hayes, J. E. Michie, D., and Mikulich, L. I. (Eds) *Machine Intelligence* (Amsterdam: Elsevier), vol. 9, pp. 149–194.

Zadeh, L. A. (1979b) Fuzzy sets and information granularity. In Gupta, M. M., Ragade, R. K., and Yager, R. R. (Eds) *Advances in Fuzzy Set Theory and Applications* (Amsterdam: North-Holland), pp. 3–18.

FUZZY LOGIC CONTROL

2
Fuzzy Logic Control: A Systematic Design and Performance Assessment Methodology

G. Vachtsevanos
School of Electrical and Computer Engineering, Georgia Institute of Technology, USA

and

S. Farinwata
Ford Research Laboratory, Dearborn, USA

2.1 INTRODUCTION

Fuzzy logic and its applications to modelling and control of dynamical systems has been at the centre of controversy over recent years. The successes and failures of this new technology have brought it to the attention of the public media, the professional societies and the trade periodicals.

Since the early 1970s, researchers recognized that a set of production rules, which may be cast as an expert system paradigm and use fuzzy arithmetic/logical notions, can emulate fairly well such classical control schemes as PID (proportional-integral-derivative) algorithms. Fuzzy logic control (FLC) thus became popular since it offered an alternative to the predominantly mathematical methods pursued by the traditional control community. Its claim to fame was based on the ease of design of the control rules (straight out of an operator's manual), the 'understandable' nature of the resulting control strategies by plant operators and other practitioners, the lack of dependence on accurate process models, and the inherent nonlinear nature of the FLC. Fuzzy rule-based strategies that mechanize the approximate reasoning of human experts found numerous applications in such diverse areas as transportation systems, consumer electronics, home appliances, industrial processes, financial systems, etc.

The performance claims that usually accompanied a handful of simple linguistic rules, derived primarily from heuristic information, have prompted the control community to publicly object to the lack of mathematical formalism and the total disregard for well-accepted assessment methodologies (see *IEEE Control Systems Magazine*, 1993, **13**, No. 3).

The fervour of the FLC controversy stems, of course, from a historical paradox: a flurry of applications has preceded any substantial theoretical developments. The question, therefore, is posed as follows: is there a systematic way to design and assess the performance of FLCs? Before outlining our own response to this query, it is instructive to review briefly the historical evolution of FLC design and assessment techniques. Mamdani and his co-workers were the first to utilize fuzzy set theory to translate human expressions of control strategies into a set of fuzzy decision rules [1]. Their fuzzy controller encoded human strategies to control a steam engine. This approach was followed by a host of other investigators. Sugeno and his Japanese co-workers pursued a design approach which supposes that the process under control exhibits a behaviour similar to that of a second-order system [2]. Completeness of the rule base was assured by an exhaustive enumeration of all possible combinations of the antecedent variables (typically the error and its rate of change) and the consequent variables. The resulting 'linguistic phase plane' thus depicts the control action to be taken when the process error and its derivative have certain (fuzzy) values as determined from actual output measurements.

It is obvious that these approaches are faced with some serious limitations arising from the assumed structure of the FLC and the hypothesis regarding the dynamics of the process to be controlled. For relatively simple systems, though, that satisfy these requirements, a rule-based controller that is heuristically derived seems to perform fairly well.

Attempts to formalize the analysis of closed-loop systems in terms of fuzzy set theory and to develop systematic design tools date back to the early years of the fuzzy literature. Tong pursued an approach that defines system operators as fuzzy relations on the state, output and control spaces [3]. Cumani considered the system quantities such as states, input, and output as fuzzy variables that obey time-evolving possibility distributions governed by Hisdal's Calculus of conditional possibility [4]. These attempts met with limited success since the fuzzy dynamical system is described in a recursive form, and the accumulated 'fuzziness' combined with the conservative nature of most inferencing schemes has a tendency to "flatten out' the distribution of fuzzy membership functions, thus masking the true dynamical behaviour of the system.

More recently, two major advances in nonlinear dynamical systems theory have motivated FLC researchers to explore parallel approaches in addressing fuzzy control design and analysis issues. The first one refers to the introduction of the concept of cell-to-cell mappings, and the second is concerned with sliding mode control and its ramifications to the dynamic evolution of nonlinear systems. Hsu argued that, because of the unavoidable accuracy limitations of both physical measurements and numerical calculations, a state variable should not be treated as a continuum of points but rather only as a collection of small intervals [5]. He therefore partitioned the state space into a finite number of disjoined sets, called cells, and provided an algorithm based an cell-to-cell mappings to analyze the global dynamic behaviour of nonlinear systems. This algorithm was used to describe the behaviour of fuzzy systems [6]. A number of investigators also proposed that fuzzy controllers work like modified sliding mode controllers [7]. The general concept here is the division of the phase plane into regions by means of invariant and switching manifolds. Such boundary divisions decide the dynamic behaviour of the system and suggest means to arrive at tracking or regulation strategies by forcing the system to follow a desired trajectory.

The lessons learned from the early successes and failures of FLC point towards a new look at the design, analysis and performance evaluation of controllers that employ fuzzy set theory. The motivating philosophy of the approach we are proposing is based on the realization that previous shortcomings may be alleviated by appropriately combining tools from fuzzy set theory, nonlinear dynamics and artificial intelligence. A hybrid methodology that utilizes both numerical and symbolic manipulations and capitalizes on the advantages of control-theoretic and AI algorithms is expected to produce a systematic and unified approach to the FLC problem. An objective of control theory is to develop control design techniques that improve system performance for uncertain dynamic systems. Resolving uncertainty in system design has been an important issue, since engineered systems are commonly subjected to such forms of uncertainty as parametric disturbances, unmodelled dynamics and external noise. It is believed that uncertainty can be most efficiently addressed via intelligence. An intelligent control system must be able to sense its environment, process ambiguous and incomplete information, and provide control actions based on imprecise or even unreliable knowledge. Since fuzzy set theory and fuzzy logic assist in arriving at some reasonable decisions with ambiguous and imprecise events, a fuzzy logic control approach that combines fuzzy set theory and control theory is considered to be a prime candidate for the development of intelligent control systems. Fuzzy logic control is cast as an expert system paradigm. This AI setting adjusts controller parameters on the basis of specified performance characteristics. We introduce a systematic design procedure for fuzzy linguistic controllers. This method can be applied to a class of nonlinear systems which are subjected to dynamic/parametric disturbances. We are addressing these issues by dividing the uncertain domain of interest into a finite number of manageable quantities and by assigning fuzzy sets to each linguistic representation; then the relations that govern the control objectives may be easily derived, since we are dealing with quantitative objects of an infinite point space in terms of qualitative reasoning that is expressed as a finite set of rules. The 'objects' referred to are vector fields of invariant or switching manifolds in a hyperspace. We detail below a systematic procedure for a generalized fuzzy control design and analysis algorithm. The paper is organized as follows: the following section introduces the fuzzy logic control problem and details the automatic rule generation procedure. Next, stability and robustness considerations are outlined, and finally application examples are used to illustrate the algorithmic developments.

2.2 THE PHASE PORTRAIT ASSIGNMENT ALGORITHM

2.2.1 Fuzzy Logic Control

The primary control objective for a complex nonlinear process is to maintain its state trajectories within allowable fuzzy boundaries in accordance with some desired profiles. Nonlinear fuzzy regulators can be used to control a system with fixed operating conditions; however, this approach is restrictive in that a large number of knowledge bases may be required, corresponding to different operating conditions. Therefore, a fine tracking controller is needed to resolve complex control situations. A framework for fuzzy control theory is established to address the systematic control rule base design problem and to investigate such important issues as convergence, stability and robustness. Consider the fuzzy logic controller design for a nonlinear process described by

$$\dot{x}(t) = f(x(t), u(t)) \qquad (2.1)$$

where $x(t) = [x_1(t), \ldots, x_n(t)]^T$ is the state and $u(t) = [u_1(t), \ldots, u_m(t)]^T$ is the control input vector and f is a smooth vector field.

As a first step, we define an error model as

$$u(t) = g(e(t), x^d(t)) \qquad (2.2)$$

where the error, $e(t) = x^d(t) - x(t)$, and the desired process profile $x^d(t)$ are defined appropriately. $f(.,.)$ is the function of the dynamic equations of the nonlinear process, $g(.,.)$ is the functional representation of the control input, and $u = g(e, x^d)$ is the resultant control mode from fuzzy inferencing.

The error system is given by

$$\dot{e} = \dot{x} - f(x^d - e, g(e, x^d))$$
$$= h(e, x^d \dot{x}^d) \qquad (2.3)$$

The objective of the fuzzy controller is to maintain closed-loop stability of the error system, i.e. make $e(t)$ converge to zero. The regulation is performed via a fuzzy hypercube using the phase portrait assignment algorithm and with the control law given as

$$u = g(e, \dot{x}^d) = g\left(\int h\, d\tau, \dot{x}^d\right) \qquad (2.4)$$

The construction of the fuzzy rule base depends on the phase portrait of the error $e(t)$, the desired profile x^d and the quantization specifications.

2.2.2 The Automatic Rule-Generation Method

We consider a nonlinear system described by the vector differential equation (2.1). We assume that the system is known but ill-defined because of vagueness uncertainty. If a mathematical model is not available, input–output data might suffice to arrive at the control design. For engineered systems we also assume the availability of heuristic plant information derived from the designer's or operator's expertise. The step-by-step procedure to synthesize a fuzzy rule base is based on data obtained from numerical simulations of a given nonlinear model of the physical system. It should be emphasized, though, that we do not require accurate mathematical representations, but an approximate or simplified model is acceptable. We introduce a design method called the 'phase portrait assignment algorithm' for the closed-loop system consisting of a multi-input nonlinear process and a fuzzy rule base controller, as shown schematically in Figure 2.1.

The algorithm proceeds as follows.

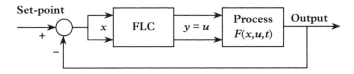

Figure 2.1 The fuzzy logic control system architecture

First, the invariant and switching manifolds are found after complete and exhaustive model simulations. The phase portrait assignment algorithm computes the vector fields of the states and finds the invariant manifolds that meet specified performance objectives. The region of interest is next divided into a finite number of smaller regions or 'fuzzy cells' in the phase space. The size of each cell is determined according to accuracy and tolerance specifications. The switching and invariant manifolds assist in the optimum design of the cell-groups and provide useful means for addressing issues of reachability and asymptotic stability.

The centre points of all cell-groups are chosen to anticipate the trajectories from one subspace to another. The term 'cell-group' is adopted to indicate that a partition is either a single cell or a grouping of cells resulting from the application of an appropriate clustering algorithm. Suppose, for illustration purposes, that we are controlling a two-state system. We denote by E_{1m} and E_{2m} the mth and nth subintervals of the x_1 and x_2 range, respectively. Linguistically, we define

$$L_{mn} = E_{1m} \times E_{2m}$$

where L_{mn} is a finite region in the state space. Figure 2.2 shows a typical space partitioning and cell-group numbering employed in the proposed scheme.

Now consider the space of two control variables u_1 and u_2. A range of permissible values is associated with each control variable. By quantizing these intervals into equal subintervals, we derive $N_{u1} + 1$ intermediate values for u_1 and $N_{u2} + 1$ intermediate values for u_2. Again, the numbers N_{u1} and N_{u2} are selected according to criteria similar to those for the states. The behaviour of the functions dictates whether a coarse or fine quantization is used. This scheme results in a set S_1 of values for the control variable u_1 and another set S_2 of values for u_2.

For each cell-group, consider the centre point. This point is the initial state for the simulation. Apply to the system the constant input (u_{1i}, u_{2j}) for all (u_1, u_2) in the set $S = S_1 \times S_2$ and perform a simulation run until the system enters into another cell-group in the state space or until the simulation time becomes larger than a fixed value t_0. For each cell-group in the state space, we perform $(N_{u1} + 1)(N_{u2} + 1)$ simulations, which result in $(N_{u1} + 1)(N_{u2} + 1)$ transitional relations of the type:

$$(u_{11}, u_{21}): L_{11} \to L_{12}$$

$$(u_{11}, u_{22}): L_{11} \to L_{21}$$

$$\vdots$$

$$(u_{13}, u_{21}): L_{11} \to L_{11}$$

44   FUZZY LOGIC CONTROL

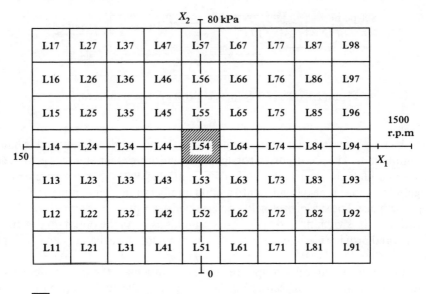

Figure 2.2   A typical space partitioning and cell-group number scheme

The first transition relation relays the following information: if the system starts at the centre of cell-group L_{11} and the constant input (u_{11}, u_{21}) is applied, then the system will end up in cell-group L_{12}. The time required for the transition is stored, as well as the energy of control input $(u^T u)$ and the Euclidean distance of the state from the equilibrium point, at the end of each simulation. Note that the ending cell-group can either be the same as the starting cell-group, in which case it is called an 'invariant manifold', or an adjacent one. In other words, the sampling period of the simulation is sufficiently small that no jumps between cell-groups are allowed. This is a rather stringent condition which may be hard to satisfy in general. However, it may be satisfied for well-behaved systems and also under a uniform cell-size construction. Still, some *ad hoc* adjustment of the sampling time may be required. It is clear that the target cell-group L_{mn} (the specified goal) must be an invariant manifold for some (u_{1i}, u_{2j})—pair. This is also necessary for convergence and asymptotic stability and is equivalent to the 'reachability condition' in modern control theory. After all pairs of control values in S have been used, we continue with the next cell-group, and we repeat this process until all cell-groups in the state space have been simulated.

The next step is to search the accumulated data file and come up with the following result: if the system starts in cell-group L_{ij}, then in order to drive it eventually to the equilibrium point, the 'best' control to be applied is (u_{1k}, u_{2l}), and such a rule has to be derived for all cell-groups in the state space. 'Best' can be in the sense of minimum time, minimum energy or minimum error. The general form of the cost of each transition is given as

$$J = \sum_{i,j} [\alpha(e^T e) + \mathcal{L}(u^T u) + \tau]t, \quad \text{for all } e \in L_{mn} \tag{2.5}$$

where u is the control input applied while in L_{mn} in order to transition to the next cell-group;

α, $\mathscr{L}$ and τ are binary coefficients to select a particular criterion as squared error, energy or time, respectively. A systematic search of all cell-groups is carried out in order to identify the optimum trajectory. The search procedure is based on a modified A^*-algorithm. The algorithm is guaranteed to terminate in a finite number of steps and to find the best path from any node to the equilibrium node, provided that such a path exists. Each cell-group becomes a node of a graph. Each transition becomes a unidirectional arc with weight equal to the transition's cost, in the sense defined above. It should also be clear that each transition (i.e. arc) is associated with a pair of constant control values. The objective is to find a path from every node to the equilibrium node, with the smallest possible sum of arc weights. The procedure begins by examining all neighbouring nodes to the equilibrium node. The total cost of each node is updated, and the search procedure is repeated until all nodes are examined and the path costs are computed and compared. Basically, this information suggests the following strategy: if the system starts in cell-group L_{mn}, apply the control corresponding to the first arc of the path connecting L_{mn} with the equilibrium cell-group, until the system enters another cell-group. Then the new path is the previous one with the first arc removed. The control associated with the new first arc is applied to the system, until we enter another cell-group. This is repeated until the system has reached the equilibrium cell-group.

Once the search procedure has generated the optimal tree, the fuzzy controller may be designed as follows: every single transition that is part of this tree, becomes a rule in the fuzzy controller. Since each transition corresponds to a node (i.e. cell-group), we can state equivalently that the number of rules is equal to the number of cell-groups from which there is a path towards the equilibrium cell-group. If the system starts in cell-group L_{mn}, we apply the control (u_{1k}, u_{2l}) for all cell-groups in the state space. The next step is to fuzzify the borders of the cell-groups and the control values. This is accomplished by introducing a linguistic term (membership function) for each dimension of a cell-group (Figure 2.3 graphically illustrates this procedure). By doing this, a point in the state space does not

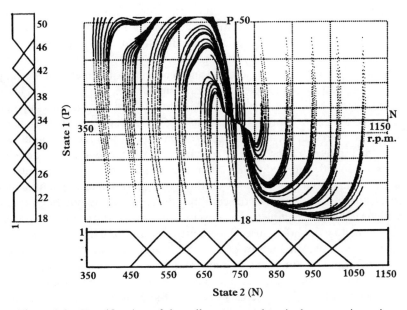

Figure 2.3 Fuzzification of the cell-groups and typical state trajectories

have to belong to only one cell-group, but it can belong to many cell-groups with different degrees of membership in each one. This implies that many rules may be fired at the same time, resulting in smooth transitions of the actual system. The states, along with the control variables, define the premises and consequence of the fuzzy rule base. In our illustrative example of the two-state two-control input system, the general form of the rules is

$$\text{IF } (x_1 \text{ is } F_1) \text{ and } (x_2 \text{ is } F_2) \text{ THEN } (u_1 \text{ is } G_1) \text{ and } (u_2 \text{ is } G_2)$$

An inferencing mechanism, such as Zadeh's compositional rule of inference, is next called upon to determine the fuzzy control action, given a set of state values usually derived from measurements. The final step involves the application of a defuzzification method, such as the mean of area, so that the crisp values of the control variables may be computed and fed as inputs to the system.

We will demonstrate the design methodology with a simple example.

Consider the following 'process':

$$\left. \begin{array}{ll} \dot{x}_1 = & x_2 \\ \dot{x}_2 = -9.25x_1 - x_2 - 0.1x_1^3 + 9.25u \\ y(t) = & x_1(t-0.1) \end{array} \right\} t \leq 5 \text{ or } t \geq 6$$

$$\left. \begin{array}{ll} \dot{x}_1 = & 1.5x_2 \\ \dot{x}_2 = -11.25x_1 - x_2 - 0.1x_1^3 + 9.25u \\ y(t) = & x_1(t-0.1) \end{array} \right\} 5 \leq t \leq 6 \quad (2.6)$$

Such systems typically arise in many engineering applications; they entail nonlinear time-delayed dynamics with random disturbances. When the set point is a step function, $5u(t)$, the closed-loop response is not satisfactory and an FLC is to be designed to address some performance issues. The $x_1 - x_2$ phase plane is quantized into 20 cells corresponding to 20 production rules. Heuristics, resolution and accuracy requirements, as well as partial knowledge of the plant dynamics, usually dictate the number of quantization levels. The automated rule generation routine allows the designer to pick the membership functions for the states and the control input. In this case, Figure 2.4 depicts the fuzzy membership functions for the states x_1, x_2 and the control input u. The invariant and switching manifolds guide the search procedure. Under minimum time conditions, the cell-to-cell trajectories are generated automatically via repeated simulations of the actual process dynamics. Next, the search procedure selects those trajectories that lead in minimum time from every cell in the state space to the equilibrium cell. The resulting fuzzy rule base is shown in Figure 2.5. This completes the development phase of the demonstration. The rule base is applied next, under closed-loop control, to the actual process. The response characteristics are shown in Figures 2.6a and b. Figure 2.6a shows the system output and the control input as functions of time, and Figure 2.6b depicts the phase plane trajectory. We should point out that input–output data could also be used to arrive at an approximate control law. Once the user provides the plant structure and membership

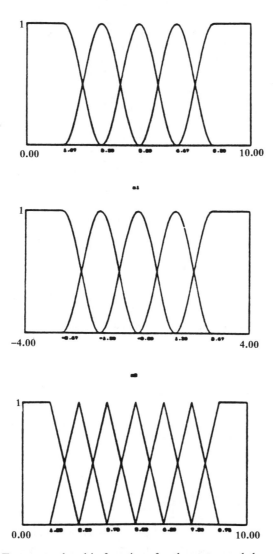

Figure 2.4 Fuzzy membership functions for the states and the control input

function information required, the program generates the 'optimum' rule base automatically. Moreover, it allows the designer to compare control designs under a variety of performance criteria. Multiple-input multiple-output systems are treated in a similar way: the phase space is now partitioned into hypercells with each cell corresponding to a rule in the rule base. Obviously, as the system dimensionality increases so does the size of the rule base. Clustering methods and dedicated fuzzy processors are then called on to limit the number of rules and to expedite the fuzzy inferencing and defuzzification operations.

48 FUZZY LOGIC CONTROL

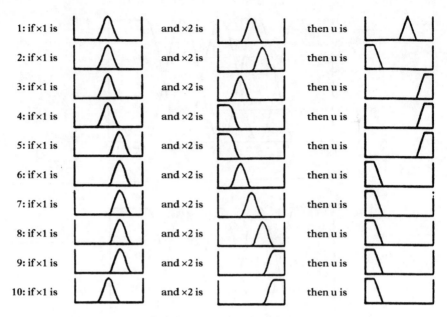

Figure 2.5 Half of the Fuzzy rule base for the test example

For this simple demonstration example, a number of other design approaches were employed, and the results were finally compared. A PID controller based on the Ziegler–Nichols method and a heuristically constructed fuzzy controller did not fare as well as the design arrived at through the phase portrait technique. The latter seems to be more robust to small time delays and disturbances, in addition to providing a smoother response without any overshoot.

2.3 PERFORMANCE ASSESSMENT

The inherent heuristic nature of most FLC design techniques points to a number of problem areas. It is difficult to ascertain the performance of such systems, especially in an objective manner. The systems' structure makes it difficult to study such performance properties as stability, robustness and controllability in a control theoretic framework. The lack of formalism in the design of these systems has given rise to numerous nonstandard ways of assessing stability, most of which are dependent on the design approach adopted. Even though in most of the stability analyses of fuzzy systems robustness is often mentioned and sometimes claimed, a concrete robustness analysis formulated in a systematic framework has been lacking. Controllability in a fuzzy sense is rarely addressed. In this section we are exploring a viable analytical means of assessing the performance of fuzzy logic control systems. Stability and robustness are considered as measures of the closed-loop system's performance. The approach is two-fold: to formulate the stability analysis of a fuzzy rule base on the basis of control theoretic concepts, and to establish the robustness problem in the framework of modern robust control and to develop conditions for robust stability of a wide class of closed-loop fuzzy systems.

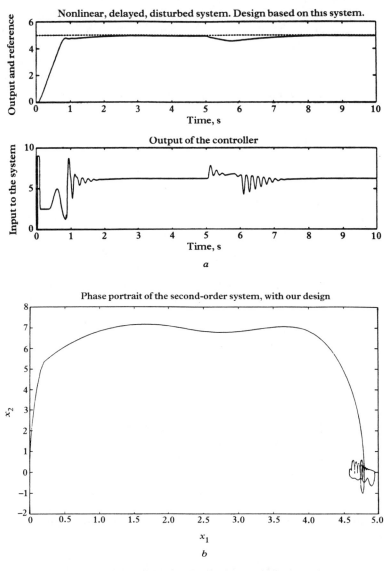

Figure 2.6 Simulation results for the test example

We address the stability of fuzzy logic control systems from an input–output standpoint. The key to the development of an appropriate theory is the formulation of an input–output mapping; as a stability requirement, the FLC should ensure that this mapping is dissipative. In conjunction with stability concerns is the issue of controllability. A controllability condition for fuzzy systems may be formulated based on fuzzy rules and fuzzy relations. As a preliminary step for robustness analysis, the concept of fuzzy sensitivity must be introduced. This allows for the following two objectives to be accomplished: the essence of fuzziness is captured via a direct utilization of membership functions and, therefore, fuzzy numbers and fuzzy rules; an approximation may be derived for the partial derivative (the sensitivity function) in terms of an equivalent expression employing membership functions of the pertinent variables, This greatly simplifies the computational task. With these preliminaries, a new robustness approach may be developed, in the framework of modern robust control, employing fuzzy sensitivity and a performance index, formulated as a Lyapunov-like function that must be minimized.

2.3.1 Stability Analysis

We begin with an observation: let us assume that a fuzzy logic controller has been designed to regulate a particular process about some set point. For typical engineered systems, in the neighbourhood of the process desired goal, or equilibrium state, very little control activity is usually required. This being the case, it is essential that the process be particularly stable in this vicinity. Elsewhere in the state space, the process may exhibit a somewhat different behaviour. For example, a greater control authority may be required to monitor far-off initial states. This is not an uncommon procedure in process set-point adjustment at the outset. We may summarize by suggesting that, in the vicinity of the equilibrium state, asymptotic convergence to the equilibrium state is of prime concern. Elsewhere in the state space, we may be content with a bounded nonlinearity and a bounded error such that the ratio of the former to the latter is vanishingly small. Under this setting, consider the following.

Let the model of the process be approximated by

$$\dot{x} = F(x, u, t), x \in R^2, u \in R^2, t \in R^+; F: R^2 \times R^2 \times R^+ \to R^2 \tag{2.7}$$

Let the desired set-point be $x_d \in R^2$. Define the error by $e = x(t) - x_d$. Suppose, for simplicity, that in the vicinity of the desired goal state, uniform control input sequences predominate for both u_1 and u_2, say

$$u_{1n} = \{NZ, NZ, NZ, \ldots, NZ\} \text{ and } u_{2n} = \{PZ, PZ, PZ, \ldots, PZ\}, \forall k$$

Let the equilibrium set of (2.7) be composed of x_d for some nominal u, determined according to

$$\{F(x, u, t) = 0, \Rightarrow x = x_d, u = u_e\} \tag{2.8}$$

The error system is

$$\dot{e} = F(e, u, t), \text{ with equilibrium state } e = 0, \text{ for some } u = u_e \tag{2.9}$$

This becomes

$$\dot{e}(t) = Ae(t) + g(e, u, t) \tag{2.10}$$

where

$$g(e, u, t) = F(e, u, t) - Ae(t)$$

Suppose that the function $g(e, u, t)$ satisfies the following conditions:

(i) $g(e, u, t)$ is continuous for $\|e\| \le a, t \in [0, \infty)$

(ii) $\lim \|g(e, u, t)\|/\|e\| = 0$ as $\|e\| \to 0$ uniformly with respect to t.

Theorem   If $\mathbb{R}_e \lambda_i(A) < 0$ and if g satisfies (i) and (ii) then the solution, $e \equiv 0$, of the system (2.9) is asymptotically stable. If a is the whole of $F(X)$, then global asymptotic stability results. Condition (i) is satisfied if F in (2.7) is differentiable at x_d (and defined at u_e) and if and only if the Jacobian $\partial F/\partial x$ at $\Omega_d = (x_d, u_e)$ is stable, where Ω_d is a specified set.

For our application, this reduces to the requirement that A given below stable.

$$A = \begin{bmatrix} \dfrac{\partial F_1}{\partial x_1} & \dfrac{\partial F_1}{\partial x_2} \\ \dfrac{\partial F_2}{\partial x_1} & \dfrac{\partial F_2}{\partial x_2} \end{bmatrix}_{\Omega_d} \tag{2.11}$$

To study the stability of the system, under the applied action of the fuzzy rule base, it suffices to show that

(iii) $A|x_d, u_e$ is stable for $u_e \in \{(NZ, NZ, \ldots, NZ), (PZ, PZ, \ldots, PZ)\}$

(iv) $\lim \|g(e, u, t)\|/\|e\| = 0$ as $\|e\| \to 0$, for $(u, x) \in \{F(U) \backslash u_e\} \times \{F(X) \backslash x_d\}$

2.3.2 Robustness Analysis

For robustness analysis, we will require: a controller K, a set of plants F, and a desired property or characteristics Θ. We will refer to the triplet

$$\mathfrak{R} = \{K, F, \Theta\} \tag{2.12}$$

as the robustness set. Any of the elements of $\mathfrak{R}$ can be generic in nature. For instance, in our case $k \in K$ is a fuzzy logic controller.

Now, given a fuzzy logic controller k and the robustness set (2.12), the objective is to achieve the following: k to stabilize the nominal plant $f \in F$, where all the parameters of f are fixed at their nominal values. k to stabilize the perturbed plant $f + \Delta f$, where $|\Delta f|$ is bounded and is due to all reasonably and practically possible perturbations in the plant's parameters of interest and the onset of bounded external disturbances. Obtain an approximate, quantitative measure of the system's robustness.

We are considering a class of nonlinear systems given by

$$F: \dot{x} = f(x', u, t; \alpha), \text{ where } x' \in \mathbb{R}^n, u \in \mathbb{R}^m, t \in \mathbb{R}^+, \text{ and } \alpha \in \mathbb{R}^r \tag{2.13}$$

where

$\alpha = [\alpha_i; i = 1, 2, \ldots, r] = $ a parameter vector with $\alpha_i \in E_\alpha = [\alpha_{\min}, \alpha_{\max}]$
$\alpha_{0i} \in E_\alpha, = $ the nominal value of α_i
$\Delta \alpha_{0i} = $ perturbation in α_i about the nominal value.

52 FUZZY LOGIC CONTROL

Assumption 1 Suppose that the parameter α_i is dominantly a characteristic of the state i dynamics, $\dot{x}_i'$, and so 'cross-coupling' between states is negligible. In other words, a perturbation in the parameter α_i has a negligible effect on the growth of error e_j, $i \neq j$. The results are extended to the case of state couplings.

Definition 1 Define a unit parameter perturbation of a state (or channel) as

$$\frac{\Delta \alpha_i}{\alpha_j} = k_{ij} \in R, i,j = 1, 2, \ldots, r \tag{2.14}$$

Note that Definition 1 allows for cross-coupling between states which we will neglect at this phase but account for below.

Definition 2 The fuzzy sensitivity of the real output variable e_i with respect to a real parameter α_j is expressed by

$$S_{e_i}^{\alpha_j} = \frac{1 - \mu_{e_i}}{1 - \mu_{\alpha_j}}, \; i,j = 1, 2, \ldots, r, \mu = \text{membership function} \tag{2.15}$$

Following Definitions 1 and 2, we have $r = 2$ and x_1 and x_2

$$\{S_{x_1}^{\alpha_1}, S_{x_1}^{\alpha_2}, S_{x_2}^{\alpha_1}, S_{x_2}^{\alpha_2}\}(t), \{k_{11}, k_{12}, k_{21}, k_{22}\} \tag{2.16}$$

We state the main result as a theorem.

Theorem Let the fuzzy controller k stabilize $f(\cdot, \alpha_{0i}) \in F$, $i = 1, 2$ (nominal stability), and suppose that Assumption 1 holds.

The fuzzy controller is robust with respect F if the matrix formed as

$$P = [p_{ij}] = [S_{x_i}^{\alpha_j} k_{ij}], \; i,j = 1, 2, \ldots, r, \text{ is negative definite, for all } t > 0 \tag{2.17}$$

We will refer to P as the total sensitivity matrix. For $r = n = 2$, and Assumption 1, this gives

$$P = \begin{bmatrix} S_{x_1}^{\alpha_1} k_{11} & 0 \\ 0 & S_{x_2}^{\alpha_2} k_{22} \end{bmatrix} < 0 \tag{2.18}$$

Proofs for the stability and robustness theorems may be found in [8]. For the robustness measure, a Lyapunov-like function $V(x)$ is defined and the differential $V(x_1, x_2)$ is required to be negative definite.

Under weak coupling assumptions, the sensitivity matrix is augmented by the coupling terms ε_{12} and ε_{21} and the main result is generalized as:

The fuzzy logic controller will stabilize the whole of F in (13), without assumption 1, if and only if:

$$S_{x_1}^{\alpha_1} k_{11} < 0 \text{ and } (S_{x_1}^{\alpha_1} S_{x_2}^{\alpha_2} - \varepsilon_{12} \varepsilon_{21}) < 0, \forall t > 0. \tag{2.19}$$

A measure of the size of the augmented sensitivity matrix may be given in terms of the supremum of its maximum singular value. As a measure of robustness, therefore, we might consider the condition number of this matrix when this is small and the smallest singular value in not zero. A fuzzy robustness measure is defined next as the ratio of the largest to the smallest value. In an interval between the smallest and largest values of the fuzzy robustness measure, the latter may now be stated linguistically (such as very robust, moderately robust, etc.) with appropriate membership functions assigned to the linguistic labels. Such performance measures are useful to the designer in the selection of optimum design parameters. Details of this approach may be found in [8].

2.4 APPLICATION EXAMPLES

The automatic rule generation routine and the performance assessment algorithms were tested on a number of examples. Among the application domains considered, the idle speed control of an automotive engine, yaw-axis control of a missile autopilot system and a simple inverted pendulum were simulated. Here we present some results from the automotive engine example in order to illustrate the feasibility of the proposed methodology.

2.4.1 The Engine Model (9)

Consider a two-state dynamic model given by [10]:

$$\dot{P} = k_p(\dot{m}_{ai} - \dot{m}_{ao})$$
$$\dot{N} = k_N(T_i - T_l) \tag{2.20}$$

where

$$\dot{m}_{ai} = (1 + 0.907\theta + 0.0998\theta^2)g(P)$$

$$\dot{m}_{ao} = -0.0005968N - 0.1336P + 0.0005341NP + 0.000001757NP^2$$

$$T_i = -39.22 + \frac{325024}{120N}\dot{m}_{ao} - 0.0112\delta^2 + 0.000675\delta N(2\pi/60)$$

$$+ 0.635\delta + 0.0216N(2\pi/60) - 0.000102N^2(2\pi/60)^2 \tag{2.21}$$

$$T_L = (N/263.17)^2 + T_d$$

$$g(P) = \begin{cases} 1 & P < 50.66 \\ 0.0197(101.325P - P^2)^{1/2}, & P \geq 50.66 \end{cases}$$

These equations are highly nonlinear. The variables in the equations are defined below.

P Manifold pressure in kPa
N engine speed in RPM.
δ Spark advance (between 10 and 45 degrees)
θ Throttle angle (between 5 and 35 degrees)

T_d Accessory load between 0 and 61 Nm.
$\dot{m}_{ai}$ Mass flow rate into the manifold.
$\dot{m}_{ao}$ Mass flow rate out of the manifold and into the cylinder.
$g(P)$ Manifold pressure influence function.
k_p Manifold dynamics constant.
k_N Rotational dynamics contant.

The dynamic equations can be transformed into the form (2.7) below with an additional scalar disturbance argument T_d, and also by defining

$$x \underline{\Delta} (x_1 \ x_2) \underline{\Delta} (P \ N) \text{ and } u \underline{\Delta} (u_1 \ u_2) \underline{\Delta} (\theta \ \delta).$$

$$\dot{x}_1 = k_p[a_0 x_2 + a_1 x_1 - a_2 x_1 x_2 - a_3 x_2 x_1^2 + g(x_1)a_4] + [k_p g(x_1)a_5]u_1 + k_p[g(x_1)a_6 u_1^2]$$
$$\underline{\Delta} F_1(x, u, T_d)$$

$$\dot{x}_2 = k_N[-a_7 + a_8 \frac{\dot{m}_{ao}}{x_2} + a_{12} x_2 - (a_{13} + a_{14})x_2^2] + [a_{10} x_2 + a_{11}]k_N u_2 + [-a_9 k_N]u_2^2 + a_{15}T_d]$$
$$\underline{\Delta} F_2(x_2, u, T_d)$$

$\theta \in [5, 35]$ degrees is the throttle angle (idling limits), $\delta \in [10, 45]$ degrees is the spark advance in degrees. The nominal (equilibrium) operating point is typically determined on the basis of operating conditions, minimum energy dissipation, etc. (in the test engine considered it is set at $N = 750$ r.p.m. at a corresponding equilibrium pressure, $P = 34.24$ Kpa).

The constants, a_i, are given bellow:

$$a_0 = 0.0005968, \ a_1 = 0.1336$$
$$a_2 = 0.0005341, \ a_3 = 0.000001757$$
$$a_4 = 1, \ a_5 = 0.907, \ a_6 = 0.0998$$
$$a_7 = 39.22, \ a_8 = 2708.533,$$
$$a_9 = 0.0112, \ a_{10} = 0.000070686$$
$$a_{11} = 0.635, \ a_{12} = 0.002306$$
$$a_{13} = 0.00000118\text{5}6, \ a_{14} = (1/263.17)^2$$
$$a_{15} = -k_N = +54.26, \ k_p = 42.4$$

The errors in P and N are defined as

$$e_p = P - P_0 \equiv x_1 - x_{1d} \text{ and } e_N = N - N_0 \equiv x_1 - x_{2d}$$

Note that since P_0 and N_0 are constant set-points, the design was conducted, without loss of generality, in $x_1 - x_2$ coordinates, instead of in the error coordinates.

We pause for a moment here to caution that the model above, as complex as it looks, is not in its most complex form. For instance, induction to power time delay and other delays inherent in the system have not been considered. Also, certain dynamics such as wall-wetting models, actuator dynamics, exhaust gas recirculation (EGR) and certain thermodynamic effects have not been considered. The model thus considered, though much simplified, maintains its essential nonlinearities in the controls and states. It is also rich in the dynamic elements necessary to produce solutions that are a large subset of those obtained from a more complex model. It is therefore appropriate for our controller design purpose. Moreover, the space spanned by the solutions provided by the simplified model is both practically reachable and non-empty. We want to emphasize that our goal in this controller design section is to introduce a systematic design methodology for fuzzy logic control systems. It is not our goal at the outset to necessarily solve the world's most complex control systems design problem, using fuzzy logic. Moreover, extensions of the methodology to systems with time delays have been completed. One example has already been provided herein. This understanding is very important, for this and any other newly introduced design methodology, in particular, fuzzy control, so as not to miss the essence of a contribution, and possibly reach unwarranted conclusions or create undue hysteria.

2.4.2 The Fuzzy Controller

The quantities of interest are the pressure and speed errors and their time derivatives. The three perturbations, namely spark advance, throttle angle and accessory load, define the premises and consequents. So, we are proposing a five-premise and two-consequent structure. The general form of the rules is given below.

IF $(e_N$ is $F_1)$ and $(e_p$ is $F_2)$ and $(\dot{e}_N$ is $F_3)$ and $(\dot{e}_p$ is $F_4)$ and $(\Delta T_d$ is $F_5)$
THEN $(D\delta$ is $G_1)$ and $(\Delta\theta$ is $G_2)$

where F_i and G_i are the consequent and premise linguistic terms, respectively. We denote these as follows:

$$\begin{array}{ll} \textit{Premise} & \textit{Consequent} \\ x_1 = F_1(e_N) & y_1 = G_1(\Delta\delta) \\ x_2 = F_2(e_p) & y_2 = G_2(\Delta\theta) \\ x_3 = F_3(\Delta T_d) & \\ x_4 = F_4(\dot{e}_N) & \\ x_5 = F_5(\dot{e}_p) & \end{array}$$

We denote by P_i, for $i = 1$ to 5, the premise matrices, and C_1 and C_2 the consequent matrices. In one simple implication of the control strategy, the derivatives of the errors are not included in the scheme and x_4 and x_5 are dropped out. This, of course, does not compromise the generality of the approach. Then

$$P_i = \begin{bmatrix} x_i^{1T} \\ \vdots \\ x_i^{kT} \end{bmatrix}, \quad \text{for } i = 1, \ldots, 3$$

$$C_j = \begin{bmatrix} y_j^{1T} \\ \vdots \\ y_j^{kT} \end{bmatrix}, \quad \text{for } j = 1, 2$$

where k is the number of rules

If we congregate the three-premise fuzzy vectors into one fuzzy set, the latter becomes a three-dimensional premise hypercube of the fuzzy inputs given by

$$x^k[1,\ldots,3] = x_1^k[1] \vee x_2^k[2] \vee x_3^k[3]$$

Similarly, the two-dimensional consequent fuzzy hypercube is formed as

$$y_k[1,2] = y_1^k[1] \vee y_2^k[2]$$

The degree of fulfillment for the ith premise is formulated as

$$d_f^i = P_i \circ \tilde{x}_i, \quad \text{for } i = 1,\ldots,3$$

The overall degree of fulfillment d_f is then formed as

$$d_f = d_f^1 \circ \cdots \circ d_f^3$$

The jth fuzzy output vector $\tilde{y}_j$ is obtained using the max–min composition with respect to the consequent matrix C_j so, $\tilde{y}_j = C_j \circ d_f$, $j = 1,2$.

The fuzzy output vector is finally converted to a crisp control input using any one of the available defuzzification strategies such as the centroid or centre of area method.

As we change (δ_i, θ_j) we compute the approximate tangential directions. For those (δ_i, θ_j), that satisfy 'min energy' or 'min time' or 'min error' trajectories, we assign membership functions μ_θ and μ_δ, respectively. One should also make sure that in addition to min time or min energy or min error, every switching curve at the boundary of the tolerance region is directed towards the origin or toward a primary attracting curve. Next, the representative values for each pair (premise/consequent) for each cell-group are stored as a rule. A search procedure is initiated by finding all possible dynamic transitions to the target cell-group L^*. We extract only one transition based on the optimal strategy. The elements of the finalized tree constitute a control role base. These rules are automatically generated on the basis of the optimal strategy chosen. Simply, the transitional relations that force the trajectories from any points in the state space to the desired goal within the prescribed tolerances are themselves the control rules for feedback regulation. The trade-offs between the number of quantization levels in L_{mn} and the smoothness in transition, or the total numbers of E_{1m} and E_{2n} and the controller performance are resolved via heuristics. Given the fuzzy sets $\tilde{E}_1$ in

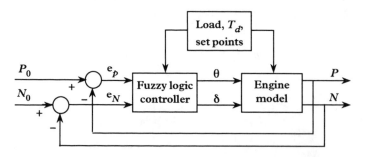

Figure 2.7 Block diagram of main fuzzy logic control loop

APPLICATION EXAMPLES 57

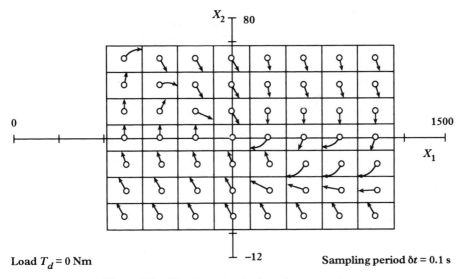

Figure 2.8 The phase portrait in the error space

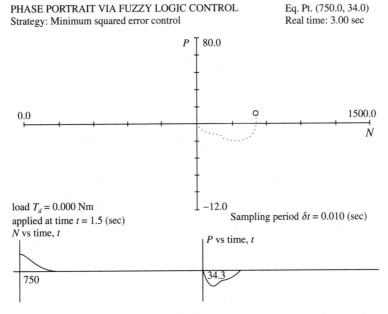

Figure 2.9 Phase plane trajectories and time response characteristics under minimum squared error control

58   FUZZY LOGIC CONTROL

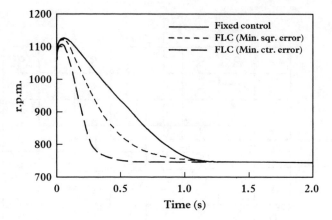

Figure 2.10   Engine rotor speed as a function of time

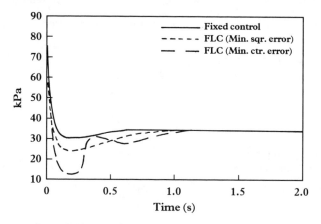

Figure 2.11   Engine pressure as a function of time

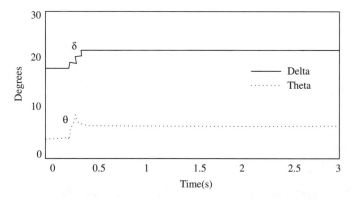

Figure 2.12   Throttle opening and spark advance as functions of time

E_{1m}, and $\tilde{E}_2$ in E_{2n}, the control input fuzzy sets $\tilde{\Delta}$ and $\tilde{\theta}$ are obtained as

$$\tilde{\Delta} = C_1^T \circ [(P_1 \circ \tilde{E}_1) \vee (P_2 \circ \tilde{E}_2)]$$

$$\tilde{\Theta} = C_2^T \circ [(P_1 \circ \tilde{E}_1) \vee (P_2 \circ \tilde{E}_2)]$$

where C_i^T is the transpose of the matrix C_i and $\circ$ is the element-wise minimum operator. The crisp results for Δ and Θ are $\delta = \text{DEFUZZ}(\tilde{\Delta})$ and $\theta = \text{DEFUZZ}(\tilde{\theta})$, where $\text{DEFUZZ}(\cdot)$ is a defuzzification operator.

2.4.3 Simulation Results

Figure 2.7 is the block diagram of the main fuzzy logic control loop employed in the simulation studies. The engine model has been tested separately under various operating conditions and was found to be in good agreement with experimental results provided by the Ford Motor Company. Figure 2.8 depicts the phase portrait of the cell-groups selected by the automatic rule generation algorithm and the associated search procedure. Trajectories of the first cell column are not used in the rule base, since they are not part of the stable regime. Optimum results are obtained when the rule base is composed of 25 rules. The simulation studies showed desirable properties namely, stability and robustness with respect to small accessory load perturbations. Figure 2.9 depicts the phase plane and time trajectories for minimum squared error control. Figures 2.10 and 2.11 show the engine speed and manifold pressure as functions of time, respectively. Comparisons are drawn for fuzzy logic control (minimum squared error and minimum control effort) and a fixed state feedback control law. It is obvious that a significant improvement is achieved via the rule based controller. Finally, Figure 2.12 shows the control action required for throttle angle and spark advance under minimum squared error conditions.

2.5 STABILITY AND ROBUSTNESS RESULTS

For the automotive engine example, the fuzzy control is uniform in the region around the equilibrium point $x = (34.250 \text{ kPa}, 750 \text{ r.p.m.})$. The sequence of control inputs in this vicinity is $u_{1n} = \{6.01\}$ and $u_{2n} = \{28.1\}$, for all time instants k. with these values, the matrix A in (2.11) for this system is determined to be

$$A = \begin{bmatrix} \dfrac{\partial F_1}{\partial x_1} & \dfrac{\partial F_1}{\partial x_2} \\ \dfrac{\partial F_2}{\partial x_1} & \dfrac{\partial F_2}{\partial x_2} \end{bmatrix}_{\{(51.0, 750)\,(6.01, 28.1)\}} = \begin{bmatrix} -15.147 & -0.8377 \\ 69.9253 & 1.3059 \end{bmatrix}$$

This Jacobian matrix is nonsingular and has eigenvalues of -9.9368 and -3.9044, thus satisfying condition (iii). The function g in this case is constructed as

$$g_1 = F_1(x, u, t) - \left(\dfrac{\alpha F_1}{\alpha x_1} e_p + \dfrac{\alpha F_1}{\alpha x_2} e_N \right)$$

$$g_2 = F_2(x, u, t) - \left(\dfrac{\alpha F_2}{\alpha x} e_p + \dfrac{\alpha F_2}{\alpha x} e_N \right)$$

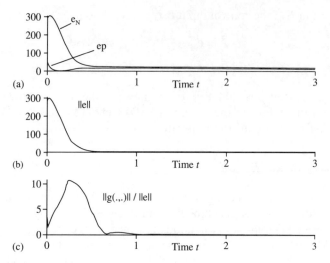

Figure 2.13 (a) Errors in N and P, (b) norm of e, (c) norm ratio

The Euclidean norm is used on g and e, that is

$$\|g(.,.)\| = (g_1^2 + g_2^2)^{1/2}$$
$$\|e(\cdot)\| = (e_P^2 + e_N^2)^{1/2}$$

Condition (iv) is satisfied for various initial conditions, in the state space, with their associated fuzzy control sequences, via simulation of the norm ratios. For relatively simple systems it may not be too difficult to evaluate, analytically, the limit of $\|g(.,.)\|/\|e(..)\|$, as $\|e(.)\| \to 0$. However, for this system, a lengthy and possibly 'brute force' algebra is avoided by computer simulation of the ratio. Where the norm of the error does tend to zero, the ratio is checked to see if it too tends to zero, thus satisfying condition (iv). One result of this procedure is provided as an example. The initial point, $(x_1, x_2) = (51.0\,\text{kPa}, 1050\,\text{r.p.m.})$, falls in a region with the fuzzy control sequence $(u_1, u_2) = (26.1, 4.01)$ degrees. These values are substituted into the norm ratio and simulated. The errors and the ratio are seen to both converge to zero, as shown in Figure 2.13. It is concluded from this that the fuzzy control values drive the system exponentially to a stable equilibrium at the origin, and that the control rules elsewhere in the state space, lead to a bounded and diminishing nonlinearity (Figures 2.13 and 2.14). The significance of satisfying (ii) is that one is now able to say that $\|g(.,.)\| = O.\|e\|$, uniformly with respect to u, as $\|e\| \to 0$. Results of this simulation show that the ratio of the norms, $0 = 4.61823 \times 10^{-5}$, is indeed of order zero. Alternatively, a necessary and sufficient condition on the nonlinear residuals is that $\lim g_i \to 0$ as $t \to \infty$, $\forall u$. This is shown to be the case in Figure 2.14. The plots of the residuals g_1 and g_2 show their uniform diminishing effect, under the fuzzy controls. This is the most crucial of the two conditions to satisfy, because it implies that $F(t, 0) = 0$, hence $e(t) = 0 \Rightarrow x_d = (34.24, 750)$, for the subset of fuzzy rules considered, is asymptotically stable. Satisfying this for all $u \in F(U) \setminus u_e$ and $x_0 \in F(X) \setminus x_d$ implies stability of the rulebase. The fact that F is differentiable at x_d ensures the uniqueness of this solution, which is the equilibrium point. This result is a statement of L_2 stability.

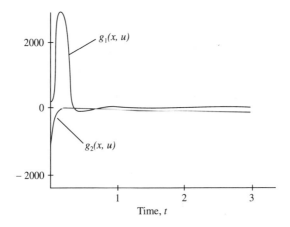

Figure 2.14 Nonlinear Residuals $g_1(x, u)$ and $g_2(x, u)$

Simulation studies were used to investigate the robustness characteristics of the automotive engine under fuzzy control. Variations of the parameters k_p and k_N in the engine model (2.20) and (2.21) were considered above and below their nominal values. The unit parameter perturbations, fuzzy sensitivity measures and cross-coupling terms were computed; moreover, estimates of the differential $dV(x_1, x_2)$ and the fuzzy robustness measure were arrived at. An optimum fuzzy robustness set was then sought such that these set-point deviations are minimized. The system designer can now use these results in order to assess the trade-offs between minimum steady-state errors and minimum fuzzy robustness measures as well as to determine the best parameter values for a particular application.

2.6 CONCLUSIONS

The proposed systematic fuzzy logic control design methodology uses an approximate state-space representation of a nonlinear dynamic system to develop the fuzzy rule base. The procedure guarantees completeness (since the cell-groups or rules cover the whole state space) and optimality in some sense (since an exhaustive search technique forms the basis for choosing the final trajectories). The simulation results indicate the effectiveness of the fuzzy logic spelling: control tool in terms of stability and robustness to external disturbances. We should note, however, that the procedure is not without some drawbacks. Complexity abounds easily as the number of inputs increases. The search can become very tasking indeed as the number of cells becomes large. Design time can also be considerably long in factoring the heuristic part of the information required for the input data. Once, though, this information has been input, the program generates the rule base over a matter of minutes on a conventional PC platform. The tool-kit is user-freiendly. The analysis algorithms provide a basis for performance assessment as well as for optimum design purposes. They integrate well with the design platform into a unified approach for fuzzy dynamic systems.

ACKNOWLEDGMENT

This research work was sponsored in part by Ford Motor Company and Georgia Tech's Manufacturing Research Center; their continued support is gratefully acknowledged.

REFERENCES

1. Mamdani, E. H. (1974) Applications of fuzzy algorithms for simple dynamic plant. *Proceedings of the IEE*, **121**, 1585–1588.
2. Sugeno, M. (1985) An introduction survey of fuzzy control. *Information Sciences*, **36**, 59–83.
3. Tong, R. M. (1980) Some properties of fuzzy feedback systems. *IEEE Transactions on Systems, Man and Cybernetics*, **10**, 327–330.
4. Cumani, A. (1982) On a possibilistic approach to the analysis of fuzzy feedback systems. *IEEE Transactions on Systems Man and Cybernetics*, **12**, 417–422.
5. Hsu, C. S. (1980) A theory of cell-to-cell mapping dynamical systems. *ASME Journal of Applied Mechanics*, **47**, 931–939.
6. Chen, Y. Y., and Tsao, T. C. (1989) A description of the dynamical behavior of fuzzy systems. *IEEE Transactions on Systems, Man and Cybernetics*, **19**, 745–754.
7. Palm, R. (1992) Sliding mode fuzzy control. *IEEE Conf. on Fuzzy Systems*, pp. 519–526.
8. Farinwata, S. (1993) Performance assessment of fuzzy logic control systems via stability and robustness measures. PhD dissertation, Georgia Institute of Technology.
9. Vachtsevanos, G., Farinwata, S., and Pirovolou, D. (1993) Fuzzy control of an automotive engine. *IEEE Control Systems Magazine*, **13**, 62–68.
10. Olbrot, A. W., and Powell, B. K. (1989) Robust design and analysis of third and fourth order time delay systems with application to automotive idle speed control. *1989 ACC*, vol. 2, pp. 1029–1039.
11. Power, B. K., and Cook, J. (1987) Non-linear low frequency phenomenological engine modeling and analysis. *1987 ACC*, pp. 332–340.
12. Crossley, P. R., and Cook, J. A. (1991) A nonlinear engine model for drivetrain system development. *IEE Control Conference*, UK.

3
On the Compatibility of Fuzzy Control and Conventional Control Techniques

R. Palm
Siemens AG Corporate Research and Development, Munich, Germany

3.1 INTRODUCTION

Fuzzy control can be divided into two groups with respect to two main aspects: the first group deals with pure fuzzy systems in which both the signals and the model of the system to be controlled are described by fuzzy sets and respective relations. Most applications, however, use system descriptions based on crisp models, e.g. differential equations, with a nonlinear control element (the fuzzy controller). With this point of view the question arose of how to determine stability, performance and robustness of such mixed crisp-fuzzy systems. Very soon one came to the conclusion that conventional linear and nonlinear control theory is able to contribute very much to deal with such mixed or hybrid systems. To solve the problems of stability, performance and robustness for such systems one is forced to create a common basis in which the fuzzy world is compatible to the crisp one. This common basis can be translation of the whole control loop into fuzzy terms. The other approach is to represent the control loop with all its elements, e.g. the fuzzy controller, as pure conventional and crisp systems, respectively.

Most approaches prefer the second way, since the conventional techniques of linear and nonlinear control (e.g. Lyapunov stability) can be adopted easily.

In the following, we want to address three major issues:

(a) The similarity between fuzzy control and VSS-systems (Variable Structure Systems with Sliding Modes)

(b) The representation of a fuzzy controller as a nonlinear control element and its interpretation as an equivalent gain.

(c) Noisy signals in the control loop and their interpretation as fuzzy signals.

For the next three sections these viewpoints show the compatibility of fuzzy control and nonlinear conventional control, where the individual issues can briefly be described as follows.

Most fuzzy controllers (FCs) for nonlinear second-order systems are designed with a two-dimensional phase plane in mind. In the first section we show that the performance and the robustness of this kind of FC stem from their property of driving the system into the sliding mode (SM) in which the controlled system is invariant to parameter fluctuations and disturbances. Additionally, the continuous distribution of the control values in the phase plane causes a behaviour similar to that of a sliding mode controller (SMC) with a boundary layer (BL) near the switching line. This gives assured tracking quality even in the presence of high model uncertainties. By tracing the FC back to the principle of an SMC, one obtains evidence about the stability of the closed-loop system. The choice of the scaling factors for the crisp inputs and outputs can be guided by the comparison of the FC with the SMC and with the modified SMC, respectively. Finally, an FC for a higher-order system is proposed. Simulation results show the practicability of the method.

The second section deals with the optimal adjustment of input scaling factors for fuzzy controllers. The method is based on the assumption that in the stationary case an optimally adjusted input scaling factor meets a specific statistical input–output dependence. A measure for the strength of statistical dependence is the correlation function and the correlation coefficient, respectively. The adjustment of input scaling factors using correlation functions is, without loss of generality, pointed out by means of a single input, single output (SISO) system. First, the section deals with the so-called equivalent gain which is closely connected to the cross-correlation of the controller input and the defuzzified controller output. The section concludes with an example with respect to a system of fuzzy rules controlling a redundant robot manipulator.

The third section deals with fuzzy signals at the controller input. Almost all theoretical results and applications of fuzzy control deal with crisp signals fed to the input of the controller. In the next steps they are fuzzified and further processed in an inference step. The result is a fuzzy set that is defuzzified by means of an appropriate defuzzification method. However, this is a simplification of the actual situation in a control loop. Normally, the system states are attainable only by means of sensors measuring the states in such a way so that the task 'reaching the desired goal x_d' can be performed. Thus, one has to take into account the negative side effects that come up with the use of sensory information like noise or spatial distribution of a signal. Much has been published about noise in the control loop (Schlitt, 1968, Kalman and Bucy, 1963). On the other hand, it would be also of interest to know how to deal with signals represented by an amplitude distribution (statistics). Signals of this type together with statistical ensembles of measurements can be interpreted as fuzzy signals. It would be advantageous if such a fuzzy signal could be processed by the controller directly instead of handling a number of parameters describing the fuzzy signal. In this section some specific operations with respect to fuzzy signals are defined:

  differentiation of a fuzzy set with respect to time
  thresholds
  the sgn-function
  the sat-function

With regard to fuzzy signals, pure sliding model, sliding mode with boundary layer, and

3.2 SLIDING MODE FUZZY CONTROL

For a large class of second-order nonlinear systems fuzzy controllers are designed with respect to a phase plane determined by error e and change of error $\dot{e}$ in relation to the states x and $\dot{x}$ (Ray and Majumder, 1984, Tang and Mulholland, 1987, Wakileh and Gill, 1988). A fuzzy value for the control variable is determined with respect to fuzzy values of error and change of error. The general approach to control design is the partition of the phase plane into two semi-planes by means of a switching line. Within the semi-planes positive and negative control outputs are produced, respectively. The magnitude of the control outputs depends on the distance of the state vector from the switching line. Although fuzzy control is very successful, especially for the control of nonlinear systems, there is a lack in design of such controllers with respect to performance and stability of FC controlled systems. On the other hand, considering the large field of FC applications one can ask why fuzzy control is so successful especially in the presence of disturbances and ill-defined knowledge about the system. The reason is that FC is similar to sliding mode control which is, for a specific class of nonlinear systems, an appropriate robust control method (Utkin, 1977, Slotine, 1985). SMC can be applied especially in the presence of model uncertainties, parameter fluctuations and disturbances, provided that the upper bounds of their absolute values are known.

3.2.1 The Principle of Sliding Mode Control

To show the relation between FC and SMC the principle of the latter will briefly be pointed out. Let

$$x^{(n)}(t) = f(x,t) + u + d(t) \tag{3.1}$$
$$x(t) = (x, \dot{x}, \ldots, x^{(n-1)})^T$$

with $x(t)$ as the state vector, $d(t)$ as the disturbances and u as the manipulated variable. Furthermore, let $f(x,t)$ be a nonlinear function of the state vector x and of time t. The control problem is to obtain the state x for tracking a desired state x_d in the presence of model uncertainties and disturbances.

We then define upper bounds of $\Delta f = f - \hat{f} < \tilde{F}, d < D, \ddot{x}_d < v$ with $\hat{f}$ as an estimate of f.
With the tracking error

$$e = x(t) - x_d(t) = (e, \dot{e}, \ldots, e^{(n-1)})^T \tag{3.2}$$

a sliding surface (switching line for second-order systems)

$$s(x,t) = 0 \tag{3.3}$$

$$s(x,t) = (d/dt + \lambda)^{n-1} e, \lambda \geq 0 \tag{3.4}$$

is defined. Starting from the initial conditions

$$e(0) = 0 \tag{3.5}$$

the tracking problem $x = x_d$ can be considered as the state vector e remaining on the sliding surface $s(x, t) = 0$ for all $t \geq 0$. The corresponding sliding condition is

$$\dot{s} \cdot \text{sgn}(s) \leq -\eta \tag{3.6}$$

with $\eta > 0$. To achieve the sliding mode of equation (3.6), we choose

$$u = (-\hat{f} - \lambda \dot{e}) - K(x, t) \cdot \text{sgn}(s) \quad \text{with } K(x, t) > 0 \tag{3.7}$$

where $(-\hat{f} - \lambda \dot{e})$ is a compensation term. By installing the upper bounds one obtains finally

$$K(x, t)_{\max} \geq \tilde{F} + D + v + \eta \tag{3.8}$$

A certain disadvantage of this method is the drastic changes of the manipulated variable which leads to high stress for the plant to be controlled. However, this can be avoided by means of a boundary layer near the switching line which smoothes out the control behaviour and ensures that the system states remain within this layer. Suppose that the upper bounds of the model uncertainties, etc., are known; stability and high performance of the controlled system are guaranteed. Therefore, we substitute the function sgn(s) by sat(s/Φ) in equation (3.7):

$$u = (-\hat{f} - \lambda \dot{e}) - K(x, t) \cdot \text{sat}(s/\Phi) \tag{3.9}$$

where

$$\text{sat}(x) = \begin{cases} x, & \text{if } |x| < 1 \\ \text{sgn}(x), & \text{if } |x| \geq 1 \end{cases}$$

This substitution corresponds to the introduction of a boundary layer (BL) $|s| \leq \Phi$, where Φ represents the width of the BL. From equations (3.1), (3.6) and (3.9) follows the filter function for a second-order system

$$\dot{s} + K(x, t) \cdot \frac{s}{\Phi} = \Delta f + d - \ddot{x}_d \tag{3.10}$$

with the input $\Delta f + d - \ddot{x}_d$. The output s of the filter is the distance to the switching line.

In principle, FCs work like modified SMCs (Palm, 1989, Kawaji, 1991, Palm, 1992a, Hwang and Li, 1992). Compared with an ordinary SMC, however, an FC gives a input–output transfer characteristic which is more adaptable to the state space of the system to be controlled. Moreover, the design of the FC can be performed by means of linguistic control rules, which is an advantage in the case of the change of control structures. In the following, the design principles of a fuzzy controller are explained. Then the relation between the membership functions and the transfer function (operating line) is discussed. Coming from the SMC principle with boundary layer, the similarities between FC and SMC are brought out concerning stability and robustness. The determination of the break frequencies of the controller is discussed. Moreover, the method is extended to the n-dimensional case where an easy way of saving rules is performed. The advantage of sliding mode fuzzy control over ordinary SMC is shown by means of the dynamical simulation of a robot effector acting on a surface.

3.2.2 The Similarity Between SMC and FC

We now discuss the high similarity between an FC whose rules have been derived from the phase plane and the SMC with boundary layer.

Similarly to the SMC, the rules are so conditioned that above the switching line $s = 0$ a negative control output is generated and a positive one below it. Let $K_{\text{Fuzz}}(e, \dot{e}, \lambda)$ be the absolute value of the control output of the FC. Then the working principle of a FC can be interpreted by

$$u = -K_{\text{Fuzz}}(e, \dot{e}, \lambda) \cdot \text{sgn}(s) \qquad (3.11)$$

Figure 3.1 shows an example of a phase plane where, close to the sliding surface (switching line) control, outputs are smaller than at a longer distance from it. Neglecting the compensation term $-\hat{f} - \lambda \cdot \dot{e}$ of the SMC with boundary layer in equation (3.9) and comparing the result with the FC (3.11) we come to the conclusion: fuzzy control, in the above-mentioned sense, is an extension of sliding mode control with boundary layer.

The FC with partial compensation has the form

$$u = -\lambda \dot{e} - K_{\text{Fuzz}}(e, \dot{e}, \lambda) \cdot \text{sgn}(s) \qquad (3.12)$$

which is a hybrid version of a conventional compensation strategy and fuzzy control. Due to the property of the fuzzy algorithm to produce a control value depending on the directional distance $s = \text{const.}$ of the state vector $e = (e, \dot{e})^{\text{T}}$ from the switching line $s = 0$

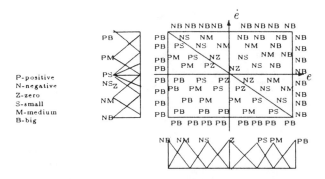

Figure 3.1 Rules in the normalized phase plane

(see equation (3.12)) one automatically also obtains a type of BL. However, this BL depends strongly on the number and shape of the membership functions used in the rules.

For a fuzzy controller whose input is an error vector e and whose output is a scalar control value $\bar{u}$ the general design rules may be formulated as follows.

Rule 1 The normalized control value u_N (controller output) should be negative above the switching line and positive below it.

Rule 2 Normalized states (controller inputs) $e_N, \dot{e}_N, \ldots$, that fall out of the phase plane should produce maximum values $|u_N|_{\max}$ with the respective sign of u_N.

Rule 3 $|u_N|$ should increase as the directional distance s between the actual state and the switching line $s = 0$ grows.

Rule 4 $|u_N|$ should increase as the distance di between the actual state and the line perpendicular to the switching line grows, for the following reasons:

normalized states $e_N, \dot{e}_N, \ldots$, that are situated at the boundaries inside the phase plane produce maximum values $|u_N|_{\max}$ with the respective sign of u_N so that discontinuities at the boundaries of the phase plane can be avoided (see Rule 2)

the central domain of the phase plane can be arrived at very quickly.

3.2.3 The Sliding Mode FC (SMFC) as a State-dependent Filter

Generally, in contrast to the SMC with BL the FC generates an approximately piecewise linear function $u = f(s)$ (see Figure 3.2). Yet, with an increasing number of membership functions $u = f(s)$ becomes linear [Tang and Mulholland, 1987]. Hence, similarly to equation (3.10) we obtain the following filter function for the ith segment

$$\dot{s} + \frac{k_i}{\phi_i} \cdot s = u_i \cdot \mathrm{sgn}(s) + f(x,t) + d - \ddot{x}_d \qquad (3.13)$$

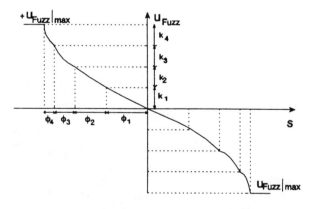

Figure 3.2 Nonlinear operating line

with

$$u_i = \begin{cases} -\sum_{v=1}^{i-1} k_v + \dfrac{k_i \cdot \sum_{v=1}^{i-1} \phi_v}{\phi_i}, & \text{if } i \geq 2 \\ 0, & \text{if } i = 1 \end{cases} \quad (3.14)$$

where $k_v, \phi_v > 0$, $i = 1$ to n, n is the number of segments.

Equation (3.13) is a state-dependent filter with different break frequencies k_i/ϕ_i. This can be utilized to choose different approach velocities for s depending on the distance between the state and the switching line. For a large directional distance s between the state vector e and the switching line $s = 0$ no unmodelled frequencies are able to cause a change of sign of the control-output variable. Therefore, for a large s we may choose a bigger control-output value than for a small s.

To produce such a transfer function the inequalities

$$\frac{k_1}{\phi_1} \leq \frac{k_2}{\phi_2} \leq \cdots \leq \frac{k_2}{\phi_2}$$

must hold.

On the other hand the different slopes of the nonlinear operating line closely correspond to the shape and position of the membership functins especially to the fixed points in the input domain, where the degrees of membership are equal to one. In the example of the last section these relations are

$$\frac{k_1}{\phi_1} = \frac{c_2}{a}, \quad \frac{k_2}{\phi_2} = \frac{c_1}{\frac{2}{3} \cdot a}$$

This also means that tuning of the transfer function in relation to a performance criterion can be performed by changing the shape of the input membership functions. This, however, can easily be carried out by appropriate fuzzy tools (SieFuzzy 2.0, 1995). Tracing back to design principles: related to SMC the following condition

$$\frac{k_i}{\phi_i} \leq \lambda \quad (3.15)$$

has to be fulfilled only in the vicinity of the origin of the phase space, i.e. for $i = 1$.

Moreover, the tracking quality is guaranteed by the maximum values $K_{\text{Fuzz}}|_{\max} = \sum_{v=1}^{i-1} k_v$ and $\Phi_{\max} = \sum_{v=1}^{i-1} \phi_v$ so long as

$$\frac{K_{\text{Fuzz}}|_{\max}}{\Phi_{\max}} \leq \lambda \quad (3.16)$$

3.2.4 Normalization and Denormalization

Concerning input normalization, normalization is a linear state-transformation. The actual processing of the fuzzy rules takes place within the normalized phase plane. The switching line $s = 0$ has to be transformed as follows: within the non-normalized phase plane we have $\lambda \cdot e + \dot{e} = 0$. In the normalized plane we obtain $\lambda_N \cdot e_N + \dot{e}_N = 0$. By means of the relationship

$$e_N = e \cdot N_e, \quad \dot{e}_N = \dot{e} \cdot N_{\dot{e}}, \quad N_e, N_{\dot{e}} \text{ are normalization factors} \tag{3.17}$$

one directly obtains

$$\lambda = \lambda_N \cdot \frac{N_e}{N_{\dot{e}}} \tag{3.18}$$

Because of the design rule $\lambda \ll v_{su}$, where v_{su} are unmodelled frequencies, one obtains

$$\frac{N_e}{N_{\dot{e}}} \ll \frac{v_{su}}{\lambda_N} \tag{3.19}$$

However, this leads only to the quotient of the normalization factors $N_e/N_{\dot{e}}$. To determine N_e or $N_{\dot{e}}$ separately we should use different methods (Palm, 1993).

Concerning output denormalization, the only structural difference between the FC of equation (3.11) and the SMC of equation (3.7) is the compensation term. If there is no sufficient model of the nonlinear part $f(x, t)$, the estimate of the upper bound of $f(x, t)$ has to be modified:

$$\tilde{F} = |f(x, t)_{\max}| \tag{3.20}$$

If we now return to (3.8), we obtain for the maximum required of K_{Fuzz}

$$K_{\text{Fuzz}}|_{\max} \geq \tilde{F} + D + v + \eta \tag{3.21}$$

where

$$K_{\text{Fuzz}}|_{\max} = \max(K_{\text{Fuzz}}(e, \dot{e}, \lambda)) \tag{3.22}$$

From this, the denormalization factor N_u can be calculated as follows.

Let $\mu_{U_{\max}}(u)$ be the membership function for the largest possible control value in the normalized phase plane (e.g. $\mu_{PB}(u)$). Let, furthermore, $u_{N_{\max}} = \text{Defuzz}(\mu_{U_{\max}}(u))$ be the corresponding defuzzified and normalized control value, and finally $K_{\text{Fuzz}_N}|_{\max} = |u_{N_{\max}}|$. The Defuzz operation uses any defuzzification method, e.g. centre of gravity. From this and $K_{\text{Fuzz}}|_{\max} = N_u \cdot K_{\text{Fuzz}_N}|_{\max}$, the denormalization factor N_u follows directly:

$$N_u = \frac{K_{\text{Fuzz}}|_{\max}}{K_{\text{Fuzz}_N}|_{\max}} \tag{3.23}$$

3.2.5 FC with Boundary Layer

In most cases where the error vector represents the controller input the corresponding control output values are symmetrical in relation to the switching line but with different sign. One can therefore save half of the rules if they are restricted to a semiplane. The resulting control value $-K_{\text{Fuzz}}$ is then multiplied by $\text{sgn}(s)$. However, this method has a disadvantage.

Because of the fact that input and output domains have exclusively positive values the defuzzification by the centre of gravity always leads to a control output greater than zero. Together with the sign of s this causes a jump from $-K_{\min}$ to $+K_{\min}$ at the origin of the transfer function of the controller. It is therefore useful to install a BL at the FC. According to equations (3.9) and (3.12), we get the modified control law for a FC with BL:

$$u = -\lambda \cdot \dot{e} - K_{\text{Fuzz}}(e, \dot{e}, \lambda) \cdot \text{sat}(s/\Phi) \qquad (3.24)$$

This gives the following advantages:

(a) at the boundary of the $(e, \dot{e})$-plane, drastic changes of u can be avoided

(b) inside the $(e, \dot{e})$ plane and close to the switching line condition (3.15) is fulfilled if Φ is chosen as follows: let $|u|_{\min}$ be the minimum value of the crisp (defuzzified) value of u. Then Φ has to be $\Phi = |u|_{\min}/\lambda$

(c) the number of rules are reduced, i.e. one can choose a minimum set of rules generating. however, a relatively rough gradation of u.

The FC with BL provided adaptive tracking quality even under changing process parameters. It achieves this both inside and outside the layer, together with a filtering of unmodelled frequencies.

3.2.6 FC of Higher Order

Concerning normalization factors, a FC of higher order can be derived from a higher order SMC with BL in analogy to the method just shown for SMC and FC. From the ith rule of an nth-order FC

$$R_i: \text{if } (e \text{ is } LE_0^{(i)}) \text{ and } (\dot{e} \text{ is } LE_1^{(i)}) \text{ and} \ldots \text{and } (e^{(n-1)} \text{ is } LE_{NM_1}^{(i)}) \text{ then } u \text{ is } LU^{(i)}$$

where $LE_k^{(i)}$ and $LU^{(i)}$ are the attributes with respect to the state variable $e^{(k)}$ and the control variable u, respectively, it is evident that also in the high-order case one has to pay attention to the right choice of normalization factors. We obtain the normalization factors $N_e, N_{\dot{e}}, \ldots, N_{e^{(n-1)}}$ through the following consideration: with respect to the non-normalized phase plane we have

$$\left(\frac{d}{dt} + \lambda\right)^{(n-1)} e = e^{(n-1)} + \binom{n-1}{1}\lambda e^{(n-2)} + \binom{n-1}{2}\lambda^2 e^{(n-3)} + \cdots + \lambda^{n-1} e = 0 \qquad (3.25)$$

and for the normalized phase plane

$$\left(\frac{d}{dt} + \lambda_N\right)^{(n-1)} e_N = e_N^{(n-1)} + \binom{n-1}{1}\lambda_N e_N^{(n-2)} + \binom{n-1}{2}\lambda_N^2 e_N^{(n-3)} + \cdots + \lambda_N^{n-1} e_N = 0$$

$$(3.26)$$

With normalization (scaling) factors, N_e, $N_{\dot e}$, ..., equation (3.26) can be rewritten as

$$e^{(n-1)} + \binom{n-1}{1}\lambda_N \cdot \frac{N_{e^{(n-2)}}}{N_{e^{(n-1)}}} e^{(n-2)} + \cdots + \lambda_N^{n-1} \frac{N_e}{N_{e^{(n-1)}}} e = 0 \tag{3.27}$$

The comparison of the coefficients of equations (3.26) and (3.27) leads to

$$\lambda = \lambda_N \cdot \frac{N_{e^{(n-2)}}}{N_{e^{(n-1)}}}, \lambda^2 = \lambda_N^2 \cdot \frac{N_{e^{(n-3)}}}{N_{e^{(n-1)}}}, \ldots, \lambda^{n-1} = \lambda_N^{n-1} \cdot \frac{N_e}{N_{e^{(n-1)}}} \tag{3.28}$$

From this we obtain the design rule for the normalization factors.

$$\frac{N_{e^{(n-2)}}}{N_{e^{(n-2)}}} = \cdots = \frac{N_e}{N_{\dot e}} = \frac{\lambda}{\lambda_N} \tag{3.29}$$

This means that after selecting of the parameters λ, λ_N and N_e all normalization factors from $N_{\dot e}$ to $N_{e^{(n-1)}}$ are already determined.

Concerning reduction of the number of rules, due to the number of input values and their related fuzzy attributes, the minimum number of rules yields

Number of rules = number of input variables × number of fuzzy attributes assigned to each input variable.

The connection of normalized inputs through their corresponding normalization factors shows that there is no reason to assign a specific rule to each combination of input variable and fuzzy attribute. The better way is to transform the elements of the state vector e_N in such a way that by forming the rules only one or two variables are taken into account.

Therefore, a simple method for to build a FC of higher order is the following: according to Rule 4 for the two-dimensional case (see the preceding section) let

$$\boldsymbol{n}_N = \frac{\left(1, \binom{n-1}{1}\lambda_N, \binom{n-1}{2}\lambda_N^2, \ldots, \lambda_N^{n-1}\right)^T}{\left|\left(1, \binom{n-1}{1}\lambda_N, \binom{n-1}{2}\lambda_N^2, \ldots, \lambda_N^{n-1}\right)^T\right|} \tag{3.30}$$

be the normal vector of the sliding surface within the normalized $(n-1)$-dimensional phase plane. Furthermore, let $s_p = \boldsymbol{e}_N^T \cdot \boldsymbol{n}_N$ be the projection of the state vector $\boldsymbol{e}_N$ on the direction of the normal vector $\boldsymbol{n}_N$ (the directional distance from $\boldsymbol{e}_N$ to the sliding surface). Hence, one obtains the shortest Euclidean distance di (the perpendicular) between $\boldsymbol{e}_N$ and the direction of $\boldsymbol{n}_N$:

$$di = \sqrt{|\boldsymbol{e}_N|^2 - s_p^2} \tag{3.31}$$

Now, the distance s_p between $\boldsymbol{e}_N$ and the sliding surface and the distance di between the direction of the normal vector and the sliding surface are evaluated by fuzzy rules. We

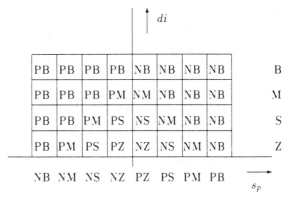

Figure 3.3 s_p-di diagram

assign a fuzzy set to each of the variables s_p, di and u_N and obtain from this the s_p – di diagram (see Figure 3.3). The attributes in the cells of this diagram describe the magnitude of the control value. They are filled in according to the design rules in the last section. In the two-dimensional case this diagram corresponds uniquely to the phase plane of the system. For higher dimensions the diagram is an aggregation of the phase space so that the correspondence between s_p and di on the one hand and the vector e_N on the other hand is not unique any more.

The ith rule of a set of rules can be stated as

$$R_i: \text{if } s_p \text{ is } L_{s_p}^{(i)} \text{ and di is } L_d^{(i)} \text{ then } u \text{ is } L_u^{(i)}$$

with linguistical terms like positive small, negative big, etc., for the variables s_p, di_i and u_i. Although the computational effort for s_p and di seems to be rather large, this controller reduces the number of rules considerably in contrast with a controller in which the states are evaluated independently. According to equations (3.12) and (3.24) the structure of an FC without and with BL and partial compensation can be defined as

$$u = -\sum_{k=1}^{n-1} \binom{n-1}{k} \lambda^k \cdot e^{(n-k)} - K_{\text{Fuzz}}(s_p, di, \lambda) \cdot \text{sgn}(s) \qquad (3.32)$$

and

$$u = -\sum_{k=1}^{n-1} \binom{n-1}{k} \lambda^k \cdot e^{(n-k)} - K_{\text{Fuzz}}(s_p, di, \lambda) \cdot \text{sat}(s/\Phi) \qquad (3.33)$$

respectively. The result is that all design rules postulated in the preceding section are also applicable to FCs of higher order.

3.2.7 Numerical Example

The following example shows the sliding mode FC compared with a SMC with BL for a second-order system.

The task is to follow a contour by means of a force adaptive manipulator which is used for robot-guided grinding of castings, welding tasks and related industrial applications. Suppose that the *a priori* knowledge of the contour is incomplete. The robot effector (tool, gripper) has to follow the surface, keeping a constant desired force. The actual force between effector and surface is measured with the help of a force sensor. The force error between desired value and actual value is given to the FC whose output (control variable) is a position correction of the robot arm. Assume for simplicity that the whole dynamics of the system are determined by the robot effector and the surface. A further simplification is the restriction to one degree of freedom which means that we have only one coordinate in our model. Let furthermore the effector be modelled as a mass–spring–damper system and the surface as a spring. The corresponding differential equation is

$$m \cdot \ddot{y} + c \cdot (\dot{y} - \dot{y}_0) - m \cdot g + K_1 \cdot (y - y_0 - \tilde{D}) + K_2 \cdot (y - y_c + r + A) = 0 \quad (3.34)$$

Equation (3.35) is a force balance equation (see Figure 3.4):

$$F_T + F_D - F_G + F_F + F_O = 0 \quad (3.35)$$

with F_T as inertial force, F_D as damping force, F_G as weighting force, F_F as spring force, and F_O as reaction force within the surface.

Let the sensor force be represented by the spring force F_F. Then if

$$F_F = K_1(y - y_0 - \tilde{D}), \quad \dot{F}_F = K_1(\dot{y} - \dot{y}_0), \quad \ddot{F}_F = K_1(\ddot{y} - \ddot{y}_0) \quad (3.36)$$

we obtain the equation for the spring (sensor) force F_F:

$$\ddot{F}_F = -\frac{c}{m} \cdot \dot{F}_F + K_1 \cdot g - \frac{K_1}{m} \cdot F_F - \frac{K_1}{m} \cdot F_O - K_1 \cdot \ddot{y}_0 \quad (3.37)$$

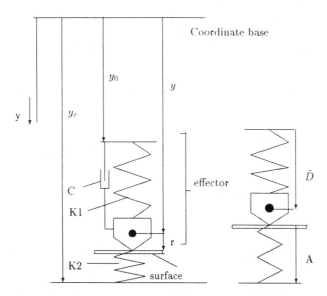

Figure 3.4 Simplified model of the robot effector acting on a surface

In this equation

$$u = -K_1 \cdot \ddot{y}_0 \tag{3.38}$$

the control variable which has to be produced by the controller. The specific parameters are

$$C = 50 \text{ kg/s}$$
$$m = 0.05 \text{ kg}$$
$$K_1 = 5000 \text{ kg/s}^2$$
$$K_2 = 10\,000 \text{ kg/s}^2$$
$$\tau_{\text{sample}} = 0.01 \text{ s}$$

Concerning a sliding mode fuzzy controller with BL, the comparison between equations (3.1) and (3.37) leads to

$$x(t) \text{ becomes } F_F(t) \tag{3.39}$$

$$x^{(n)}(t) \text{ becomes } \ddot{F}_F \tag{3.40}$$

$$f(x,t) \text{ becomes } -\frac{c}{m} \cdot \dot{F}_F + K_1 \cdot g - \frac{K_1}{m} \cdot F_F - \frac{K_1}{m} \cdot F_O \tag{3.41}$$

Furthermore we have

$$s = \lambda e + \dot{e} \tag{3.42}$$

$$e = F_F - F_{F_d} \tag{3.43}$$

The controller corresponding to equation (3.24) has the form

$$u = -\lambda \cdot \dot{e} - K_{\text{Fuzz}}(s_p, d, \lambda) \cdot \text{sat}(s/\Phi) \tag{3.44}$$

The $\tilde{F}$ in equation (3.21) is equal to

$$\tilde{F} = -\frac{c}{m} \cdot \dot{F}_F + K_1 \cdot g - \frac{K_1}{m} \cdot F_F - \frac{K_1}{m} \cdot F_O \tag{3.45}$$

Within the normalized phase plane the directional distance s_{p_N} and the distance di_N described in the last section are

$$s_{p_N} = (e_N + \dot{e}_N)/\sqrt{2} \tag{3.46}$$

and

$$di_N = |(-e_N + \dot{e}_N)/\sqrt{2}| \tag{3.47}$$

For the normalized values s_{p_N} and di_N and the normalized control variable u_N the following fuzzy sets are defined:

PSS: s_{p_N} is positive small
PSB: s_{p_N} is positive big
NSS: s_{p_N} is negative small
NSB: s_{p_N} is negative big
DS: di_N is small
DB: di_N is big
PUS: u_N is positive small
PUB: u_N is positive big
NUS: u_N is negative small
NUB: u_N is negative big

Figure 3.5 shows the corresponding membership functions. The respective s_p-di diagram is shown in Figure 3.6.

From this diagram we obtain the rules for the calculation of the control u_N:

For $s_{p_N} > 0$

if PSS and DS then NUS
if ((PSS and DB) OR PSB) then NUB

For $s_{p_N} \leq 0$

if NSS and DS then PUS
if ((NSS and DB) OR NSB) then PUB

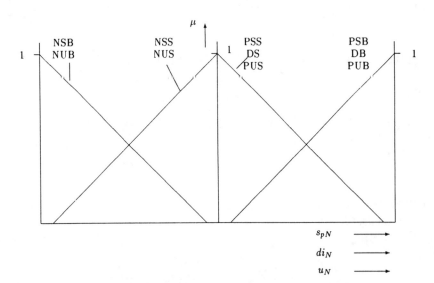

Figure 3.5 Membership functions for s_{p_N}, $|di_N|$ and u_N

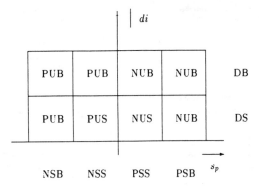

Figure 3.6 $s_p - di$ diagram for the contour tracking problem

The following FC parameters have been chosen:

$$\lambda = 50 \, 1/s$$
$$N_e = 20 \, s^2/\text{kg m}$$
and from λ and N_e
$$N_{\dot{e}} = 0.4 s^3/\text{kg m}$$
$$N_u = 100 \text{kg m}/s^4$$
$$\Phi = 2.5 \, \text{m/s}$$
$$F_d = 10 \text{N}$$
$$v = 0.2 \, \text{m/s (feed rate)}$$

Concerning sliding mode controller with BL, the ordinary SMC with BL evaluates only the sign of s. The maximum output and the denormalization factor are the same as those with the SMFC with BL. Also, the BL has the same width.

The corresponding control law is

$$u = -\lambda \cdot \dot{e} - K \cdot \text{sat}(s/\Phi)$$

where

$$K = 3900 \, \text{kg m}/s^4$$
$$\Phi = 2.5 \, \text{m/s}$$

Concerning simulation results, the surface along which the robot effector should slide has three different slopes: zero, positive and negative. Both controllers have been tested with respect to the different types of slopes (see Figure 3.7). Because of their similar control parameters, both controllers work in a similar way. However, the results coming from the ordinary SMC with BL tend to chattering around the force setpoint. This is due to the relatively high slope of the s-u-operating line around its origin. On the other hand the slope of the operating line of the SMFC with BL is low around its origin and steep far from it, which causes a fast and also smooth control action.

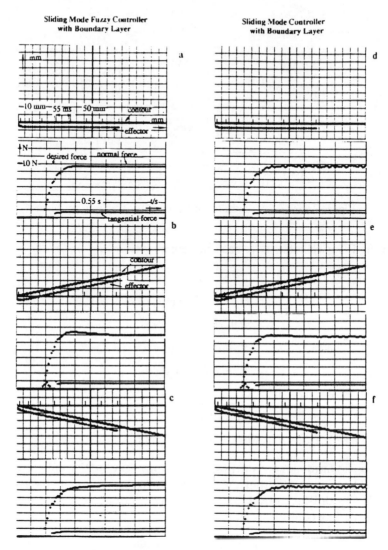

Figure 3.7 Simulation results of the contour tracking

3.3 SCALING OF FUZZY CONTROLLERS USING THE CROSS-CORRELATION

In most technical applications fuzzy controllers (FC) work in such a way that, on the basis of crisp desired inputs and crisp actual outputs, the system (plant) is controlled also by crisp manipulated variables. In this wide-spread case fuzziness is restricted only to the controller which is said to be more robust than conventional controllers (Lee, 1990, Driankov *et al.*, 1994, Palm, 1992a). Figure 3.8 shows the general FC structure. The operation of a FC of this type requires fuzzification of the inputs (e.g. error and change of error): each crisp input is assigned to a subset of grades of membership, depending on the

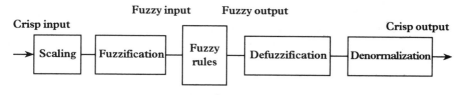

Figure 3.8 FC structure

a priori chosen subset of membership functions. The design of a FC requires information about the system to be controlled such as operating areas of the FC inputs and the manipulated variable of the FC output. For simplicity, in most cases the membership functions of the input and output variables are defined within normalized intervals (universes of discourse). In the case of normalized universes of discourse an appropriate choice of specific operating areas requires respective scaling or denormalization factors. An input scaling factor transforms a physical signal into the normalized universe of discourse of the controller input, whereas the output denormalization factor provides a transformation of the defuzzified output signal from the normalized universe of discourse of the controller output into a physical manipulated variable. The importance of an optimal choice of input scaling factors is evidently shown by the fact that ill-scaling results in either shifting the operating area to the boundaries or utilizing only a small area of the normalized universe of discourse. On the other hand, the adjustment of the output denormalization factor affects the closed-loop gain which has a direct influence on system stability. The behaviour of the system controlled finally depends on the choice of the normalized transfer characteristics (control surface) of the controller. In the case of a predifined set of rules the control surface is mainly determined by the shape and location of the input and output membership functions. When taking these influences into account for controller design one should pay attention to the following priority list.

(a) The output denormalization factor has the greatest influence on the stability and oszillation tendency. Because of its strong impact on stability this factor is assigned to the first priority in the design process.

(b) Input scaling factors have the greatest influence on the basic sensitivity of the controller with respect to optimal choice of the operating areas of the input signals. Therefore, input scaling factors are assigned to the second priority.

(c) Shape and location of input and output membership functions and, with this, the transfer function of the normalized controller positively influence the behaviour of the systems controlled in different areas of the state space, provided that the operating areas of the signals are optimally chosen through a well-adjusted input scaling factor. Therefore, this aspect is getting the third priority.

Once, by means of some system analysis, the scaling factors and the parameters of the membership functions have been chosen, the next task is to tune them in order to improve the systems's behaviour with respect to some optimization criteria. In this context tuning should consider the same priority list as for the design process.

Most of the reports on tuning refer to membership functions in order to change the transfer characteristics of the controller (Smith and Comer, 1993, Zheng, 1993, Boscolo, 1992). Examples for gain tuning can be found in Katayama *et al.* (1991), Maeda (1992), Braae (1979) and Zheng (1992). Tuning of rules has been considered by Proczyk and Mamdani (1979) and Peng (1990). Many reports deal with integral criteria with respect to particular test signals such as step, pulse, and random functions (Smith, 1993, Zheng, 1993). This chapter deals with the second-level of the tuning hierarchy, namely with the appropriate scaling of controller inputs which has the most influence on the sensitivity of the controller. It is assumed that both the rule set and the membership functions are predefined and kept constant during the tuning process. The input data are assumed to be Gaussian distributed signals whose parameters are assumed to be unknown. A poor knowledge of the distribution parameters can be explained by slowly-varying plant parameters or drift in the sensory used for state observation. Here, we distinguish three relevant cases:

(*a*) known mean (e.g. mean = 0) and unknown deviation

(*b*) unknown mean and known deviation

(*c*) mean and deviation are unknown.

Considering a controller with multi-input and single-output, the three cases can be processed on-line by measuring the linear dependence between each input and output signal of the controller. A measure for linear input–output dependence of a transfer element is the cross-correlation function and the cross-correlation coefficient, respectively. Firstly, the shift of the signal's mean value along the universe of discourse ensures that the signal meets the relevant operating area of the control surface. Secondly, once the relevant region has been reached the tuning procedure keeps on changing the particular input scaling factor until the goal, the cross-correlation coefficient to be a certain value near its maximum, is reached. It is shown that for Gaussian input signals a given FC can be imaginarily replaced by an equivalent gain which strongly depends on the nonlinear transfer characteristics of the FC (Schlitt, 1968). This method allows the utilization of linear system theory even in the case of nonlinear elements within the control loop. Therefore, an appropriate choice of the equivalent gain has a great influence on the behaviour of the closed-loop system. The equivalent gain can be expressed in terms of the standard deviation of the input and the input–output cross-correlation function. The claim is that for a stationary input a certain amount of signal amplitudes around the operating point of the FC should be linearily transmitted by the FC. As already mentioned, a measure for linear input–output dependence of a transfer element is the cross-correlation function. Hence, if a specific linearity between input and output is required, one has to adjust the standard deviation in such a way that a corresponding cross-correlation coefficient is met. For a given SISO FC structure the only parameter to influence the equivalent gain is the scaling factor for the input signal. This result can easily be extended to the multi-input/multi-output case (MIMO) if the individual states of the plant to be controlled are not correlated with each other. The method presented deals with the optimal adjustment of scaling factors for FCs with the help of the input–output cross-correlation (Palm, 93). If the distribution of the signal is *a priori* known the method can be characterized as a design approach only by consideration of the nonlinear FC without closing the control loop.

3.3.1 Input–Output Correlation for an FC

3.3.1.1 Equivalent gain: SISO case

Fuzzy controllers with crisp inputs and outputs can be considered as multidimensional nonlinear transfer elements with upper and lower limits. Let the system to be controlled be linear or, within the operating area, linearizable with lowpass characteristic (see Figure 3.9). Furthermore, let the desired value w include Gaussian noise. We then obtain non-Gaussian noise at the output of the FC because of its nonlinear transfer characteristic. On the other hand, we suppose the system to filter out all frequencies causing a non-Gaussian distribution. In this way, we expect to have Gaussian noise at the output of the system and, with this, at the adder where the desired value w is compared with the actual value x. With this assumption the scaled signal $e_s = (w - x) \cdot s_c$ is also of Gaussian type. From nonlinear system theory we know the terms 'describing function' for sinusoidal signals and 'equivalent gain' for signals with noise (Schlitt, 1968). The aim of this method is to replace the nonlinear element in a closed-loop system by a linear one whose gain depends on the amplitude e_0 (for sinusoidal inputs) or variance σ_e^2 (for noise) of the controller input.

Let e be a stationary and ergodic GAUSSian process. Furthermore, let

$$\bar{u} = K(\sigma_e^2) \cdot e + v \tag{3.48}$$

where $K(\sigma_e^2)$ is the equivalent gain corresponding to a specific FC transfer characteristic, and v is noise which is not correlated with e_s. Let

$$R_{e,\bar{u}}(\sigma_e) = E[(e(t) - E[e(t)]) \cdot (\bar{u}(t) - E[\bar{u}(t)])] \tag{3.49}$$

be the linear cross-correlation function for $\tau = 0$, and

$$E[x(t)] = \lim_{t \to \infty} \frac{1}{2T} \int_{-T}^{+T} x(t) \, dt$$

the expected value.

For simplicity, the mean values of e and $\bar{u}$ are equal to zero:

$$E[e] = 0, \quad E[\bar{u}] = 0$$

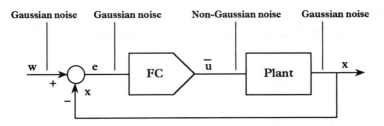

Figure 3.9 Closed-loop system with a nonlinear FC

By multiplying both sides of (3.48) with e and computing the expected values, one obtains

$$E[e \cdot \bar{u}] = K(\sigma_e^2) \cdot E[e^2] + E[e \cdot v] \qquad (3.50)$$

with $E[e \cdot \bar{u}]$ as the cross-correlation $R_{e\bar{u}}$ and $E[e^2]$ as the variation σ_e^2.
Because of $E[e \cdot v] = 0$ one obtains for the equivalent gain

$$K(\sigma_e^2) = \frac{R_{e\bar{u}}(\sigma_e)}{\sigma_e^2} \qquad (3.51)$$

With regard to scaled input signals e_s with noise, we then obtain the equivalent gain

$$K(\sigma_{e_s}) = \frac{R_{e_s\bar{u}}(\sigma_{e_s})}{\sigma_{e_s}^2} \qquad (3.52)$$

3.3.1.2 Equivalent gain: MIMO case

For the MIMO case let

$$\bar{u} = K(\sigma_e^2) \cdot e + v \qquad (3.53)$$

with

$$\bar{u} = (\bar{u}_1, \bar{u}_2, \ldots, \bar{u}_m)^T$$
$$e = (e, \dot{e}, \ddot{e}, \ldots, e^{(n-1)})^T$$
$$v = (v_1, v_2, \ldots, v_m)^T$$

and

$$K(\sigma_e^2) = \begin{pmatrix} K_1(\sigma_e^2) & K_1(\sigma_{\dot{e}}^2) & \cdots & K_1(\sigma_{e^{(n-1)}}^2) \\ K_2(\sigma_e^2) & K_2(\sigma_{\dot{e}}^2) & \cdots & K_2(\sigma_{e^{(n-1)}}^2) \\ \vdots & \vdots & \ddots & \vdots \\ K_m(\sigma_e^2) & K_m(\sigma_{\dot{e}}^2) & \cdots & K_m(\sigma_{e^{(n-1)}}^2) \end{pmatrix}$$

Moreover, we assume zero mean for all elements of e, $\bar{u}$ and v:

$$\forall i,k \quad E[e^{(k)}] = 0, \quad E[\bar{u}_i] = 0, \quad E[v_i] = 0 \qquad (3.54)$$

Finally, we assume the following cross-correlations to be zero:

$$\forall i \neq k, \quad E[e^{(i)} \cdot e^{(k)}] = 0, \quad E[v_i \cdot v_k] = 0 \qquad (3.55)$$
$$\forall i = k, \quad E[e^{(i)} \cdot v_k] = 0$$

with

$$i = 1, \ldots, m, \quad k = 1, \ldots, n-1$$

Rewriting equation (3.53) yields

$$\bar{u}_1 = K_1(\sigma_e^2) + K_1(\sigma_{\dot{e}}^2) + \cdots + K_1(\sigma_{e^{(n-1)}}^2) + v_1$$
$$\bar{u}_2 = K_2(\sigma_e^2) + K_2(\sigma_{\dot{e}}^2) + \cdots + K_2(\sigma_{e^{(n-1)}}^2) + v_2$$
$$\vdots \quad \vdots \quad F \quad \ddots \quad \vdots$$
$$\bar{u}_i = K_i(\sigma_e^2) + K_i(\sigma_{\dot{e}}^2) + \cdots + K_i(\sigma_{e^{(n-1)}}^2) + v_i \qquad (3.56)$$
$$\vdots \quad \vdots \quad F \quad \ddots \quad \vdots$$
$$\bar{u}_m = K_m(\sigma_e^2) + K_m(\sigma_{\dot{e}}^2) + \cdots + K_m(\sigma_{e^{(n-1)}}^2) + v_m$$

Multiplication of equation (3.56) with $e, \dot{e}, \ldots, e^{(n-1)}$, and computing the expected values leads, with conditions (3.54) and (3.55), for the ith row to

$$K_i(\sigma_e^2) = \frac{R_{e,\bar{u}_i}(\sigma_e)}{\sigma_e^2}$$

$$K_i(\sigma_{\dot{e}}^2) = \frac{R_{\dot{e},\bar{u}_i}(\sigma_{\dot{e}})}{\sigma_{\dot{e}}^2} \qquad (3.57)$$

$$\vdots$$

$$K_i(\sigma_{e^{(n-1)}}^2) = \frac{R_{e^{(n-1)},\bar{u}_i}(\sigma_{e^{(n-1)}})}{\sigma_{e^{(n-1)}}^2}$$

From equation (3.57) we obtain the advantageous result that the individual equivalent gains can be derived independently from each other. Therefore, the next steps regarding input scaling can be performed by considering only the SISO case without loss of generality.

3.3.1.3 Input scaling

The input e is scaled by means of

$$e_s = e \cdot s_c \qquad (3.58)$$

where

$\quad s_c$ is the scaling factor, e_s is the scaled input $\qquad (3.59)$

The scalar output signal $\bar{u}$ is computed by the centre of gravity:

$$\bar{u} = \int_A^B \frac{\mu_u \cdot u}{\int_A^B \mu_u \, du} \, du \qquad (3.60)$$

where

$\quad \mu_u \in (0, 1)$ is the degree of membership, $u \in (A, B)$ is the universe of discourse $\qquad (3.61)$

84 COMPATIBILITY OF FUZZY CONTROL AND CONVENTIONAL CONTROL TECHNIQUES

For simplicity we assume the denormalization factor of $\bar{u}$ to be one. Furthermore, let $e(t)$ and $\bar{u}(t)$ be stationary and ergodic processes. The standard deviations of the scaled signal $e_s(t)$ and non-scaled signal $e(t)$ are connected in the same way as the signals e_s and e are:

$$\sigma_{e_s} = s_c \cdot \sigma_e \tag{3.62}$$

From (3.52) and (3.62) we obtain

$$K(\sigma_e) = \frac{R_{e,\bar{u}}(s_c \cdot \sigma_e)}{s_c^2 \cdot \sigma_e^2} \tag{3.63}$$

Both gain K and cross-correlation $R_{e,\bar{u}}$ reach their maximum values in the case of maximum linear input–output connection. The normalized correlation coefficient

$$\tilde{R}_{e,\bar{u}} = \frac{R_{e,\bar{u}}}{\sigma_{e_s} \cdot \sigma_{\bar{u}}} \tag{3.64}$$

reaches its maximum at $\tilde{R}_{e,\bar{u}}|_{max} = 1$. $\tilde{R}_{e,\bar{u}}$ reaches its minimum at

$$\tilde{R}_{e,\bar{u}}|_{min} = \sqrt{\frac{2}{\pi}}$$

if the transfer characteristic is symmetrical with respect to the mean value $\bar{e}_s = E[e_s(t)]$ (see Figure 3.10). This is related to a relay transfer characteristic, since for a very large standard deviation compared with the width $2A$ of the interval considered, every symmetrical transfer characteristic acts as a relay function. Moreover, it is evident that a shift of the mean $\bar{e}_s$ of the distribution to the limits of the transfer characteristic of the FC leads to decreasing input–output correlation of the controller. The extreme point is reached when the largest part of the distribution is covered by one of the branches at which the control output is always a constant value. For this case we obtain $\tilde{R}_{e,\bar{u}} \approx 0$.

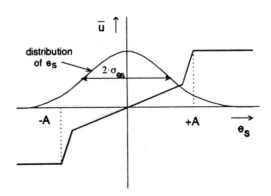

Figure 3.10 Symmetrical nonlinear transfer characteristic of an FC with limits

Optimal input scaling actually means searching for an optimal σ_{e_s} with respect to the interval $[-A, +A]$ of the FC between its limits. We assume the optimal scaling for the following case.

We start the searching procedure with a large s_c (which means a large $\sigma_{e_s} = \sigma_{e_s}|_{max}$). Then, keeping σ_{e_s} constant, we change $\bar{e}_s$ stepwise by adding $\Delta \bar{e}_s$ to $\bar{e}_s$ which corresponds to a shifting of the distribution of the input signal along the $\bar{e}_s$ axis. The result is a curve $\tilde{R}(\bar{e}_s, \sigma_{e_s})$ with a maximum $\tilde{R}(\bar{e}_s, \sigma_{e_s})|_{max}$ that is supposed to be located at only one point with

$$\tilde{R}(\bar{e}_s + \Delta \bar{e}_s, \sigma_{e_s}) \neq \tilde{R}(\bar{e}_s, \sigma_{e_s}), \quad \forall \tilde{R}(\bar{e}_s, \sigma_{e_s}) \tag{3.65}$$

and

$$\sigma_{e_s} = \text{const.}$$

After that we decrease σ_{e_s} by $\Delta \sigma_{e_s}$ and change the result $\tilde{R}(\bar{e}_s, \sigma_{e_s} - \Delta \sigma_{e_s})$ in the same way. Because of the monotone function $\tilde{R}(\bar{e}_s, \delta_{e_s})|\bar{e}_s = \text{const.}$ We obtain a higher maximum $\tilde{R}(\bar{e}_s, \sigma_{e_s})|_{max}$ as before:

$$\tilde{R}(\bar{e}_s, \sigma_{e_s} - \Delta \sigma_{e_s})_{max} > \tilde{R}(\bar{e}_s, \sigma_{e_s})_{max} \tag{3.66}$$

We stop the searching procedure at

$$\tilde{R}(\bar{e}_s, \sigma_{e_s}) \geq 1 - \alpha$$

with condition (3.65) where $\alpha \in (0, 1)$ is a free parameter. If condition (3.65) is not fulfilled we obtain a plateau. In this case the domain of the FC is assumed to be insufficient with respect to the given standard deviation σ_{e_s}. Hence, one has to increase the scaling factor s_c until (3.65) is met. Figure 3.12 shows a typical $\tilde{R}(\bar{e}_s, \sigma_{e_s})$ plot and Figure 3.13 shows the corresponding block scheme.

We choose the free parameter α such that for a linear FC characteristic between the upper and lower limit (see Figure 3.11) the standard deviation σ_{e_s} of the scaled signal e_s is

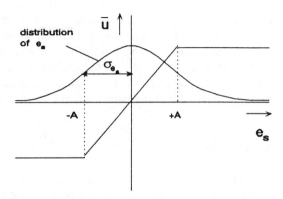

Figure 3.11 Linear transfer characteristic with limits

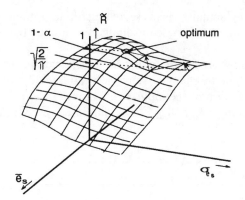

Figure 3.12 $\tilde{R}(\bar{e}_s, \sigma_{e_s})$ plot

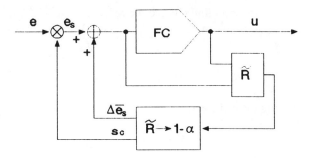

Figure 3.13 Block scheme for tuning of scaling factors

set equal to the interval A of the controller:

$$\alpha = 1 - \tilde{R}(\bar{e}_s, \delta_{e_s})_{\delta_{e_s} = A} \tag{3.67}$$

This means that we have an input signal probability $P = 0.68$ for the linear region of the FC. However, if the characteristic of the controller between its limits is nonlinear (see Figure 3.10) then one obtains automatically an s_c such that $\sigma_{e_s} < A$. So it is clear that the cross-correlation $\tilde{R}$ as a linear operation meets its maximum value when for a given standard deviation σ_{e_s} the function between the limits of the transfer characteristic of the FC is linear. If the function between the limits is nonlinear one obtains for the same σ_{e_s} a lower value for $\tilde{R}$. This, however, corresponds to a smaller standard deviation σ_{e_s} and, with this, a smaller scaling factor s_c if a linear function between the limits is assumed.

3.3.1.4 Numerical example

Suppose a standard deviation $\sigma_e = 1$ of the input signal $e(t)$. For an FC with a linear characteristic between the limits as shown in Figure 3.14a) we obtain $\tilde{R}(\bar{e}_s, \sigma_{e_s})_{\sigma_{e_s} = 10} = 0.95$ and $\alpha = 0.05$. This corresponds to a scaling factor $s_c = 10$ and the scaled standard

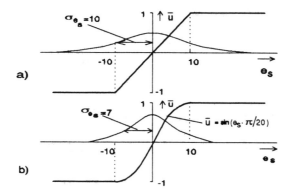

Figure 3.14 FCs with (a) linear and (b) nonlinear characteristics between the limits

deviation $\sigma_{e_s} = 10$. For an FC with a sinusoidal characteristic as shown in Figure 3.14b) we obtain for the same $\alpha = 0.05$ a scaling factor $s_c = 7$ and the scaled standard deviation $\sigma_{e_s} = 7$.

3.3.2 Application to a Redundant Manipulator Arm

The application sample of this subsection deals with the control of the kinematic of a redundant robot arm (the inverse task) whose static transfer characteristic is highly nonlinear but, in contrast to this, whose internal dynamics consist of a simple dead time element or a delay. This section will not, however, describe the whole problem of solving the inverse task in the case of kinematical redundancy. This has already been discussed by (Palm, 1992b) in detail. In the following, only the aspect of appropriate tuning of the input scaling factors required for kinematical control of the robot arm is considered.

It has been shown that a shift of the mean value of an input signal leads to a reduction of the correlation coefficient. The following example deals with input signals whose signs do not change during the control process. Moreover, the input scaling factor does not only affect the standard deviation, but also the mean value. The task is the optimal choice of scaling factors so that the distribution of the corresponding input signal is located within the corresponding operating area of the FC. The basic assumption is that an optimal scaling factor is obtained when the statistical input–output dependence meets its maximum:

$$|\tilde{R}_{\tau=0}| \to \max$$

To be independent of any change of sign in the control loop the absolute value of $\tilde{R}$ has been chosen. The problem of kinematical control of a redundant manipulator arm can be simply described as follows: the effector (gripper, tool) of the planar robot arm is supposed to follow a predefined path (see Figure 3.15). The robot kinematic is constructed in such a way that the individual links of the manipulator are able to avoid both external obstacles and internal restrictions (e.g. boundaries of the links).

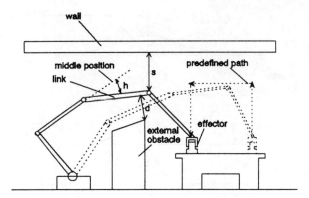

Figure 3.15 Motion of a redundant robot arm

In the special case the motion of each link is, in addition to the given effector task, determined by: the distance h between link and middle position, the distance s between link and wall, and the distance d between link and obstacle.

The distances h, s and d are evaluated by fuzzy attributes (e.g. s = positive small) and their membership functions. For each link a fuzzy rule base provides the corresponding correction of the joint angle. The actions z (angle corrections) of each link are evaluated also by fuzzy attributes (e.g. z = positive big angle correction). Although the process to be controlled is highly nonlinear in the large, it can be considered as linear for small angle corrections. The distances h, s and d are scaled so that they fit the predefined normalized universes of discourse.

For the internal restriction 'scaled distance between the ith link and its middle position' $h_{iN} = h$ yields

$$h \text{ negative big:} \quad \text{HNB} = (\mu_{\text{HNB}}(h)/h)$$
$$h \text{ negative small:} \quad \text{HNS} = (\mu_{\text{HNS}}(h)/h)$$
$$h \text{ positive big:} \quad \text{HPB} = (\mu_{\text{HPB}}(h)/h)$$
$$h \text{ positive small:} \quad \text{HPS} = (\mu_{\text{HPS}}(h)/h), \quad \forall h \in H$$

For the external restriction 'scaled distance between the ith link and some wall' $s_{iN} = s$ yields

$$s \text{ big:} \quad \text{SIB} = (\mu_{\text{SIB}}(s)/s)$$
$$s \text{ small:} \quad \text{SIS} = (\mu_{\text{SIS}}(s)/s), \quad \forall s \in S$$

For the external restriction 'scaled distance between the ith link and some obstacle' $d_{iN} = d$ yields

$$d \text{ big:} \quad \text{DIB} = (\mu_{\text{DIB}}(s)/s)$$
$$d \text{ small:} \quad \text{DIS} = (\mu_{\text{DIS}}(s)/s), \quad \forall d \in D$$

For the 'scaled output' $z_i = z$ regarding the ith link yields

z negative big: $\text{ZNB} = (\mu_{\text{ZNB}}(h)/h)$
z negative small: $\text{ZNS} = (\mu_{\text{ZNS}}(h)/h)$
z negative zero: $\text{ZNZ} = (\mu_{\text{ZNZ}}(h)/h)$
z positive big: $\text{ZPB} = (\mu_{\text{ZPB}}(h)/h)$
z positive small: $\text{ZPS} = (\mu_{\text{ZPS}}(h)/h)$
z positive zero: $\text{ZPZ} = (\mu_{\text{ZPZ}}(h)/h), \quad \forall z \in Z$

All membership functions μ vary only witin a predefined standard interval h, s, d, $z \in [\text{MAX}, \text{MIN}]$. Outside this interval the value of μ is either 0 or 1. Furthermore, all fuzzy sets are normal, i.e. there is an h, s, d or z with a corresponding $\mu = 1$.

To obtain an appropriate motion for each link the following set of rules has been applied:

IF (SIS AND DIS AND (HNS OR HPS)) OR (SIS AND HNB AND DIB) THEN ZNZ
IF (SIS AND HPB AND DIS) OR (SIB AND HPS AND DIB) THEN ZNS
IF (SIS AND DIB AND (HNS OR HPS)) OR ((SIS OR SIB) AND HPB AND DIB)
   THEN ZNB
IF SIS AND HNB AND DIS THEN ZPZ
IF (SIB AND DIS AND (HPS OR HPB)) OR (SIB AND HNS AND DIB) THEN ZPS
IF (SIB AND DIS AND (HNS OR HNB)) OR (SIB AND HNB AND DIB) THEN ZPB

The scalar output value has been computed by the centre of gravity:

$$z_N = \frac{\int_{z_{\min}}^{z_{\max}} z \cdot \mu_z \, dz}{\int_{z_{\min}}^{z_{\max}} \mu_z \, dz}$$

Within the rules for the operations AND and OR, the MAX and MIN operator, respectively, have been chosen.

The correlation coefficient $\tilde{R}$ for discrete points of time has been applied concerning the distance s between each link and the wall:

$$\tilde{R}[s,z] = \frac{\sum_{i=1}^{n} s_i z_i - \frac{1}{n}\left(\sum_{i=1}^{n} s_i \cdot \sum_{i=1}^{n} z_i\right)}{\sqrt{\sum_{i=1}^{n} s_i^2 - \frac{1}{n}\left(\sum_{i=1}^{n} s_i\right)^2} \cdot \sqrt{\sum_{i=1}^{n} z_i^2 - \frac{1}{n}\left(\sum_{i=1}^{n} z_i\right)^2}} \tag{3.68}$$

Figures 3.16 and 3.17 show the change of the correlation coefficient $\tilde{R}$, where s_s is the scaling factor for distance s.

The other scaling factors are $s_h = 20$ and $s_d = 120$ (see Figure 3.16). The peak of $\tilde{R}$ is lying at about $s_s = 80$. Figure 3.17 shows a similar situation for $s_h = 100$ and $s_d = 60$. The peak of $\tilde{R}$ lies at $s_s = 80$ again. The result is that although there is a certain change in the curvature

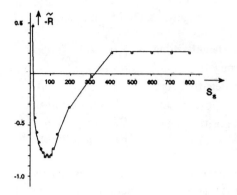

Figure 3.16 Simulation results fo s_s with $s_h = 20$ and $s_d = 120$

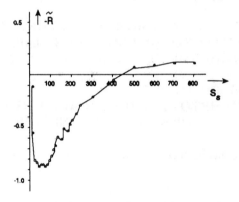

Figure 3.17 Simulation results for s_s with $s_h = 100$ and $s_d = 60$

of $R(s_s)$, depending on a different choice of the other scaling factors s_d and s_h, the abscissa of the maximum value of $\tilde{R}$ does not change. This finally illustrates the independence of the location of the maxima of different correlation coefficient curves.

3.4 FUZZY INPUTS

The scope of pure fuzzy systems including fuzzy signals has been studied extensively by Tong (1980), Pedrycz (1992), Gupta et al. (1986) in the past. However, one also should pay attention to the mixed case where some signals are crisp and some are fuzzy (Palm and Driankov, 1994). This is the case when the objective x_d is crisp and the output y, fed back via a sensor, is fuzzy (see Figure 3.18). Then, the error signal e is also fuzzy. The error e is fed to the input of a fuzzy controller without any fuzzification block since the input signal e is already fuzzy. The output u of the controller is crisp since the system to be controlled requires crisp inputs. The crisp states x of the system are measured by means of a sensor providing fuzzy outputs y.

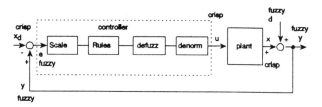

Figure 3.18 General block scheme of a control loop

Although some methods exist to prove the stability of fuzzy controlled systems (de Glas, 1984, Tanaka and Sano, 1993, Palm, 1992a) all of these methods deal with crisp signals throughout the control loop.

Therefore, it is of interest to find corresponding methods for investigating stability and robustness of fuzzy controlled systems, even in the case of fuzzy signals at the input of the controller.

One of the most important control methods for a large class of nonlinear systems is sliding mode control (Utkin, 1977, Slotine, 1985). Hence, it would be interesting how stability and robustness are determined for sliding mode control and related methods in the presence of fuzzy signals.

Before dealing with control methods some specific opeations with respect to fuzzy signals are defined. According to this, some operations on fuzzy numbers need to be defined: differentiation of a fuzzy set with respect to time, thresholds and the sgn function and the sat function.

Furthermore, with regard to fuzzy signals the following topics are discussed: pure sliding mode, sliding mode, sliding mode with boundary layer and sliding mode fuzzy control.

3.4.1 Some Useful Operations on Fuzzy Sets

3.4.1.1 Difierentiation of a fuzzy set with respect to time

The definition of the differentiation of a fuzzy set is a matter of fuzzy analysis and has already been proposed by Dubois and Prade (1980) and Zimmermann (1991). However, this definition does not completely satisfy the problems arising for dynamical fuzzy sets with time variable parameters. Therefore, a different definition of differentiating a fuzzy set with respect to time is proposed.

Normally, differentiation of a crisp function $x(t)$ with respect to time is defined by

$$\dot{x} = \lim_{\Delta t \to \infty} \frac{x(t + \Delta t) - x(t)}{\Delta t}$$

In order to define the differentiation of a fuzzy set we start with its continuous description. Let, for example, a fuzzy set $X(t)$ be described by a bell-shaped membership function

$\mu_X(x(t))$ similar to a Gaussian probability distribution.

$$\forall t, \quad \mu_X(x(t)) = e^{-(x-\bar{x}(t))^2/2\cdot\sigma(t)^2} \tag{3.69}$$

where

$\bar{x}(t)$ is the time variable mean and $\sigma(t)$ is the time variable standard deviation (width).

Similarly to a probability distribution we characterize the width of the membership function by a scaled deviation $\sigma(t)$. The fuzzy set is normal which means $\mu_X(x(t) = \bar{x}(t)) = 1$.

Since $x(t)$ is a function of time the fuzzy set X moves along its universe of dicourse in relation to the velocity $\dot{\bar{x}}(t)$ of the mean $\bar{x}(t)$ and the velocity $\dot{\sigma}(t)$ of the deviation $\sigma(t)$. Thus, the dynamics of the membership function only depend on the two parameters $\bar{x}(t)$ and $\sigma(t)$. The representation of a time variable fuzzy set and its derivatives with respect to time by a finite number of parameters (e.g. $\bar{x}(t)$ and $\sigma(t)$) is very useful to bridge some gaps between conventional and fuzzy system theory.

Example Let $X(t)$ be a fuzzy set whose membership function $\mu_X(x(t))$ is defined as the normalized bell-shaped function

$$\mu_X(x(t)) = e^{-(x-\bar{x}(t))^2/2\cdot\sigma_x(t)^2} \tag{3.70}$$

$$\sigma_x(t) = \sigma_{x_0} = \text{const.}$$

where x_0 is the initial value of $x(t)$. Additionally, let the mean $\bar{x}(t)$ satisfy the differential equation

$$\dot{\bar{x}} + k\cdot\bar{x} = 0 \tag{3.71}$$

with the solution

$$\bar{x} = \bar{x}_0 \cdot e^{-k\cdot t} \tag{3.72}$$

where $\bar{x}_0$ is the initial value of $\bar{x}(t)$.
Then the resulting membership function yields

$$\mu_X(x(t)) = e^{-(x-x_0\cdot e^{-k\cdot t})^2/2\cdot\sigma_{x_0}^2} \tag{3.73}$$

Figure 3.19 shows the dynamics of the fuzzy set $X(t)$ together with the corresponding membership function $\mu_X(x(t))$ in a qualitative way.

A similar procedure can always be performed if the membership function can be represented by an analytical function.

However, the representation of a time variable fuzzy set $X(t)$ in terms of its parameters is not a fuzzy set any more. To be more specific, let the time variable fuzzy set be predefined. What is the velocity of the given fuzzy set in terms of a fuzzy set? Without loss of generality this problem shall be discussed by means of the bell-shaped membership function.

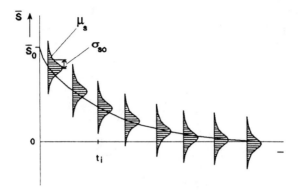

Figure 3.19 Dynamics of the fuzzy set with respect to time

Consider a fixed pair $\langle \mu_X(x^i(t)), x^i(t) \rangle$. The behaviour of this pair in time is based on the following conditions:

$$\forall \Delta t, x^i(t + \Delta t) = x^i(t) + \Delta x^i(t)$$

and

$$\mu_X(x^i(t + \Delta t)) = \mu_X(x^i(t))$$

The fuzzy set $\dot{X}(t)$ is then defined as follows:

$$\forall i, \mu_{\dot{X}}(\dot{x}^i(t)) = \max_k \{\mu_X(x^k(t))\} \tag{3.74}$$

where

$$\forall k, \dot{x}^k(t) = \dot{x}^i(t)$$

This means, in the case of several points $x^k(t)$ with the same velocity $\dot{x}^i(t)$ but different degrees of membership $\mu_X(x^k)$ we choose the maximum degree of membership $\max_k \{\mu_X(x^k(t))\}$ for $\dot{x}^i(t)$. This is justified because the fuzzy set $\dot{X}$ should be a normal set like $X(t)$.

To show how to form a fuzzy set $\dot{X}(t)$ given a fuzzy set $X(t)$ let us now apply definition (3.74) to a bell-shaped membership function (see Figure (3.20)). Let $\mu_X(x^i(t))$ and $\mu_X(x^i(t + \Delta t))$ the membership functions for point x^i at times t and $t + \Delta t$, respectively:

$$\mu_{x^i(t)} = e^{-(x^i(t) - \bar{x}(t))^2 / 2 \cdot \sigma(t)^2} \tag{3.75}$$

$$\mu_{x^i(t + \Delta t)} = e^{-(x^i(t + \Delta t) - \bar{x}(t + \Delta t))^2 / 2 \cdot \sigma(t + \Delta t)^2}$$

where

$$\mu_X(x^i(t)) = \mu_X(x^i(t + \Delta t)) \tag{3.76}$$

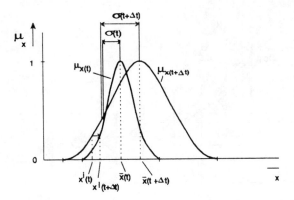

Figure 3.20 Motion of a time variable bell-shaped membership function along the x-axis of the universe of discourse

The velocity or the derivative of the fuzzy set with respect to time is determined by

$$\forall i, \dot{x}^i(t) = \lim_{\Delta t \to 0} \frac{x^i(t+\Delta t) - x^i(t)}{\Delta t} \tag{3.77}$$

From equations (3.75) and (3.76) follows

$$e^{-(x^i(t) - \bar{x}(t))^2 / 2 \cdot \sigma(t)^2} = e^{-(x^i(t+\Delta t) - \bar{x}(t+\Delta t))^2 / 2 \cdot \sigma(t+\Delta t)^2}$$

and further

$$\frac{x^i(t) - \bar{x}(t)}{\sigma(t)} = \frac{x(t+\Delta t) - \bar{x}(t+\Delta t)}{\sigma(t+\Delta t)}$$

From the linear approximations

$$\begin{aligned} x^i(t+\Delta t) &\approx x^i(t) + \dot{x}^i \cdot \Delta t \\ \bar{x}(t+\Delta t) &\approx \bar{x}(t) + \dot{\bar{x}} \cdot \Delta t \\ \sigma(t+\Delta t) &\approx \sigma(t) + \dot{\sigma} \cdot \Delta t \end{aligned} \tag{3.78}$$

one obtains the velocity

$$\dot{x}^i(t) = \dot{\bar{x}}(t) + (x^i(t) - \bar{x}(t)) \cdot \frac{\dot{\sigma}(t)}{\sigma(t)} \tag{3.79}$$

According to our definition (3.74) the corresponding membership degree for $\dot{x}^i(t)$ can be obtained as follows.

For $\dot{\sigma} = 0$ one obtains $\forall i, \dot{x}^i(t) = \dot{\bar{x}}(t)$. Since $\mu_{\bar{X}}(\bar{x}(t)) = 1$ we obtain, according to our definition of $\dot{X}$

$$\forall i, \mu_{\dot{X}}(\dot{x}^i(t)) = 1$$

(see Figure 3.21a)

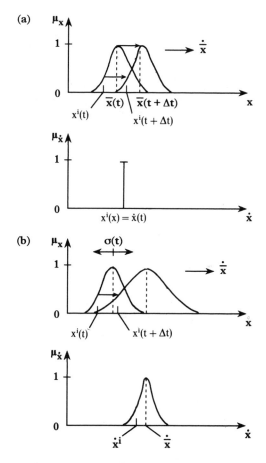

Figure 3.21 Fuzzy set for the velocity of a fuzzy set

For $\dot{\sigma} > 0$ one obtains $\forall i, \mu_{\dot{x}}(\dot{x}^i(t)) = \mu_x(x^i(t))$. If $\forall i, \dot{x}^i(t) = \dot{\bar{x}}(t)$, as a special case, we obtain $\forall i, \mu_{\dot{x}}(\dot{x}^i(t)) = \mu_{\dot{x}}(\dot{\bar{x}}(t)) = \mu_x(\bar{x}(t))$ (see Figure 3.21b).

We demonstrate the formation of the velocity of a bell-shaped fuzzy set in a three-dimensional way (see Figure 3.22).

The points $x^i(t)$ and $x^i(t + \Delta t)$ with the same degree of membership meet at the diagonal, which is described by

$$x(t + \Delta t) = \bar{x}(t + \Delta t) + (x(t) - \bar{x}(t)) \cdot \frac{\sigma(t + \Delta t)}{\sigma(t)} \qquad (3.80)$$

It is a little more complicated to deal with functions that are asymmetrical to the maximum. If such a curve can be described by a continuous function in a closed form, the velocity of this function can be obtained by doing the same procedure as in the symmetrical case. In the general case, however, points with the same degree of membership do not lie on a diagonal. The mapping $\mu_X(x(t)) \to \mu_X(x(t + \Delta t))$ is then described by

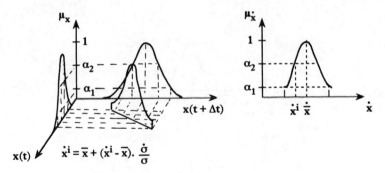

Figure 3.22 Mapping between $\mu_X(x(t))$ and $\mu_X(x(t + \Delta t))$ for the bell-shaped case

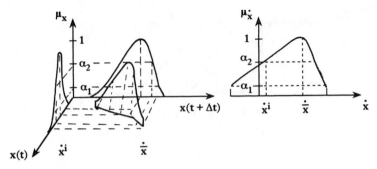

Figure 3.23 Mapping between $\mu_X(x(t))$ and $\mu_X(x(t + \Delta t))$ in the case of an asymmetrical membership function

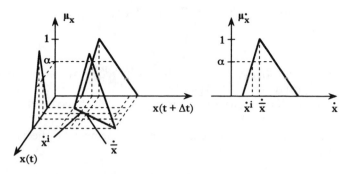

Figure 3.24 Mapping between $\mu_X(x(t))$ and $\mu_X(x(t + \Delta t))$ in the piecewise continuous case

a more complicated but still continuous curve which the corresponding velocities $\dot{x}^i$ belong to (see Figure 3.23).

The approach holds even if the membership function is only piecewise continuous, like symmetrical or asymmetrical triangles (see Figure 3.24).

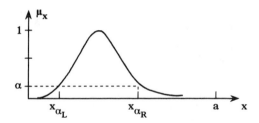

Figure 3.25 Checking of a fuzzy set X on a threshold a

3.4.1.2 Thresholds

Let X be a normal and convex fuzzy set with $X = \{\mu_X(x), x\}$. Its α-cut is defined as (see Fig. 3.25)

$$X_\alpha = \{\mu_X(x), x \mid \mu_X(x) \geq \alpha\}$$

Then let

$$x_{\alpha L} = \{x \mid \alpha \leq \mu_X(x) = \min_{x \in X}\{\mu_X(x)\}\}$$
$$x_{\alpha R} = \{x \mid \alpha \leq \mu_X(x) = \max_{x \in X}\{\mu_X(x)\}\}$$

Obviously $x_{\alpha L} \leq x_{\alpha R}$. Then the fuzzy set X is said to be smaller than a crisp threshold a iff

$$x_{\alpha R} < a \tag{3.81}$$

3.4.2 The sgn-function

Let X be a normal and convex fuzzy set. Then let the sgn-function be defined as (see Figure 3.26)

$$\text{sgn}(X) = \text{sgn}(\text{defuzz}(X)) \tag{3.82}$$

where defuzz(X) is the result of the defuzzification operation on the fuzzy set X, e.g. the centre of gravity (c.o.g.).

3.4.2.1 The sat-function

Let $X = \{\mu_X(x(t)), x(t)\}$ be a normal and convex fuzzy set; then the division of X by a crisp number Φ is defined by

$$\frac{X}{\Phi} = \left\{\mu_X\left(\frac{x(t)}{\Phi}\right), \frac{x(t)}{\Phi}\right\}$$

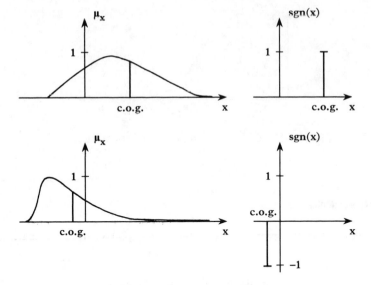

Figure 3.26 Definition of sgn(X)

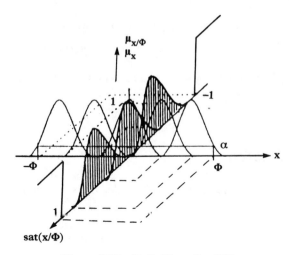

Figure 3.27 Definition of sat(X)

Then the sat function is defined as follows. For a crisp value $\Phi > 0$ we define

$$\text{sat}\left(\frac{X}{\Phi}\right) = \frac{X}{\Phi} \quad \text{for } -\Phi \leq \text{defuzz}(X) \leq +\Phi \tag{3.83a}$$

$$\text{sat}\left(\frac{X}{\Phi}\right) = \text{sgn}(X) \text{ else} \tag{3.83b}$$

where equation (3.83a) is fuzzy and equation (3.83b) is crisp.

3.4.3 Sliding Mode Control and Related Control Strategies

3.4.3.1 Pure sliding mode control

Let us again start with the state equation of an, in general, nonlinear system:

$$x^{(n)}(t) = f(x,t) + u + d \tag{3.84}$$

with

$$x(t) = (x, \dot{x}, \ldots, x^{(n-1)})^T$$

where $x(t)$ is a crisp state vector, d is a fuzzy disturbance and u is a crisp manipulated variable. Further, let

$$y = C \cdot x + \bar{d} \tag{3.85}$$

be the observation equation (sensory equation) with y as the fuzzy output vector, C as the $(n-1) \times (n-1)$ diagonal matrix and d as the fuzzy vector of uncertainties of the sensory.

Equation (3.84) means that $x^{(n)}$ is a fuzzy value since d is a fuzzy scalar. In equation (3.85) the vector y is also a fuzzy signal which originates from the fuzzy interpretation of sensor uncertainties. Figure 3.28 shows the corresponding block scheme.

It has to be pointed out that there are two different sources of uncertainty in the control loop: disturbances and uncertainties that affect the control value u and the output vector y. x_d is a crisp vector of desired states and

$$e = y - x_d = C \cdot x + \tilde{d} - x_d \tag{3.86}$$

is a fuzzy vector of errors. The vector e is fuzzy since $\tilde{d}$ is a fuzzy vector, too. For the sake of simplicity we set $C = E$ (E is a unity matrix).

Further, let $f(x,t)$ be a nonlinear function of the state vector x and of time t.

Now, we formulate

$$s = (d/dt + \lambda)^{(n-1)} e \tag{3.87}$$

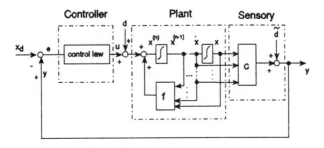

Figure 3.28 Block scheme regarding a control loop with fuzzy signals

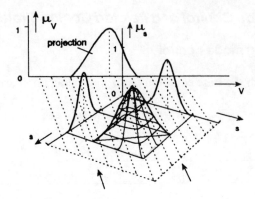

Figure 3.29 Lyapunov function regarding a fuzzy value s

with $e = x + \tilde{d} - x_d$, and e is a fuzzy scalar error, x is a crisp scalar position, $\tilde{d}$ is a fuzzy scalar of position uncertainty, x_d is a crisp scalar desired position, λ is a positive scalar value and s is a scalar fuzzy value.

With the help of the fuzzy value s, the cross product $s \times s$ and a subsequent projection (see Figure 3.29) we define a fuzzy Lyapunov-like function

$$V = \tfrac{1}{2} \cdot s^2 \tag{3.88}$$

Further, let the condition for stability be

$$\dot{V} \leq -\eta \cdot s \cdot \operatorname{sgn}(s) \tag{3.89}$$

where η is a crisp positive value.

Condition (3.89) means a decrease of energy within the system. The derivative of the fuzzy function V with respect to time can be obtained with the help of the procedure of forming the derivative of a fuzzy values mentioned above. The resulting fuzzy value $\dot{V}$ is added to the fuzzy value $\eta \cdot s \cdot \operatorname{sgn}(s)$

$$A = \dot{V} + \eta \cdot s \cdot \operatorname{sgn}(s) \tag{3.90}$$

by taking into account the cross product $\dot{V} \times \eta \cdot s \cdot \operatorname{sgn}(s)$ and its projection onto the diagonal as shown in Figure 3.30.

The test

$$A = \dot{V} + \eta \cdot s \cdot \operatorname{sgn}(s) \leq 0 \tag{3.91}$$

is performed by means of the above-mentioned test of a fuzzy value with regard to a predefined threshold. The task is to find a crisp manipulated variable u that satisfies equation (3.89). In the following we refer without loss of generality to a second order system. From equations (3.88) and (3.89)

$$\dot{V} = \frac{1}{2} \cdot \frac{d}{dt}(s^2) = s \cdot \dot{s} \leq -\eta \cdot s \cdot \operatorname{sgn}(s) \tag{3.92}$$

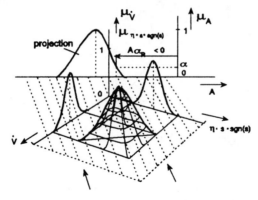

Figure 3.30 Forming of $A = \dot{V} + \eta \cdot s \cdot \text{sgn}(s)$

Equation (3.87) becomes

$$s = \lambda \cdot e + \dot{e} \quad (3.93)$$

and with

$$\dot{s} = \lambda \cdot \dot{e} + \ddot{e} \quad (3.94)$$

$$\ddot{e} = \ddot{x} + \ddot{d} - \ddot{x}_d$$

we obtain

$$s \cdot \dot{s} = s \cdot (\lambda \cdot \dot{e} - \ddot{x}_d + \ddot{d} + f(\dot{x}, x, t) + d + u) \leq -\eta \cdot s \cdot \text{sgn}(s) \quad (3.95)$$

in which only $\ddot{x}_d$ and u are crisp values.

We now introduce the following upper bounds $E, v, F, D1$ and $D2 > 0$:

$$\lambda \cdot \dot{e} \cdot \text{sgn}(\dot{e}) < E$$

$$-\ddot{x}_d \cdot \text{sgn}(-\ddot{x}_d) < v$$

$$f(\dot{x}, x, t) \cdot \text{sgn}(f(\dot{x}, x, t)) < F \quad (3.96)$$

$$\ddot{d} \cdot \text{sgn}(\ddot{d}) < D1$$

$$d \cdot \text{sgn}(d) < D2$$

This means, the upper bounds have to be greater than a value whose grade of membership μ is less then a particular α (see Figure 3.31): $A > x_{\alpha\max}$. By rewriting equation (3.95) we obtain

$$\text{sgn}(s) \cdot (\lambda \cdot \dot{e} - \ddot{x}_d + \ddot{d} + d + f(\dot{x}, x, t)) + \eta + \text{sgn}(s) \cdot u \leq 0 \quad (3.97)$$

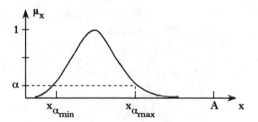

Figure 3.31 Upper bounds of fuzzy sets

With the upper bounds $E, v, F, D1$ and $D2$ we chose the control law in the same way as in the crisp case:

$$u = -K \cdot \text{sgn}(s) \tag{3.98}$$

where K has to satisfy the condition

$$K > E + v + F + D1 + D2 + \eta \tag{3.99}$$

According to the motion along a switching line in the crisp case, one obtains for fuzzy errors the differential equation (see equation (3.87))

$$\dot{e} + \lambda \cdot e = 0 \tag{3.100}$$

Let error e be a fuzzy set described by the membership function

$$\mu_e(t) = f(e, \bar{e}, p_1, \ldots, p_n) \tag{3.101}$$

where $\bar{e}$ is the c.o.g. of the fuzzy set depending on time t and $p_1, \ldots, p_n$ are parameters depending on time t.

Assume, finally, that $\bar{e}, p_1, \ldots, p_n$ do not depend on each other.

Now, one has to choose a representative of the fuzzy set in order to solve a differential equation like equation (3.100). To do this we replace e by $\bar{e}$ in equation (3.100). This is quite reasonable because in many cases the fuzzy set e can be interpreted as a superposition of a time variant crisp signal and a fuzzy signal with other time variant parameters but with a constant mean. Thus, from equation (3.100) we obtain

$$\dot{\bar{e}} + \lambda \cdot \bar{e} = 0 \tag{3.102}$$

with the solution

$$\bar{e} = \bar{e}_0 \cdot e^{-\lambda \cdot t} \tag{3.103}$$

where $\bar{e}_0$ is the starting value of $\bar{e}(t)$. Equivalent to the sliding mode in the crisp case the mean $\bar{e}$ of the error signal e tends to zero with a velocity depending on the slope λ of the switching line.

FUZZY INPUTS

Although the inputs of the controller are fuzzy values similar to a sliding mode with crisp controller inputs, drastic changes of the control output still occur. This is due to the fact that control law (3.98) makes a crisp decision regarding the sign of the mean (centre of gravity) of s.

However, regarding the tracking precision of the controlled system there is a difference between crisp and fuzzy inputs: for crisp inputs error e goes (theoretically) exponentially to zero. For fuzzy inputs this can only be shown for the mean $\bar{e}$ (see equation (3.103)). This means that, in contrast to the crisp case, there remains a finite tracking precision depending on the width of the deviation σ of error e. We define the tracking precision to be

$$T^*_{prec} = 2\sigma \tag{3.104}$$

which means that for Gaussian noise the probability of a crisp measurement e^* falling into the interval $[-2\sigma, 2\sigma]$ is $P(e^*) = 0.95$.

3.4.3.2 Sliding mode with boundary layer

To avoid drastic changes of the control output control law (3.98) is changed into

$$u = -K \cdot \text{sat}(s/\Phi) \tag{3.105}$$

where $\text{sat}(s/\Phi)$ has already been explained in equation (3.83). Within the boundary layer from equations (3.105) and (95), follows formally the filter equation

$$\dot{s} + \frac{K}{\Phi} \cdot s = G \tag{3.106}$$

for

$$G = \lambda \cdot \dot{e} - \ddot{x}_d + \dot{\tilde{d}} + d + f(\dot{x}, x, t)$$

as a fuzzy input of the 'filter'. By taking the means of s, $\dot{s}$ and G we obtain the filter equation

$$\dot{\bar{s}} + \frac{K}{\Phi} \cdot \bar{s} = \bar{G} \tag{3.107}$$

with the solution

$$\bar{s}(t) = e^{-\int_0^t \frac{K}{\Phi} dt'} \left[\bar{s}_0 - \int_0^t \bar{G}(t) \cdot e^{\int_0^t \frac{K}{\Phi} dt'} dt \right] \tag{3.108}$$

Equation (3.108) describes the behaviour of the mean $\bar{s}$ of the fuzzy set s approaching the line

$$\bar{s} = \lambda \cdot \bar{e} + \dot{\bar{e}} = 0 \tag{3.109}$$

Figure 3.32 shows the two approaching components regarding equations (3.103) and (3.108).

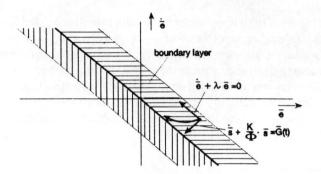

Figure 3.32 Behaviour of the mean of a fuzzy state vector in the phase space

Following equation (3.83) the control law (3.105) still provides a fuzzy control output which does not make sense for the input of a plant. Therefore, the fuzzy control law (3.105) is changed into the crisp control law

$$u = -K \cdot \text{sat}(\text{defuzz}(s)/\Phi) = -K \cdot \text{sat}(\bar{s}/\Phi) \qquad (3.110)$$

In contrast to the pure sliding mode, this control law does not produce high changes in the control variable. However, by using a boundary layer the tracking precision of the controlled system needs to be extended to

$$T_{\text{prec}} = T^*_{\text{prec}} + \Theta = (2 \cdot \sigma + \Theta) \qquad (3.111)$$

where $\Theta = \Phi/\lambda$ is the $\bar{e}$-component of the width of the boundary layer.

3.4.3.3 Sliding mode fuzzy control

It has been shown that a specific class of fuzzy controllers, e.g. PI and PD fuzzy controllers, work in a similar way as a sliding mode controller with boundary layer. Palm (1992a) presented a fuzzy control method which evaluates the distances of the state vector from the switching line in the following way: within the normalized phase plane the distance

$$s_N = e_N + \dot{e}_N \qquad (3.112)$$

is evaluated by means of the following control rules:

$$
\begin{array}{llll}
\text{IF} & s_N = NB & \text{THEN} & u_N = PB \\
\text{IF} & s_N = NM & \text{THEN} & u_N = PM \\
\text{IF} & s_N = NS & \text{THEN} & u_N = PS \\
\text{IF} & s_N = Z & \text{THEN} & u_N = Z \\
\text{IF} & s_N = PS & \text{THEN} & u_N = NS \\
\text{IF} & s_N = PM & \text{THEN} & u_N = NM \\
\text{IF} & s_N = PB & \text{THEN} & u_N = NB \\
\end{array}
\qquad (3.113)
$$

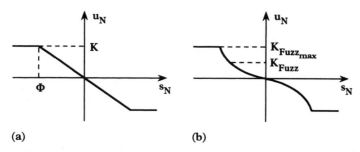

Figure 3.33 Comparison of the transfer functions of a sliding mode controller with boundary layer (*a*) and a sliding mode fuzzy controller (*b*)

where NB is negative big, NM is negative medium, NS is negative small, Z is zero, PS is positive small, PM is positive medium, PB is positive big, and u_N is the normalized control output. Although rather simple, this set of rules leads to a very good control behaviour, even if the system to be controlled includes nonlinearities. As already pointed out, this comes from the strong relationship between this kind of fuzzy controller and the sliding mode controller with boundary layer. Figure 3.33 shows an example of the transfer functions of the two controllers.

These results have been obtained using crisp controller inputs. According to rule set (3.113) the related denormalized version of the control law is

$$u = -K_{\text{Fuzz}} \cdot \text{sgn}(s) \qquad (3.114)$$

where K_{Fuzz} denotes the denormalized absolute value of the defuzzified result obtained from the rules (3.113).

With respect to fuzzy inputs control law (3.98) does not change very much, although the computation of K_{Fuzz} in (3.114) is more complicated. One method of computing u_N from fuzzy inputs is as follows. According to (3.113) the membership functions $\mu_{s_{\text{NB}}}, \mu_{s_{\text{NM}}}, \mu_{s_{\text{NS}}}, \mu_{s_Z}, \mu_{s_{\text{PS}}}, \mu_{s_{\text{PM}}}, \mu_{s_{\text{PB}}}$ for s_N and $\mu_{u_{\text{NB}}}, \mu_{u_{\text{NM}}}, \mu_{u_{\text{NS}}}, \mu_{u_Z}, \mu_{u_{\text{PS}}}, \mu_{u_{\text{PM}}}, \mu_{u_{\text{PB}}}$ for u_N are defined. The inference step is performed by means of the max-min-composition which, by using the Mamdani relation, can be performed very easily rule-by-rule (Hellendoorn, 1990, Driankov, 1993).

(a) Take the maximum degree of membership of the crossing points between the input fuzzy set and the reference set according to the rule considered.

(b) Clip the corresponding output reference set at the level calculated in step 1.

The complete fuzzy output set is then calculated through the union of the fuzzy output set obtained from each rule.

Figure 3.34 shows a corresponding example of several fuzzy inputs with different shapes but with identical mean values resulting in different output sets and, after defuzzification, different control outputs, too.

This example shows that the output set of the controller reflects, to a certain degree, the shape and location of the fuzzy input. On the other hand, it is obvious that a reduction of

106 COMPATIBILITY OF FUZZY CONTROL AND CONVENTIONAL CONTROL TECHNIQUES

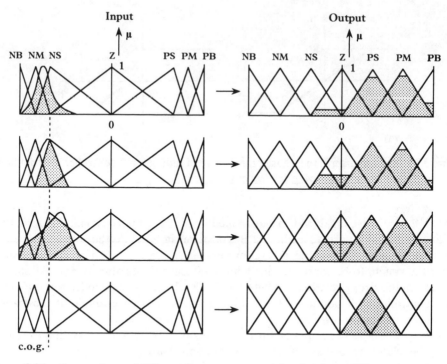

Figure 3.34 Comparison of different fuzzy outputs originating from different fuzzy inputs

a fuzzy input set to its mean value leads to a loss of information about the confidence about the input signal.

Control surface The transfer function (control surface) of a fuzzy controller with fuzzy inputs and crisp outputs cannot be determined in the same easy way as for crisp inputs and outputs because there is no point-to-point mapping between input and output any more. Figure 3.35 shows the transfer function $u_N = f(\bar{e}_N, \sigma_{e_N})$ of the fuzzy controller described above for a bell-shaped fuzzy input with different standard deviations σ_{e_N}. An essential

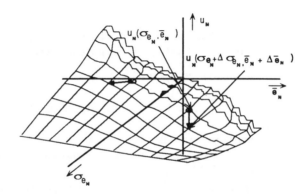

Figure 3.35 Control surface of a fuzzy controlle with fuzzy inputs

result is that the transfer function becomes flatter, the larger the standard deviation of the input signal is going to be. The consequence for the system's behaviour is that model uncertainties, unmodelled disturbances and parameter fluctuations are filtered more strongly than in the case of small deviations. Moreover, the approach phase towards the origin of the phase plane is slowing down, which can be clearly understood from filter equation (3.107). Although the filter for s is more complicated in the case of a fuzzy controller in contrast to a SMC with boundary layer (see Driankov (1993)), it can be seen that the maximum value of K_{Fuzz} directly affects the filtering process. This is quite reasonable because measurements with little confidence (large deviation) are less weighted than those with high confidence (small deviation). This has its counterpart in Kalman filtering, where the filter coefficient increases with rising confidence of the measurement.

Stability In pure sliding mode control and sliding mode with boundary layer the width of the standard deviation affects only the tracking quality (tracking precision) but not the stability since the gain K, which is responsible for stability, is a fixed value. In the case of sliding mode fuzzy control $K_{Fuzz_{max}}$ changes its value in relation to the deviation σ. Hence, gain K_{Fuzz} in control law (3.114) depends on shape and location of the fuzzy input set. Therefore, knowing the upper bounds in equation (3.96) and the relationship between σ and $K_{Fuzz_{max}}$ one is able to decide for which standard deviation σ the system becomes unstable. Let the relationship between $K_{Fuzz_{max}}$ and σ be modelled by

$$K_{Fuzz_{max}} = \frac{C1}{C2 + C3 \cdot \sigma} \quad (3.115)$$

where $C1$, $C2$ and $C3$ are constant. Then, from equation (3.115) and condition (3.96) one obtains the stability condition

$$\sigma < \frac{C1 \cdot (E + v + F + D1 + D2 + \eta) - C2}{C3} \quad (3.116)$$

3.4.4 Simulation Results

The combination of fuzzy inputs with sliding mode fuzzy control has been tested with the help of a simple linear model of a second-order system

$$\ddot{x} + \dot{x} = u \quad (3.117)$$
$$y = x + n$$
$$u = -K_{Fuzz} \cdot \text{sgn}(s)$$

with the error

$$e = y - x_d$$

The noise n has been produced by using a pseudo-random generator providing a mixture of uniform distributions which are predefined on several intervals:

$$n = 0.5 \cdot \text{random}(4) - 0.5 \cdot \text{random}(4) + 0.1 \cdot \text{random}(2) - 0.1 \cdot \text{random}(2)$$

108 COMPATIBILITY OF FUZZY CONTROL AND CONVENTIONAL CONTROL TECHNIQUES

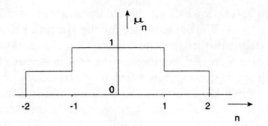

Figure 3.36 Membership function for the pure noise

This results in a normalized distribution with respect to 1 (the membership function of n) shown in Figure 3.36

The controller is the SMFC described in the last subsection. Firstly, in order to construct the fuzzy set S_N the calculation of a single crisp value s_{i_N} is performed by means of six steps:

(a) measurement of e_i at time t_i

(b) filtering of e_i

$$e_{\text{new}_i} = e_{\text{old}_i} \cdot (1-f) + f \cdot e_i$$

where $f \in [0, 1]$ is a constant filter factor

(c) calculation of the estimate

$$\hat{\dot{e}}_i = \frac{(e_{\text{new}_i} - e_{\text{old}_i})}{\Delta t}$$

(d) scaling of e_i

$$e_{i_N} = N_e \cdot e_i$$

(e) scaling of $\hat{\dot{e}}_i$

$$\hat{\dot{e}}_{i_N} = N_{\dot{e}_i} \cdot \hat{\dot{e}}_i$$

(f) calculation of s_{i_N}

$$s_{i_N} = e_{i_N} + \hat{\dot{e}}_{i_N}$$

After that, s_{i_N} is classified with respect to 22 classes within the interval $[-100, +100]$. In the classification process a predefined number of s_{i_N} values (e.g. 10 or 100 values) are gathered from which one obtains a histogram related to the actual distribution. Then the histogram is normalized with respect to the highest likelihood in order to obtain a normal membership function.

It should be noted that the way of calculating the derivative of e is different from the method of differentiation of a fuzzy set with respect to time mentioned above. In this approach we firstly filter the error e and then make an estimation of $\dot{e}$.

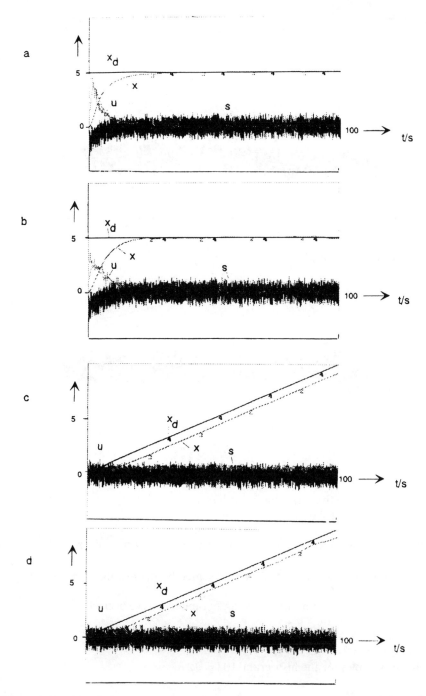

Figure 3.37 Simulation results for crisp (a, c, e) and fuzzy (b, d, f) controller inputs

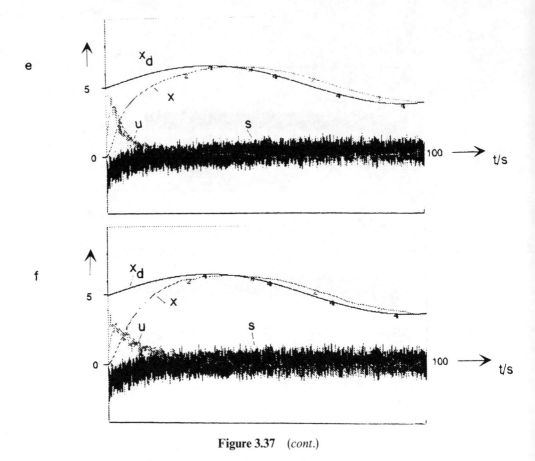

Figure 3.37 (*cont.*)

In order to compare the performance with respect to crisp or fuzzy controller inputs the calculation of the controller input has been done in two ways:

calculation of the mean value $\bar{s}_N$ of the distribution with its degree of membership $\mu_{\bar{s}_N} = 1$
calculation of the whole distribution represented by the membership function μ_{s_N}

For both ways of calculation the following parameters have been chosen:

scaling factors: $N_e = 25$, $N_{\dot{e}} = 0.75$
sampling time: $\Delta t = 0.02\text{s}$
number of samples for the histogram: $10 \, (= 0.2\text{s})$

The test functions are

step function $x_d = 5$
ramp function $x_d = 0.1t$
sinusoidal function $x_d = 5 + 1.5 \cdot \sin(0.05 \cdot t)$

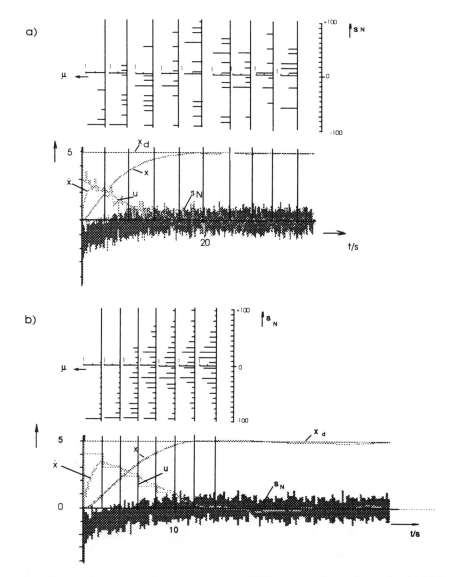

Figure 3.38 Step responses for fuzzy inputs and different numbers of samples: (*a*) 10 samples, (*b*) 100 samples

The result is as follows.

For the step function sample the state reaches the objective faster in the fuzzy input case (see Figure 3.37*b*) than for crisp inputs (see Figure 3.37*a*).

For the ramp function sample the remaining deviation $x - x_d$ is smaller in the fuzzy input case (see Figure 3.37*d*) than for crisp inputs (see Figure 3.37*c*).

For the sinusoidal function one obtains a better tracking quality for the fuzzy inputs (see Figure 3.37*f*) than for crisp inputs (see Figure 3.37*e*).

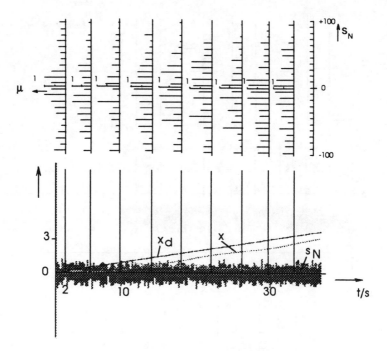

Figure 3.39 Ramp response for fuzzy inputs

The better quality of the fuzzy input samples is caused by the fact that, provided the distribution is well scaled, more rules contribute to the end result than for the crisp input case. These additional rules lead to a higher rate of change of the control output compared with the rate of change which would have resulted from using crisp inputs alone. This means that the system becomes more sensitive with respect to disturbances and desired input changes.

However, this advantage is lost if the deviation of noise increases without updating the corresponding scaling factors. For this reason the scaling factors should be adjusted (or tuned) either on-line or off-line with respect to the deviation of noise.

Finally, by way of three examples we show how the input membership function for the normalized s_N internally behaves. Figures 3.38, 3.39 and 3.40 show the step response, ramp response and a mixture of step response and sinusoidal response of the system to be controlled. The sample times are $T = 2\,\text{s}$ (concerning 100 gathered samples) and $T = 0.2\,\text{s}$ (concerning 10 gathered samples), respectively. The time plots are accompanied by the input membership functions defined in a normalized universe of discourse.

Step response (10 and 100 samples)   The distribution (membership function) is strongly nonsymmetrical if the state x is far from the set point. A comparison between 10 and 100 samples shows a faster reaction for 10 samples (because of the short delay). The membership function has more gaps in the 10 sample case. However, the information about the characteristics of the original distribution (deviation, asymmetry) can be recognized..

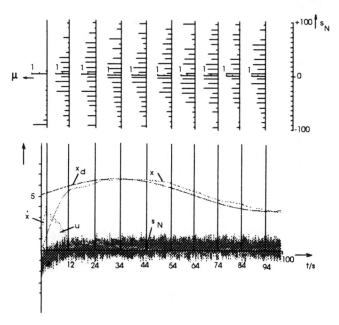

Figure 3.40 Mixture of step response and sinusoidal response for fuzzy inputs

Ramp response (100 samples) Here we observe a constant asymmetry of the distribution coming from the remaining deviation between the desired value x_d and the actual state x.

Step and sinusoidal response (100 samples) In this example we observe the largest asymmetries in the starting phase and at the turning points of the desired curve.

REFERENCES

Boscolo, A., and Drius, F. (1992) Computer aided tuning and validation of fuzzy systems. *1st IEEE Int. Conf. on Fuzzy Systems*, San Diego, pp. 605–614.

Braae, M., Rutherford, D. A. (1979) Selection of parameters for a fuzzy logic controller. *Fuzzy Sets and Systems*, **2**, 185–199.

de Glas, M. (1984) Invariance and stability of fuzzy systems. *Journal of Mathematical Analysis and Applications*, **199**, 428–447.

Driankov, D., Hellendoorn, H., and Reinfrank, M. (1993) *An Introduction to Fuzzy Control* (Berlin: Springer Verlag).

Driankov, D., Palm, R., and Hellendoorn, H. (1994) *Fuzzy Control With Fuzzy Inputs: The Need for New Rule Semantics, FUZZ IEEE'94* Orlando, June 26–29, 1994, Proceedings, pp. 13–21.

Dubois, D., and Prade, H. (1990) *Fuzzy Sets and Systems: Theory and Applications* (New York, London, Toronto, Academic Press, Inc.).

Gibson, J. (1963) *Non-linear Automatic Control* (New York: McGraw-Hill).

Gupta, M. M., Kiszka, J. B., and Trojan, G. M. (1986) Multivariable structure of fuzzy control systems. *IEEE Transactions on Systems Man and Cybernetic*, **16**, 638–656.

Hellendoorn, H. (1990) Reasoning with fuzzy logic. PhD. thesis, Delft University of Technology.

Hwang, G.-C., and Li, S.-C. (1992) A stability approach to fuzzy control design for nonlinear systems. *Fuzzy Sets and Systems*, **48**, 279–287.

Kalman, R. E., and Bucy, R. S. (1963) New results in linear filtering and prediction theory. *Transactions of ASME*, Series D, **83**, 95–108.

Kawaji, S. and Matsunaga, N. (1991) Fuzzy control of VSS type and its robustness. *IFSA'91, Brussels*, pp. 81–88.

Katayama, R., Kajitani, Y., Matsumotu, K., Watanabe, M. and Nishida, Y. (1991) An automatic knowledge acquisition and fast self tuning method for fuzzy controllers based on hierarchical fuzzy inference mechanisms. *IFSA '91*, Brussels, pp. 105–108.

Chuen Chien Lee (1990) Fuzzy logic in control systems: fuzzy logic controller—Part I. *IEEE Transactions on Systems, Man and Cybernetics*, **20**, 404–418.

Maeda, M., Murakami, S., (1992) A self-tuning fuzzy controller. *Fuzzy Sets and Systems*, **51**, 29–40.

Palm, R. (1989) Fuzzy controller for a sensor guided robot manipulator. *Fuzzy Sets and Systems*, **29**, 133–149.

Palm, R. (1992a) Sliding mode fuzzy control. *IEEE International Conference on Fuzzy Systems*, San Diego, pp. 519–526.

Palm, R. (1992b) Control of a redundant manipulator using fuzzy rules. *Fuzzy Sets and Systems*, **45**, 279–298.

Palm, R. (1993) Tuning of scaling in scaling factors in fuzzy controllers using correlation functions. *IEEE International Conference on Fuzzy Systems*, San Francisco, 691–696.

Palm, R., and Driankov, D. (1994) Fuzzy inputs. *FUZZ IEEE'94*, Orlando.

Pedrycz, W. (1992) *Fuzzy control and fuzzy systems*, 2nd revised edition, Research Studies Publication.

Peng, Xian-Tu (1990) Generating rules for fuzzy logic controllers by functions. *Fuzzy Sets and Systems*, **36**, 83–89.

Procyk, T. J., Mamdani, E. H. (1979) A linguistic self-organizing process controller. *Automatica*, **15**, 15–30.

Ray, K. S., Majumder, D. D. (1984) Application of circle criteria for stability analysis with fuzzy logic controller *IEEE Transactions on Systems, Man and Cybernetics*, **14**, 345–349.

Schlitt, H. (1968) *Stochastische Vorgaenge in linearen und nichtlinearen Regelkreisen* (Vieweg, Braunschweig).

Slotine, J. E. (1985) The robust control of robot manipulators. *The International Journal of Robotics Research*, **4**, 49–64.

Smith, S. M. and Comer, D. J. (1992) An algorithm of automated fuzzy logic controller tuning. *1st IEEE International Conference on Fuzzy Systems*, San Diego, pp. 615–622.

Tanaka, K., and Sano, M. (1993) concept of stability margin for fuzzy systems and design of robust fuzzy controllers. *IEEE Int. Conf. Fuzzy Systems*, San Francisco, California, pp. 29–34.

Tang, K. L. and Mulholland, R. J. (1987) Comparing fuzzy logic with classical controller designs. *IEEE Transctions on Systems, Man and Cybernetics*, **17**, 1085–1087.

SieFuzzy 2.0 (1995) Siemens AG, Munich Germany, Sie Fuzzy 2.0 mannal.

Tong, R. M. (1980) Some properties of fuzzy feedback systems. *IEEE Transactions on Systems, Man and Cybernetics*, **10**, 327–330.

Utkin, V. I. (1977) Variable struxture systems with sliding mode: a survey. *IEEE Transactions on Automatic Control*, **22**, 212–222.

Wakileh, B. A. M. and Gill, K. F. (1988) Use of fuzzy logic in robotics. *Computers in Industry*, **10**, 35–46.

Viljamaa, P., and Koivo, H. K. (1993) Tuning of multivariable fuzzy logic controller. *2nd IEEE International Conference on Fuzzy Systems*, San Francisco, pp. 697–701.

Li Zheng (1992) A practical guide to tune PI-like fuzzy-controllers. *1st IEEE International Conference on Fuzzy Systems. Fuzz-IEEE'92*, San Diego, pp. 633–640.

Li Zheng (1993) A practical computer-aided tuning technique for fuzzy control. *2nd IEEE International Conference on Fuzzy Systems*, San Francisco, pp. 702–707.

Zimmermann, (1991), *Fuzzy set theory and its applications* (Kluwer, Boston, Dortrecht, London).

4
On the Crisp-type Fuzzy Controller: Behaviour Analysis and Improvement

Wu Zhi Qiao and **Masaharu Mizumoto**
*Department of Management Engineering,
Osaka Electro-Communication University, Japan*

4.1 INTRODUCTION

Many fuzzy controller structures based on various inference methods have been presented [1–14]. Among them the most widely used methods in practice are: the Mamdani method proposed by Mamdani and his associates [2] who adopted the min–max compositional rule of inference based on an interpretation of a control rule as a conjunction of the antecedent and consequent; and the product–sum method proposed by Mizumoto [10–12] who suggested introducing the product and arithmetic mean aggregation operators to replace the logical AND (minimum) and OR (maximum) calculations in the min–max compositional rule of inference. The product–sum method greatly simplifies the algorithm of the fuzzy controller. In the algorithm of a fuzzy controller, the defuzzification calculation is also a complicated and time-consuming task. Using different defuzzification methods would not give a distinct improvement of performance. Hans and Christoph [22] have given a review on the defuzzyfication. Takagi and Sugeno proposed a crisp-type model in which the consequent parts of the fuzzy control rules are crisp functional representation or crisp real numbers in the simplified case instead of fuzzy sets [13, 14]. With this model of crisp real number output, the fuzzy set of the inference consequence will be a discrete fuzzy set with a finite number of points, this can greatly simplify the calculation of the defuzzification algorithm. The product–sum inference method and the crisp output model are often applied in a mixed manner. The mixed product–sum crisp model has a fine performance and the simplest algorithm that is very easy to be implemented in hardware system and converted into a fuzzy neural network model. Unsurprisingly, it has been much appreciated by the Japanese industrial engineering. In this chapter we will take account of the product–sum crisp model and the min–max crisp model.

There is not yet a sound theoretical method available to analyze a fuzzy controller, whereas the conventional control theory is highly developed. It is natural for the researchers to apply the conventional theory, mostly linear systems theory, to solve the nonlinear problem of fuzzy controller and much work has been done in this direction [15, 17–20]. Wang and his colleagues [20] first compared the linear rule type fuzzy controller [20, 33] with a classical PID controller in their input–output relationship. Buckley [18, 19] proved that a fuzzy controller of the crisp type can approximate any real-valued continuous function to any degree of accuracy. Mizumoto [10] shows that under some special control rules, a product–sum crisp type fuzzy controller will generate exactly the same input–output relation as a regular PID controller. In this chapter, the authors try to analyze the behaviour of a crisp type fuzzy controller with both inference methods of min–max compositional rule and the product–sum inference, by relating the fuzzy controller to the conventional PID controller. We will investigate the influence of the membership functions, the control rules, as well as the scaling parameters to the performance of a fuzzy controller according to the well-known classical design method of a PID controller. Furthermore, the authors work out two new fuzzy controller structures, the PID type fuzzy controller and the parameter adaptive method, which promises to improve considerably the performance of the fuzzy controller.

4.2 THE CRISP-TYPE FUZZY CONTROLLER

We briefly describe the crisp-type fuzzy controller mixed with both of product–sum and min–max inference methods as follows [2, 11–14].

Suppose that the fuzzy controller in consideration is a two-input and one-output one. The two inputs to the fuzzy controller are error e and change rate of error $\dot{e}$, and the output of the fuzzy controller (that is, the input to the controlled process) is u. The universes of discourses of $e, \dot{e}$ and u are $E \subset R, \dot{E} \subset R$ and $U \subset R$, respectively. Denote the linguistic values of e and $\dot{e}$ by A_i ($i \in I = [-m, \ldots, -2, -1, 0, 1, 2, \ldots, m]$) and B_j ($j \in J = [-n, \ldots, -2, -1, 0, 1, 2, \ldots, n]$), respectively. The fuzzy control rules are given in the form

$$\text{if } e \text{ is } A_i \text{ and } \dot{e} \text{ is } B_j \text{ then } u \text{ is } u_{ij}$$

where $u_{ij} \in U$ ($i \in I$, $j \in J$) is a crisp value instead of a fuzzy subset. The fuzzy controller with such kind of control rules is called a crisp-type fuzzy controller [13, 14]. This fuzzy controller can be shown in Figure 4.1.

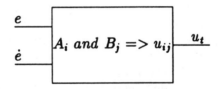

Figure 4.1 The crisp type fuzzy controller

If the number of control rules is equal to $I \times J$, the fuzzy control rule base is said to be complete [20]. In the following discussion we assume that the fuzzy control rule base is complete.

Suppose that the membership functions A_i and B_j are $A_i(e)$ and $B_j(\dot{e})$. In a certain control instance t, the observation values of e and $\dot{e}$ are e_t and $\dot{e}_t$, respectively, then the truth values of A_i and B_j are $A_i(e_t)$ and $B_j(\dot{e}_t)$ ($i\in I, j\in J$). The truth value of the antecedent part of a fuzzy control rule will be

$$f_{ij} = \begin{cases} A_i(e_t)B_j(\dot{e}_t), & \text{product-sum method} \\ \text{Min}\{A_i(e_t), B_j(\dot{e}_t)\}, & \text{min-max method} \end{cases} \quad (i\in I, j\in J) \qquad (4.1)$$

The reasoning from the antecedent part to the consequent part will generate a conclusion fuzzy subset which we denote as C. C will be discrete fuzzy subset with finite number of points. In the case of the product-sum method, $C = \{f_{ij}/u_{ij}, | i\in I, j\in J\}$, where u_{ij} are not necessarily different from each other. Using the centre of gravity method to defuzzify the fuzzy set C, the real output of the controller u_t is given by

$$u_t = \frac{\sum_{i,j} f_{ij} u_{ij}}{\sum_{i,j} f_{ij}} \qquad (4.2)$$

Although in the case of min-max method

$$C = \sup\left\{\frac{f_{ij}}{u_{ij}}, | i\in I, j\in J\right\}$$

If u_{ij} are different from each other, obviously

$$C = \left\{\frac{f_{ij}}{u_{ij}}, | i\in I, j\in J\right\}$$

and the real output of the fuzzy controller will be the same as given by equation (4.2). If not, there will be many different situations. For example, considering the extreme case, suppose that $\forall (i\in I, j\in J)$, $u_{ij} = u_0$, C becomes a single point subset, that is $C = \{f_0/u_0\}$, where $f_0 = \max[f_{ij} | i\in I, j\in J]$, the real output of the fuzzy controller will be u_0. For convenience of discussion, we only consider the case that the u_{ij} are different from each other so that equation (4.2) can apply for both the product-sum and min-mix methods.

4.3 THE DYNAMIC ANALYSIS OF THE CRISP-TYPE FUZZY CONTROLLER

In our study we will employ the triangular membership function for each fuzzy linguistic value of the error e and the change rate of error $\dot{e}$ as shown in Figure 4.2. Denote the cores of fuzzy set A_i as e_i and those of B_j as $\dot{e}_j$, the interval $[e_i, e_{i+1}]$ as Δ_i, and the interval

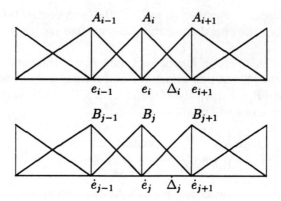

Figure 4.2 The membership functions of A_i and B_j

$[\dot{e}_j, \dot{e}_{j+1}]$ as $\dot{\Delta}_j$. We design the membership functions so that the support sets of A_i are equal to $[e_{i-1}, e_{i+1}]$ and those of B_j are equal to $[\dot{e}_{j-1}, \dot{e}_{j+1}]$, as illustrated in Figure 4.2. We have

$$A_i(e) = 1 - \frac{e - e_i}{\Delta_i}, \quad A_{i+1}(e) = \frac{e - e_i}{\Delta_i}, \quad A_k(e) = 0 \quad (k \neq i \in I)$$

for $e \in [e_i, e_{i+1}]$, and

$$B_j(\dot{e}) = 1 - \frac{\dot{e} - \dot{e}_j}{\dot{\Delta}_j}, \quad B_{j+1}(\dot{e}) = \frac{\dot{e} - \dot{e}_j}{\dot{\Delta}_j}, \quad B_t(\dot{e}) = 0 \quad (t \neq j \in J)$$

for $\dot{e} \in [\dot{e}_j, \dot{e}_{j+1}]$.

Obviously, $A_i(e) + A_{i+1}(e) = 1$ ($e \in [e_i, e_{i+1}]$), and $B_j(\dot{e}) + B_{j+1}(\dot{e}) = 1$ ($\dot{e} \in [\dot{e}_j, \dot{e}_{j+1}]$).

On the $e - \dot{e}$ plane, we call the set $\{e, \dot{e} | e = e_i, \dot{e} = \dot{e}_j, i \in I, j \in J\}$ a NET [20] and the points $(e_i, \dot{e}_j)$ the NODE of the NET. The NET is illustrated in Figure 4.3. In the following description we will analyze the output of the crisp-type fuzzy controller in the NET and the NODE on an $e - \dot{e}$ plane.

Under the conditions in Figure 4.3, in each control cycle of the fuzzy control system, only four rules are fired. For instance, if the inputs of the fuzzy controller are located in the NET lattice areas $S = [e_i, e_{i+1}] \times [\dot{e}_j, \dot{e}_{j+1}]$ of the e–$\dot{e}$ plane, the four fired control rules will be as follows:

if e is A_i and $\dot{e}$ is B_j then u is u_{ij}
if e is A_{i+1} and $\dot{e}$ is B_j then u is $u_{(i+1)j}$
if e is A_i and $\dot{e}$ is B_{j+1} then u is $u_{i(j+1)}$
if e is A_{i+1} and $\dot{e}$ is B_{j+1} then u is $u_{(i+1)(j+1)}$

In the lattice area S of the NET, we denote the membership degree $A_i(e) = \mu$, $A_{i+1}(e) = 1 - A_i(e) = 1 - \mu$, and $B_j(\dot{e}) = \dot{\mu}$, $B_{j+1}(\dot{e}) = 1 - B_j(\dot{e}) = 1 - \dot{\mu}$.

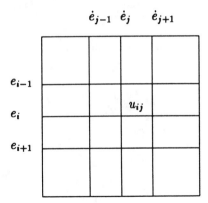

Figure 4.3 The NET on the e–$\dot{e}$ plane

We will discuss the product–sum method and the min–max method.
For the product–sum method, by combining (4.1), equation (4.2) can be written as

$$u_t = \frac{\mu\dot{\mu}u_{ij} + (1-\mu)\dot{\mu}u_{(i+1)j} + \mu(1-\dot{\mu})u_{i(j+1)} + (1-\mu)(1-\dot{\mu})u_{(i+1)(j+1)}}{\mu\dot{\mu} + (1-\mu)\dot{\mu} + \mu(1-\dot{\mu}) + (1-\mu)(1-\dot{\mu})}$$

It is easy to check that the denominator of the right-hand side of the above equation is

$$\mu\dot{\mu} + (1-\mu)\dot{\mu} + \mu(1-\dot{\mu}) + (1-\mu)(1-\dot{\mu}) = 1$$

Therefore

$$u_t = \mu\dot{\mu}u_{ij} + (1-\mu)\dot{\mu}u_{(i+1)j} + \mu(1-\dot{\mu})u_{i(j+1)} + (1-\mu)(1-\dot{\mu})u_{(i+1)(j+1)} \qquad (4.3)$$

where μ and $\dot{\mu}$ are functions of arguments e and $\dot{e}$ respectively.

Let us examine the dynamic input–output behaviour of the product–sum crisp-type fuzzy controller. Assume that at time instant t, the observation value of error and the change rate of error are e and $\dot{e}$, respectively.

Obviously, when $e = e_i$ and $\dot{e} = \dot{e}_j$, that is at the NODE $(e_i, \dot{e}_j)$ of the NET, $A_i(e) = \mu = 1, A_k(e) = 0$ $(k \neq i \in I)$, $B_j(\dot{e}) = \dot{\mu} = 1$, $B_t(\dot{e}) = 0$ $(t \neq j \in J)$, from equation (4.3), the output of the product–sum crisp type fuzzy controller will be

$$u_t = u_{ij} \qquad (4.4)$$

On the NET of e–$\dot{e}$ plane, on the line of $\dot{e} = \dot{e}_j$, $B_j(\dot{e}) = \dot{\mu} = 1$ and $B_t(\dot{e}) = 0$ $(t \neq j \in J)$, we further suppose the error e is within the interval $[e_i, e_{i+1}]$, the membership degree of the linguistic values subject to e will be $A_i(e) = 1 - (e - e_i)/\Delta_i = \mu$, $A_{i+1}(e) = 1 - A_i(e) = 1 - \mu$.

122 CRISP-TYPE FUZZY CONTROLLER

Then from (4.3), we have

$$u_t = \mu u_{ij} + (1-\mu)u_{(i+1)j}$$
$$= u_{(i+1)j} + (u_{ij} - u_{(i+1)j})\mu$$
$$= u_{ij} + \frac{u_{(i+1)j} - u_{ij}}{\Delta_i}(e - e_i)$$
$$= a + pe \qquad (4.5)$$

where $a = u_{ij} - pe_i$ and

$$p = \frac{u_{(i+1)j} - u_{ij}}{\Delta_i}$$

In the same manner, we can derive the input–output relation of the fuzzy controller on the line of $e = e_i$, and $\dot{e}_j \in [\dot{e}_j, \dot{e}_{j+1}]$. Now, $A_i(e) = \mu = 1$, $A_k(e) = 0$ $(k \neq i \in I)$, $B_j(\dot{e}) = 1 - (\dot{e} - \dot{e}_j)/\dot{\Delta}_j = \dot{\mu}$, and $B_{j+1}(\dot{e}) = 1 - B_j(\dot{e}) = 1 - \dot{\mu}$. Then from (4.3), we have

$$u_t = \dot{\mu} u_{ij} + (1 - \dot{\mu})u_{i(j+1)}$$
$$= u_{i(j+1)} + (u_{ij} - u_{i(j+1)})\dot{\mu}$$
$$= u_{ij} + \frac{u_{i(j+1)} - u_{ij}}{\dot{\Delta}_j}(\dot{e} - \dot{e}_i)$$
$$= a + d\dot{e} \qquad (4.6)$$

where $b = u_{ij} - d\dot{e}_j$ and

$$d = \frac{u_{i(j+1)} - u_{ij}}{\dot{\Delta}_j}$$

Now let us look at the min–max crisp-type fuzzy controller. In this case, by combining (4.1), equation (4.2) can be written as

$$u_t = \frac{(\mu \wedge \dot{\mu})u_{ij} + ((1-\mu) \wedge \dot{\mu})u_{(i+1)j} + (\mu \wedge (1-\dot{\mu}))u_{i(j+1)} + ((1-\mu) \wedge (1-\dot{\mu}))u_{(i+1)(j+1)}}{\mu \wedge \dot{\mu} + (1-\mu) \wedge \dot{\mu} + \mu \wedge (1-\dot{\mu}) + (1-\mu) \wedge (1-\dot{\mu})}$$

(4.7)

Unlike the product–sum method, the denominator of the right-hand side of the above equation is not always equal to 1. Obviously, at the NODE $(e_i, \dot{e}_j)$ of the NET, $A_i(e) = \mu = 1$, $A_k(e) = 0$ $(k \neq i \in I)$, $B_j(\dot{e}) = \dot{\mu} = 1$, $B_t(\dot{e}) = 0$ $(t \neq j \in j)$, from the above equation, the output of the min–max crisp-type fuzzy controller will be $u_t = u_{ij}$, which is the same as the output of the product–sum crisp type fuzzy controller.

On the NET of e–$\dot{e}$ plane, on the line of $\dot{e} = \dot{e}_j$ (obviously $B_j(\dot{e}) = \dot{\mu} = 1$ and $B_t(\dot{e}) = 0$ $(t \neq j \in J)$ as in the above analysis of the product–sum crisp-type fuzzy controller) we suppose the error e is within the interval $[e_i, e_{i+1}]$, the membership degree of the linguistic values subject to e will be $A_i(e) = 1 - (e - e_i)/\Delta_i = \mu$, $A_{i+1}(e) = 1 - A_i(e) = 1 - \mu$. Then the denominator of the right-hand side of equation (4.7) is

$$\mu \wedge \dot{\mu} + (1 - \mu) \wedge \dot{\mu} + \mu \wedge (1 - \dot{\mu}) + (1 - \mu) \wedge (1 - \dot{\mu}) = 1$$

and

$$u_t = (\mu \wedge \dot{\mu})u_{ij} + ((1 - \mu) \wedge \dot{\mu})u_{(i+1)j} + (\mu \wedge (1 - \dot{\mu}))u_{i(j+1)}$$
$$+ ((1 - \mu) \wedge (1 - \dot{\mu}))u_{(i+1)(j+1)}$$
$$= \mu u_{ij} + (1 - \mu)u_{(i+1)j}$$
$$= u_{(i+1)j} + (u_{ij} - u_{(i+1)j})\mu$$
$$= u_{ij} + \frac{u_{(i+1)j} - u_{ij}}{\Delta_i}(e - e_i)$$
$$= a + pe$$

where $a = u_{ij} - pe_i$ and

$$p = \frac{u_{(i+1)j} - u_{ij}}{\Delta_i}$$

In the same manner, on the line of $e = e_i$, and $\dot{e}_j \in [\dot{e}_j, \dot{e}_{j+1}]$, the membership degrees are, $A_i(e) = \mu = 1$, $A_k(e) = 0$ ($k \neq i \in I$), $B_j(\dot{e}) = 1 - (\dot{e} - \dot{e}_j)/\dot{\Delta}_j = \dot{\mu}$, and $B_{j+1}(\dot{e}) = 1 - B_j(\dot{e}) = 1 - \dot{\mu}$. Again, the denominator of the right-hand side of equation (4.7) is equal to 1. We have

$$u_t = (\mu \wedge \dot{\mu})u_{ij} + ((1 - \mu)) \wedge \dot{\mu})u_{(i+1)j} + (\mu \wedge (1 - \dot{\mu}))u_{i(j+1)}$$
$$+ ((1 - \mu) \wedge (1 - \dot{\mu}))u_{(i+1)(j+1)}$$
$$= \dot{\mu}u_{ij} + (1 - \dot{\mu})u_{i(j+1)}$$
$$= u_{i(j+1)} + (u_{ij} - u_{i(j+1)})\dot{\mu}$$
$$= u_{ij} + \frac{u_{i(j+1)} - u_{ij}}{\dot{\Delta}_j}(\dot{e} - \dot{e}_i)$$
$$= a + d\dot{e}$$

where $b = u_{ij} - d\dot{e}_j$ and

$$d = \frac{u_{i(j+1)} - u_{ij}}{\dot{\Delta}_i}$$

As we can see, on the NET of the $e - \dot{e}$ plane, the output of the min–max crisp fuzzy controller is the same as that off the product–sum crisp-type fuzzy controller, as given by (4.5) and (4.6).

At an arbitrary point in the area $E \times \dot{E}$–NET on the e–$\dot{e}$ plane, as can be seen from equations (4.3) and (4.7), the output of the fuzzy controller is a nonlinear function of the arguments of e and $\dot{e}$. There is no known analytical solution available to deal with such kind of nonlinearity. A complete treatment of this nonlinearity is impossible. Nevertheless, we can adopt the linearization method and carry out a analysis of small deviations from nominal just like the conventional or modern control theory usually does [23, 24].

We will look at the neighbourhood of the NODE of the e–$\dot{e}$ plane. Denote the input–ouput relation of the fuzzy controller with the following nonlinear model:

$$u_t = f(e, \dot{e}, t) \tag{4.8}$$

As shown in the above text of this section, at the NODE $(e_i, \dot{e}_j)$ of the e–$\dot{e}$ plane, the nominal solution is known and given by (4.4), that is

$$u_t = f(e_i, \dot{e}_j, t) = u_{ij}$$

The difference between these nominal values and some slightly perturbed functions e, $\dot{e}$ and u_t can be defined by

$$\delta e = e - e_i$$
$$\delta \dot{e} = \dot{e} - \dot{e}_j$$
$$\delta u_t = u_t - u_{ij}$$

Then equation (4.8) can be written as

$$u_{ij} + \delta u_t = f(e_i, \dot{e}_j, t) + \left[\frac{\partial f}{\partial e}\right]_n \delta e + \left[\frac{\partial f}{\partial \dot{e}}\right]_n \delta \dot{e} + \text{higher order terms}$$

where $[\cdot]_n$ means that the derivatives are evaluated on the nominal solutions, that is, the solutions at the NODE point of the NET. Since the nominal solutions satisfy equation (4.8), the first terms in the preceding Taylor series expansions cancelout. For sufficiently small δe, $\delta \dot{e}$ and δu_t perturbations, the higher-order terms can be neglected, leaving the linear equations

$$\delta u_t = \left[\frac{\partial f}{\partial e}\right]_n \delta e + \left[\frac{\partial f}{\partial \dot{e}}\right]_n \delta \dot{e} \tag{4.9}$$

A neighbourhood of each NODE (or nominal point) will be divided into four different quadrants by the two NET lines that across at the NODE. For simplicity, we only take consider of the case of the first quadrant where $\delta e \geq 0$ and $\delta \dot{e} \geq 0$, that is to say

$$(e_i + \delta e, \dot{e}_j + \delta \dot{e}) \in [e_i, e_{i+1}] \times [\dot{e}_j, \dot{e}_{j+1}]$$

We can calculate the derivatives in equation (4.9) at the NODE point $(e_i, \dot{e}_j)$ simply from (4.5) and (4.6).

From (4.5) we have

$$\frac{\partial f}{\partial e} = \frac{u_{(i+1)j} - u_{ij}}{\Delta_i}$$

and from (4.6) we have

$$\frac{\partial f}{\partial \dot{e}} = \frac{u_{i(j+1)} - u_{ij}}{\dot{\Delta}_j}$$

so

$$\left[\frac{\partial f}{\partial e}\right]_{(e_i, \dot{e}_j)} = \frac{u_{(i+1)j} - u_{ij}}{\Delta_i}$$

$$\left[\frac{\partial f}{\partial \dot{e}}\right]_{(e_i, \dot{e}_j)} = \frac{u_{i(j+1)} - u_{ij}}{\dot{\Delta}_i}$$

Then

$$\delta u_t = \left[\frac{\partial f}{\partial e}\right]_n \delta e + \left[\frac{\partial f}{\partial \dot{e}}\right]_n \delta \dot{e}$$

$$= \frac{u_{(i+1)j} - u_{ij}}{\Delta_i} \delta e + \frac{u_{i(j+1)} - u_{ij}}{\dot{\Delta}_j} \delta \dot{e}$$

that is

$$u_t - u_{ij} = \frac{u_{(i+1)j} - u_{ij}}{\Delta_i}(e - e_i) + \frac{u_{i(j+1)} - u_{ij}}{\dot{\Delta}_j}(\dot{e} - \dot{e}_j)$$

therefore

$$u_t = \left[u_{ij} - \frac{u_{(i+1)j} - u_{ij}}{\Delta_i} e_i - \frac{u_{i(j+1)} - u_{ij}}{\dot{\Delta}_j} \dot{e}_j\right] + \frac{u_{(i+1)j} - u_{ij}}{\Delta_i} e + \frac{u_{i(j+1)} - u_{ij}}{\dot{\Delta}_j} \dot{e}$$

$$= A + Pe + D\dot{e} \tag{4.10}$$

where

$$A = u_{ij} - \frac{u_{(i+1)j} - u_{ij}}{\Delta_i} e_i - \frac{u_{i(j+1)} - u_{ij}}{\dot{\Delta}_j} \dot{e}_j$$

$$= u_{ij} - Pe_i - D\dot{e}_j$$

$$P = \frac{u_{(i+1)j} - u_{ij}}{\Delta_i}$$

$$D = \frac{u_{i(j+1)} - u_{ij}}{\dot{\Delta}_j}$$

Because we can see that the crisp-type fuzzy controller with the min–max reference method and that with product–sum reference method have the same output on the NET of the e–$\dot{e}$ plane and the neighbourhood of the NODE of the NET, so the performance difference between them is minor. It is obvious that the denser the NET, the smaller the difference.

The above analysis also shows that the crisp-type fuzzy controller behaves like a P controller on the NET lines $\dot{e} = \dot{e}_j$ of the e–$\dot{e}$ plane, as shown by equation (4.5), and like a D controller on the NET lines $e = e_i$ of the $e - \dot{e}$ plane, as shown in equation (4.6). Although in the neighbourhood of the NODE point of the NET of $e - \dot{e}$ plane, the crisp-type fuzzy controller approximately performs a PD controller, as illustrated by equation (4.10). Actually, we can regard such a kind of fuzzy controller as a parameter time-varying PD controller. This is very useful in designing the fuzzy controller according to the conventional PID control theory.

To further explore the behaviour of the fuzzy controller under discussion by relating it to the conventional PID controller, we present here a brief review of the property of the PID controller.

The input–output relation of a PID controller is

$$u = K_p e + K_I \int e \, \mathrm{d}t + K_D \frac{\mathrm{d}e}{\mathrm{d}t}$$

According to the conventional automatic control theory [23], the performance of a PID (proportional-plus-integral-plus-derivative) controller is determined by its proportional parameter K_P, integral parameter K_I and derivative parameter K_D. The proportional control law can guarantee the fast response of the control system, the integral control law can eliminate the steady-state error of the control system, and the derivative control law can increase the damping of the system, thus reducing the overshoot and oscillating times of the system response.

A P or PD controller will yield a steady-state error for the system step response if the controlled plant is a type 0 system. The steady-state error is inversely proportional to K_P; if K_P is too large, the stability of the system may be adversely affected. One obvious effect of the integral control is that it increases the order of the system by one. Therefore, the steady-state error of the original system without integral control is improved by one order; that is, if the original system is type 0, the integral control reduces the steady-state error to zero. However, because the system order is increased, it may become less stable. Thus, we see that the PID controller when designed properly could yield a system with fast rise time, small overshoot and non-steady-state error.

As it can be seen from (4.5), (4.6) and (4.10) that the parameters P and D of the fuzzy controller at a specific NET lattice are decided by the control values of two neighbouring NODEs and the distances between the two neighbouring NODEs. The P and D parameters are proportional to the difference between the control values of two neighbouring NODEs: the P parameter is inversely proportional to Δ_i, and the D parameter is inversely proportional to $\dot{\Delta}_j$. Therefore, in designing a fuzzy controller, we can carefully tune the values of u_{ij} in each NODE (thus the fuzzy control rules are tuned), and the Δ_i and $\dot{\Delta}_j$ of the NET (thus the membership functions are tuned) to obtain an optimal combinations of the P and D parameters in the different areas in the e–$\dot{e}$ plane. For example, for some process control system, in the area of the e–$\dot{e}$ plane where the error is large, a larger P may be wanted to speed up the system response, and in the neighbourhood of zero, a larger P parameter may also be wanted to reduce the steady-state error.

4.4 THE STATIC ANALYSIS OF A CRISP-TYPE FUZZY CONTROL SYSTEM

We have mentioned that the min–max crisp type and the product–sum crisp-type fuzzy controllers have the same output properties on the NET of e–$\dot{e}$ plane and the neighbourhood of the NODE. The difference between their performance is small. All the following analyses and experiments will be made on the product–sum type fuzzy controller. The results also apply for the min–max type fuzzy controller.

As analyzed above, the crisp-type fuzzy controller can be regarded as a parameter time-varying PD controller. It definitely yields a steady-state error when used to control a type 0 plant. The error analysis of the fuzzy control system will be similar to that of the conventional PID controller. Since the fuzzy controller is nonlinear, some special problems for the fuzzy control system need to be dealt with.

Consider the fuzzy control system of Figure 4.4, where K_1 and K_2 are the scaling factors for e and $\dot{e}$, respectively, and α is the output gain. In the preceding sections, we did not take into account the scaling factors or the output gain in the fuzzy controller. In this section we concentrate on the control rules and membership functions, so we still set these scaling parameters and output gain to unit.

In all the following derivations of this section it is assumed that the system is stable. If the system is unstable, none of the conclusions derived in this section have meaning. We study the steady-state error of the system due to the step input. The plant considered is type 0 system, because for a type 1 plant the step response of the system is steady-state error free, the analysis becomes insignificant.

Since the system is assumed to be stable, when the time is sufficiently long, the system's error will become constant; that is to say, any order derivative of error is zero. Suppose that the equilibrium point of the system is zero and the control value in the NODE $(0, 0)$ of the e–$\dot{e}$ plane is $u_{00} = 0$; $e_0 = 0$ and $\dot{e}_0 = 0$. The input–output relation of the fuzzy controller is decided by (4.5). Denoting the steady-state error by e_{ss}, we next discuss the cases when e_{ss} is located within the interval $[0, e_{20}]$. The results can be generalized to the other areas. From (4.5), we have the input–output relation of the fuzzy controller as follows:

$$u_{t0} = \begin{cases} \dfrac{u_{10}}{e_{10}} e_{ss}, & \text{when } e_{ss} \in [0, e_{10}] \\ u_{10} + \dfrac{u_{20} - u_{10}}{e_{20} - e_{10}}(e_{ss} - e_{10}), & \text{when } e_{ss} \in [e_{10}, e_{20}] \end{cases} \quad (4.11)$$

as can be shown in Figure 4.5.

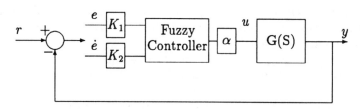

Figure 4.4 The PD-type fuzzy control system

CRISP-TYPE FUZZY CONTROLLER

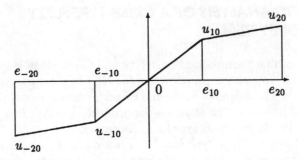

Figure 4.5 The steady-state input/output relation of fuzzy controller

Denote the transform function of the plant to be controlled by $G(S)$, and its steady-state character by K_G; then K_G is defined as

$$K_G = \lim_{s \to 0} G(S)$$

Therefore, at the steady-state of the system, to maintain the system in balance, it the following equations must hold:

$$e_{ss} = r - y = 1 - y$$

$$y = u_{t0} K_G$$

By combining the equation (4.11) with the above equations

$$e_{ss0} = \frac{1}{1 + K_G \dfrac{u_{10}}{e_{10}}} \qquad (4.12)$$

$$e_{ss1} = e_{10} + \frac{1 - K_G \dfrac{u_{10} - e_{10}}{e_{20} - e_{10}}}{1 + K_G \dfrac{u_{20} - u_{10}}{e_{20} - e_{10}}} \qquad (4.13)$$

that is to say, the e_{ss} may take the value of either e_{ss0} or e_{ss1}. When $e_{ss} \in [e_{10}, e_{20}]$, the steady-state error e_{ss} is denoted by (4.12), and when $e_{ss} \in [0, e_{10}]$, e_{ss} is denoted by (4.13). There may be different cases where the steady-state error may occur. We can carry out the following approximate analysis.

If $e_{ss0} < e_{10}$ and $e_{ss1} < e_{10}$, this means that when $e \in [e_{10}, e_{20}]$ the system tends to drive the error into the interval $[0, e_{10}]$, where the steady-state error is denoted by e_{ss0}; the system settles down finally at $e_{ss} = e_{ss0}$.

If $e_{ss0} > e_{10}$ and $e_{10} < e_{ss1} < e_{20}$, the system cannot settle down in the interval $[0, e_{10}]$, the steady-state error is denoted by e_{ss1}, so $e_{ss} = e_{ss1}$.

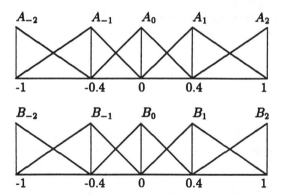

Figure 4.6 The membership functions of $A_i B_j$ for simulation

If $e_{ss0} < e_{10}$ and $e_{10} < e_{ss1} < e_{20}$, e_{ss} can settle down in either e_{ss0} or e_{ss1}, decided by the transient response of the system.

If $e_{ss0} > e_{10}$ and $e_{ss1} < e_{10}$, $e_{ss} = e_{ss1}$, the system will be unstable. When e is about to settle down within $[0, e_{10}]$, the system error tends to the point beyond e_{10}, whereas when e is about to settle down within $[e_{10}, e_{20}]$, the system tends to a balance point less than e_{10}; thus a limit loop may occur.

From (4.12) and (4.13) we can see that the value of u_{10} plays an important role in the static behaviour of the fuzzy control system. We can choose a proper u_{10} so that the static behaviour of the fuzzy control system becomes satisfactory. We will illustrate the influence of u_{10} on the static behaviour of the fuzzy control system by simulation experiments.

Before presenting the simulation, we give a description of the simulation model. The fuzzy control system is as shown in Figure 4.4. The plant model is a second-order and type 0 system with the following transform function:

$$G(s) = \frac{K}{(T_1 S + 1)(T_2 S + 1)}$$

where $K = 8$, $T_1 = 1$, and $T_2 = 0.5$. In our simulation experiments we use the discrete simulation method; the results would be slightly different from that of a continuous system. The sampling time of the system is set to be 0.1 second.

For the fuzzy controller, the fuzzy subsets of e and $\dot{e}$ are defined as shown in Figure 4.6. Their cores are

$$\{e_i\} = \{e_{-2}, e_{-1}, e_0, e_1, e_2\} = \{-1, -0.4, 0, 0.4, 1\}$$
$$\{\dot{e}_j\} = \{\dot{e}_{-2}, \dot{e}_{-1}, \dot{e}_0, \dot{e}_1, \dot{e}_2\} = \{-1, -0.4, 0, 0.4, 1\}$$

The fuzzy control rules are represented as Table 4.1. In the fuzzy control rules table we let u_{-10} and u_{10} be free to examine how the performance changes when they vary.

Figure 4.7 presents the simulation results. We assume that control rules table is anti-symmetric, that is to say u_{-10} and u_{10} have the same absolute value. The simulation

Table 4.1 The fuzzy control rules

	$\dot{e}_{-2}$	$\dot{e}_{-1}$	$\dot{e}_0$	$\dot{e}_1$	$\dot{e}_2$
e_{-2}	-1	-0.7	-0.5	-0.3	0
e_{-1}	-0.7	-0.4	u_{-10}	0	0.3
e_0	-0.5	-0.2	0	0.2	0.5
e_1	-0.3	0	u_{10}	0.4	0.7
e_2	0	0.3	0.5	0.7	1

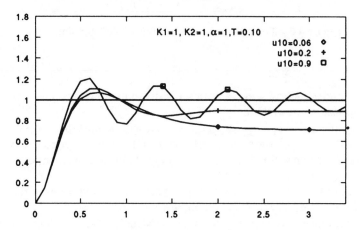

Figure 4.7 The step response of PD-type fuzzy control system

shows that when u_{10} is small, the steady-state error of the system is large. Increasing the value of u_{10} can reduce the steady-state error. With too large a value of u_{10}, oscillation occurs, thus resulting in an unstable system.

4.5 AN IMPROVEMENT: PID-TYPE FUZZY CONTROLLER STRUCTURE

As mentioned above, the fuzzy controller approximately behaves like a parameter time-varying PD controller. Since the mathematical models of most industrial process systems are type 0, obviously there would be steady-state error if they are controlled by this kind of fuzzy controller. This characteristic has been stated in the brief review of the PID controller in the preceding section. If we want to eliminate the steady-state error of the control system, we can imagine substituting the input $\dot{e}$ (the change rate of error or the derivative of error) of the fuzzy controller with the integration of error. This will result in the fuzzy controller behaving like a parameter time-varying PI controller; thus the steady-state error is expelled by the integration action. However, a PI-type fuzzy controller will have a slow rise time if the P parameters are chosen to be small, and have a large overshoot if the P or I parameters are chosen to be large. So there may be a time

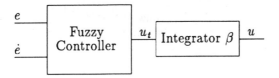

Figure 4.8 The PI-type fuzzy controller

when one might want to introduce not only the integral control but also the derivative control to the fuzzy control system, because the derivative control can reduce the overshoot of the system's response, to improve the control performance. Of course, this can be realized by designing a fuzzy controller with three inputs: the error, the change in error and the integration of error. However, these methods will be hard to apply in practice because of the difficulty of constructing fuzzy control rules. Usually fuzzy control rules are constructed by summarizing the manual control experience of an operator who has been controlling the industrial process skilfully and successfully. The operator intuitively regulates the executor to control the process by watching the error and the change rate of the error between the system's output and the set-point value given by the technical requirement. It is not the practice for the operator to observe the integration of the system's error. Therefore it is not possible to explicitly abstract fuzzy control rules from the operator's experience. Moreover, the three-inputs one-output fuzzy controller will have far more control rules than the two-inputs one-output one; the construction of fuzzy control rules is an even more difficult task and the calculation will be more complicated. Hence, it is better to design a fuzzy controller that possesses the fine characteristics of the PID controller by using only the error and the change rate of the error as its inputs. In the following statement we propose a method that can introduce the good features of the PID controller without changing the structure of the basic fuzzy controller.

One way is to have an integrator serially connected to the output of the fuzzy controller, as shown in Figure 4.8. From (4.10), the control input to the plant can be approximated by

$$u = \beta \int u_t \, dt = \beta \int (A + Pe + D\dot{e}) \, dt = \beta A t + \beta D e + \beta P \int e \, dt \quad (4.14)$$

where β is the integral constant.

Considering the scaling factors, substitute e and $\dot{e}$ with $K_1 e$ and $K_2 \dot{e}$, respectively, (4.14) can be written as

$$u = \beta \int (A + PK_1 e + DK_2 \dot{e}) \, dt = \beta A t + \beta K_2 D e + \beta K_1 P \int e \, dt \quad (4.15)$$

Hence the fuzzy controller becomes a parameter time-varying PI controller. We call this fuzzy controller the PI-type fuzzy controller, and the fuzzy controller without the integrator the PD-type fuzzy controller. Figure 4.9 shows a PI-type fuzzy control system.

In a PI-type fuzzy control system, the steady-state error is zero, but when the integral factor β is small the system's response is slow, and when it is too large there is a high

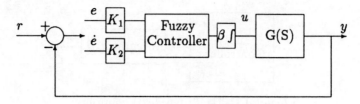

Figure 4.9 The PI-type fuzzy control system

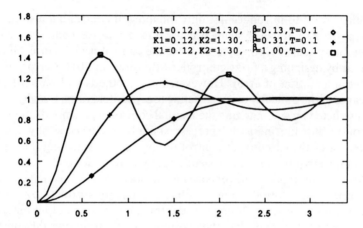

Figure 4.10 The step response of the PI-type fuzzy control system

overshoot and serious oscillation. These properties are illustrated by simulation in Figure 4.10. In this section and the following sections the basic fuzzy controller (control rules and membership functions) and the plant model used for simulation are the same as used in the preceding section, and the value of u_{10} of the conntrol rules Table 4.1 is fixed at 0.2.

Therefore, we may want to introduce the derivative control law into the fuzzy controller to overcome the overshoot and instability. We propose a controller structure that simply connects the PD-type and the PI-type fuzzy controller together in parallel. We have the equivalent structure by connecting a PI device with the basic fuzzy controller serially, as shown in Figure 4.11. Where α is the weight on the PD-type fuzzy controller and β is that on the PI-type fuzzy controller, a larger α/β means more emphasis on the derivative control and less emphasis on the integral control, and vice versa.

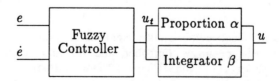

Figure 4.11 The PID-type fuzzy controller

AN IMPROVEMENT: PID-TYPE FUZZY CONTROLLER STRUCTURE

Combine (4.10) and (4.14); then the output of the whole controller is approximated by

$$u = \alpha u_t + \beta \int u_t \, dt$$

$$= \alpha(A + Pe + D\dot{e}) + \beta \int (A + Pe + D\dot{e}) \, dt$$

$$= \alpha A + \beta A t + (\alpha P + \beta D)e + \beta P \int e \, dt + \alpha D\dot{e} \quad (4.16)$$

If we consider the scaling factors, the above equation can be rewritten as

$$u = \alpha(A + PK_1 e + DK_2 \dot{e}) + \beta \int (A + PK_1 e + DK_2 \dot{e}) \, dt$$

$$= \alpha A + \beta A t + (\alpha K_1 P + \beta K_2 D)e + \beta K_1 P \int e \, dt + \alpha K_2 D\dot{e} \quad (4.17)$$

Thus the fuzzy controller behaves like a time-varying PID controller; its equivalent proportional control, integral control and derivative control components are $\alpha K_1 P + \beta K_2 D$, $\beta K_1 P$ and $\alpha K_2 D$, respectively. We call this new controller structure a PID-type fuzzy controller. The PID-type fuzzy control system is shown in Figure 4.12.

To demonstrate the good property of the proposed fuzzy controller structure, we carry out simulation on such a fuzzy control system.

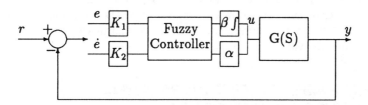

Figure 4.12 The PID-type fuzzy control system

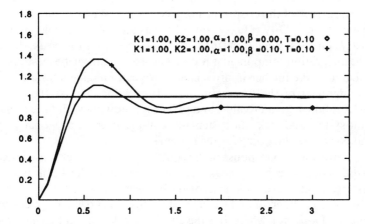

Figure 4.13 Comparison of PD-, PI- and PID-type fuzzy control system (I)

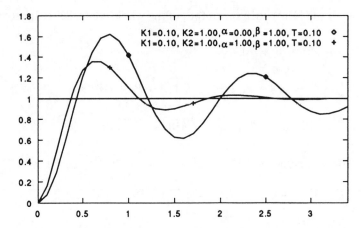

Figure 4.14 Comparison of PD-, PI- and PID-type fuzzy control system (II)

Figure 4.13 and 4.14 are the simulation results of the system's step response. The influence of α and β on the system performance is demonstrated. When $\alpha > 0$ and $\beta = 0$, meaning that the fuzzy controller behaves like a parameter time-varying PD controller, there exists a steady-state error. When $\alpha = 0$ and $\beta > 0$, meaning that the fuzzy controller behaves like a parameter time-varying PI controller, the steady-state error of the system is eliminated but there is an large overshoot and serious oscillation. When $\alpha > 0$ and $\beta > 0$, the fuzzy controller becomes a parameter time-varying PID controller and the overshoot is substantially reduced. It is possible to obtain a comparatively good performance by carefully choosing the value of α and β.

4.6 FURTHER IMPROVEMENT: THE PARAMETER ADAPTIVE FUZZY CONTROLLER

The PID-type fuzzy controller proposed in the preceding section has substantially improved the performance of the crisp-type fuzzy controller. We also find that the integration component of the PID-type fuzzy controller has an important role on the performance of the fuzzy control system. If the integration component is too weak, the response is slow, and if the integration component is too strong, the system will become unstable. So it is still desirable to make further improvements. We can imagine letting the equivalent integration component of the fuzzy controller vary with time. At the early stage of response, we let it take a greater value, and reduce it gradually with time to increase the damping of the system and make the system more stable. In this way we can hope to have a fast rise and a short settling time for the system's response.

Figure 4.15 shows the step response of the control system. The response process can be divided into different phases by the peak vaue times. From the start time 0 to the time t_1 when the first peak value occurs, the error of the system covers the whole universe of discourse. After time t_1, the error of the system's response will no longer go beyond the belt area of the interval $[-\delta_1, \delta_1]$, where δ_1 is the absolute peak value at time t_1. At t_2 another peak value occurs. After t_2 the error of the system's response will never go beyond the belt

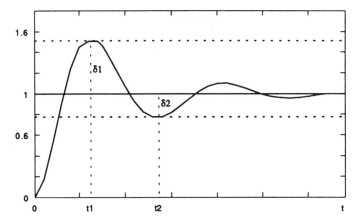

Figure 4.15 Different phases of the step response of a control system

area of the interval $[-\delta_2, \delta_2]$. And so on. We can decrease the integral control component at each peak value time, according to the absolute value of each peak.

Let us examine the equivalent proportional control, integral control and derivative control components of a PID-type fuzzy controller from (4.17). These equivalent control components are repeated as follows:

$$\begin{aligned} \text{proportional:} \quad & \alpha K_1 P + \beta K_2 D \\ \text{integral:} \quad & \beta K_1 P \\ \text{derivative:} \quad & \alpha K_2 D \end{aligned}$$

If we decrease the parameter β gradually, the integral control component will be decreased so that the damping of the system is increased and the system is more stable. Note that the proportional component includes the term of the production of β and K_2. Although decreasing the value of β will decrease the proportional control component, the reaction of the control system against the error will be slowed down. If while decreasing β we increase K_2 in the same rate as β is decreased, the equivalent proportional control will remain unchanged and the system can always keep rapid reaction against the error. We can also see that when K_2 is increased, the equivalent derivative control component will be increased at the same time; this would do no harm to the system's performance, because the derivative control law can increase the resistance against the overshoot and oscillation of the system.

Motivated by this idea, we design a parameter adaptive fuzzy controller. The parameter adaptive fuzzy controller is composed of a PID-type fuzzy controller, a peak observer and a parameter regulator. Figure 4.16 is the block diagram of the parameter adaptive fuzzy controller.

The basic PID-type fuzzy controller is as described in the preceding section. The peak observe keeps watching on the system's output, gives a signal at each peak time and measures the absolute peak value. The parameter regulator tunes the controller's parameters K_2 and β simultaneously at each peak time signal and according to the peak value at that time. The algorithm of tuning the scaling constants and the integral gain is as

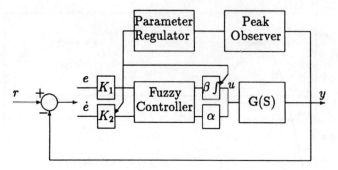

Figure 4.16 Block diagram of the parameter adaptive fuzzy controller

follows.

$$K_2 = \frac{K_{2s}}{\delta_k}, \beta = \delta_k \beta_s$$

where K_{2s} and β_s are the initial values of K_2 and β, respectively. δ_k is the absolute peak value at the peak time t_k ($k = 1, 2, 3, \ldots$).

Figure 4.17–4.19 present some simulation results of the parameter adaptive fuzzy control system with different initial values of the scaling parameters and output gains. In Figures 4.17 and 4.18, when the adaptive mechanism does not come into action the system oscillates seriously, and when the adaptive mechanism comes into action the oscillations are resisted strongly. In Figure 4.19, the system without adaptive mechanism has a slight oscillation, and the parameter adaptive control yields a non-oscillating system. Generally speaking, the simulation results demonstrate that the parameters of the adaptive fuzzy controller greatly improve the performance of the control system. It can greatly reduce the oscillating times and shorten the settling time of the system. So in

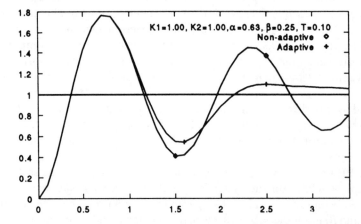

Figure 4.17 Comparison of fuzzy control system with and without adaptive mechanism (I)

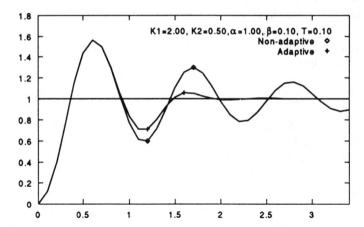

Figure 4.18 Comparison of fuzzy control system with and without adaptive mechanism (II)

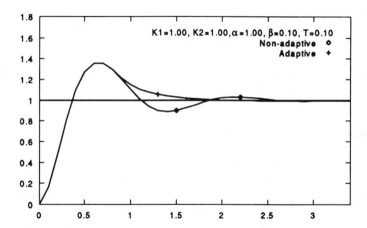

Figure 4.19 Comparison of fuzzy control system with and without adaptive mechanism (III)

practice it is possible to choose a large initial value for β to have a fast rising of response, but no instability.

4.7 CONCLUSIONS

We have studied the input output behaviour of the crisp-type fuzzy controller with the min–max reference method and the product–sum reference method. The fuzzy controller with both reference methods have the same input–output characteristics on the NET and the NODE of the e–$\dot{e}$ plane. The difference between the performance of the two methods is minor. The crisp–type fuzzy controller can be regarded as a parameter time-varying PD controller. Therefore, the analysis and designing of a fuzzy control system can take advantage of the conventional PID control theory. According to the conventional PID control theory, we have been able to propose some improvement methods for the crisp-type fuzzy controller.

It has been illustrated that the PD-type fuzzy controller yields a steady-state error for the type 0 system, the PI-type fuzzy controller can eliminate the steady-state error. We proposed a controller structure that combine the features of both PD-type and PI-type fuzzy controller, obtaining a PID-type fuzzy controller which allows the control system to have a fast rise and a small overshoot, as well as a short settling time.

To further improve the performance of the proposed PID-type fuzzy controller, the authors designed a parameter adaptive fuzzy controller. The PID-type fuzzy controller can be decomposed to the equivalent proportional control, integral control and the derivative control components. The proposed parameter adaptive fuzzy controller gradually decreases the equivalent integral control component of the fuzzy controller with the system response process time, to increase the damping of the system when the system is about to settle down, while keeping the proportional control component unchanged to guarantee the rapid reaction against the system's error. With the parameter adaptive fuzzy controller, the oscillation of the system is strongly restrained and the settling time is considerably shortened.

We have presented the simulation results to demonstrate the fine performance of the proposed PID-type fuzzy controller and the parameter adaptive fuzzy controller structure.

The analysis results and the proposed improvement methods can be used as the guide to the design of fuzzy control systems.

REFERENCES

1 Zadeh, L. A. (1973) Outline of a new approach to the analysis of complex systems and decision processes. *IEEE Transactions on Systems, Man and Cybernetics*, **3**, 24–28.
2 Mamdani, E. H., and Assilian, S. (1975) An experiment in linguistic synthesis with a fuzzy logic controller. *International Journal of Man–Machine Studies*, **7**, 1–13.
3 Koczy, L. T., and Hirota, K. (1993) Interpolative reasoning with insufficient evidence in sparse fuzzy rules bases. *Information Sciences*, **71**, 169–201.
4 Yager, R. R. (1992) Expert systems using fuzzy ets. In *An Introduction to Fuzzy Logic Applications in Intelligent Systems*, edited by R. R. Yager and L. A. Zadeh (Kluwer: Norwell), pp. 27–44.
5 Yager, R. R., and Filev, D. P. (1993) On the reasoning on fuzzy logic control and fuzzy expert systems. *Proceedings of Second IEEE International Conference on Fuzzy Systems*, San Francisco, pp. 839–844.
6 Mizumoto, M., and Zimmermann, H. J. (1982) Comparison of fuzzy reasoning methods. *Fuzzy Sets and Systems*, **8**, 253–283.
7 Mizumoto, M. (1992) Fuzzy reasoning. *Journal of Japan Society for Fuzzy Theory and Systems* (in Japanese), **1**, 256–264.
8 Mizumoto, M. (1989) Pictorial representations of fuzzy connectives, Part I: Cases of t-norms, t-conorms and averaging operators. *Fuzzy Sets and Systems*, **31**, 217–242.
9 Mizumoto, M. (1989) Pictorial representations of fuzzy connectives, Part II: Cases of compensatory operators and self-dual operators. *Fuzzy Sets and Systems*, **32**, 45–79.
10 Mizumoto, M. (1992) Realization of PID controls by fuzzy control methods. *Proceedings of IEEE International Conference on Fuzzy Systems*, San Diego, pp. 709–715.

11 Mizumoto, M. (1991) Min–max gravity method versus product–sum-gravity method for fuzzy controls. *Proceedings of IV IFSA Congress*, Brussels, Part E, pp. 127–130.
12 Mizumoto, M. (1993) Fuzzy controls under product–sum-gravity methods and new fuzzy control methods. In *Fuzzy Control Systems*, edited by A. Kandel and G. Langholz (CRC Press), pp. 276–294.
13 Takagi, T., and Sugeno, M. (1983) Derivation of fuzzy control rules from human operator's control actions. *Proc. of the IFAC Conf. on Fuzzy Information*, Marseille, France, pp. 55–60.
14 Takagi, T., and Sugeno, M. (1985) Fuzzy identification of fuzzy systems and its application to modeling and control. *IEEE Transactions on Systems, Man and Cybernetics*, **15**, 116–132.
15 Kosko, B. (1992) Fuzzy systems as universal approximators. *Proceedings of IEEE International Conference on Fuzzy Systems*, San Diego, pp. 1153–1162.
16 Buckley, J. J. (1992) Theory of fuzzy controllers: an introduction. *Fuzzy Sets and Systems*, **51**, 249–258.
17 Buckley, J. J. (1990) Fuzzy controllers: further limit theorems for linear control rules, *Fuzzy Sets and Systems*, **36**, 225–233.
18 Buckley, J. J., and Hayashi, Y. (1993) Fuzzy input–output controllers are universal approximators, *Fuzzy Sets and Systems*, **58**, 273–278.
19 Buckley, J. J. (1993) Sugeno type controllers are universal controllers, *Fuzzy Sets and Systems*, **53**, 299–303.
20 Wang Pei Zhuang, Zhang Hong Min, and Xu Wei (1990) PAD-analysis of stability o fuzzy control systems. *Fuzzy Sets and Systems*, **38**, 27–42.
21 Hu Jia Yao, Wu Zhi Qiao, and Soon Shuo Shang (1988) Static analysis of fuzzy controllers. *Journal of Beijing Light Industry College* (in Chinese), 12–18.
22 Hans, H., and Christoph, T. (1993) Defuzzification in fuzzy controllers. *Journal of intelligent and Fuzzy Systems*, **1**, 109–123.
23 Benjamin, C. K. (1987) *Automatic Control Systems* (Prentice-Hall, London).
24 William, L. B. (1991) *Modern Control Theory* (Prentice-Hall, London).
25 Kickert, W. J. M., and Mamdani, E. H. (1978) Analysis of a fuzzy logic controller. *Fuzzy Sets and Systems*, **1**, 29–44.
26 Procyk, T. J., and Mamdani, E. H. (1979) A linguistic self-organizing process controller. *Automatica*, **15**, 15–30.
27 Tong, R. M. (1976) Analysis of fuzzy control algorithms using the relation matrix. *International Journal of Man–Machine Study*, **8**, 679–686.
28 Yamazaki, T. (1982) An improved algorithm for a self-organizing controller and its experimental analysis. PhD thesis, University of London.
29 Shao, S. (1988) Fuzzy self-organizing controller and its application for dynamic processes. *Fuzzy Sets and Systems*, **26**, 151–164.
30 Tanscheit, R., and Scharf, E. M. (1988) Experiments with the use of a rule based self-organizing controller for robotics applications. *Fuzzy Sets and Systems*, **26**, 195–214.
31 Graham, B. P., and Newell, R. B. (1989) Fuzzy adaptive control of a first order process. *Fuzzy Sets and Systems*, **31**, 47–65.
32 Ollero, A., and Garcia-Cerezo, A. J. (1989) Direct digital control, auto-turning and supervision using fuzzy logic. *Fuzzy Sets and Systems*, **30**, 135–153.
33 Wu Zhi Qiao, *et al.* (1992) A rules self regulating fuzzy controller. *Fuzzy Sets and Systems*, **47**, 13–21.
34 Shi-Zhong He, *et al.* (1993) Control of dynamical processes using an on-line rule-adaptive fuzzy control system. *Fuzzy Sets and Systems*, **54**, 11–22.

FUZZY LOGIC HARDWARE IMPLEMENTATIONS

5
Design Considerations of Digital Fuzzy Logic Controllers

M. J. Patyra
*Department of Electrical and Computer Engineering,
University of Minnesota, USA*

5.1 INTRODUCTION

This paper presents an overview of the design issues for digital fuzzy logic control circuits. Digital fuzzy logic controllers (FLCs) are the most commercially successful implementations of fuzzy logic circuits to date. The origins of this success can be found in several domains, including the psychological, economical and technical ones. These are sophisticated and usually touchy problems, therefore they will not addressed in this paper. Interested readers may check a recent publication dealing with these matters [15].

Since the development of the first functional FLC by Togai and Watanabe in 1985 [31–33] there have been many successful developments of digital-based implementations of FLC hardware systems [3–5, 7–14, and 18–41]. Also, some improvements to these implementations have been proposed [1, 2]. Due to the wide variety of such achievements, it is hard to find the common 'line' of reference when trying to characterize and compare various FLC devices. This is often confusing when different data are provided, especially in terms of overall FLC performance, compatibility and capability.

This drawback motivated the research presented in this paper. The purpose of this paper is twofold. First, to establish a set of common parameters useful for fuzzy logic controller comparisons. Second, to present the analytical formulation of various controllers and the different forms of their implementations, as well as discussing these implementations in terms of hardware cost and performance.

The organization of this paper is as follows. First, the general foundation regarding digital FLC is laid out. Second, the set of major characteristics is defined, followed by the analytical formulation of different FLC configurations. Third, the analytical FLC models are mapped into hardware representations. These hardware implementations are then analyzed using the characteristics established in this paper. Finally, a summary and concluding remarks are provided.

5.2 DIGITAL-BASED FUZZY LOGIC HARDWARE

This section reviews some basic concepts behind the digital implementations of fuzzy logic circuits. It is therefore an introduction to the FLC implementations analyzed in Section 5.3.

For the purpose of this paper we assume the control scheme illustrated in Figure 5.1.

The theory behind fuzzy control is not outlined in this paper. For theoretical details regarding the fuzzy logic control and fuzzy logic controllers, the reader is referred to tutorial papers such as [16] and [17], or the book [6].

Let us briefly present the main stages of the control scheme delineated in Figure 5.1.

5.2.1 Digital Fuzzification

Fuzzification is a transformation of the crisp data into a corresponding fuzzy set. Before the data can be fuzzified, however, they should be first normalized to meet the range of universe of discourse suitable for the controller input.

Typically, the state of the process is examined by means of measurements of its parameters. Such measurements generate analogue-type data available for the control. Assume that the analogue output from the process under control is fed back to the fuzzy controller. First, the analogue-to-digital conversion needs to be performed, followed by the input data normalization. Note that both procedures have their own accuracy and resolution. This fact suggests some fuzzification techniques that can compensate for the unavoidable loss of accuracy. It is important to stress that these procedures are usually responsible for the introduction of a systematic error to the input data.

As a result, the fuzzification can follow three basic strategies.

Input crisp data are converted into a fuzzy singleton in an appropriate universe of discourse. Precisely, the fuzzy singleton is a crisp value and hence no fuzziness is introduced in this case, but this strategy has been widely used for fuzzy control applications because of its computational simplicity.

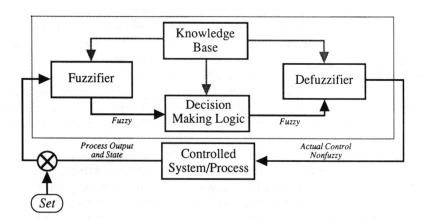

Figure 5.1 The general scheme of the fuzzy control

Input crisp data are converted into a fuzzy vector based on expert knowledge of the characteristics of measurement instruments, A/D conversion and normalization. In this case, all 'vagueness' associated with the measurement and transformation is included in the resulting fuzzy vector.

Input crisp data are randomly distributed and therefore can be converted into fuzzy vector with arbitrary shape (e.g. triangle, trapezoidal, Gaussian, etc.). However, the distribution of the fuzzy vector may be determined by the parameters of the probability distribution available for the specific measurement process. This case is impractical, and therefore is rarely implemented in the hardware.

Figure 5.2 illustrates the example of an arbitrary and experimentally measured variable $\mathcal{A}$; its value is normalized, quantized and finally discretized into a fuzzy set A. The variable $\mathcal{A}$ measured at the continuous universe of discourse can be represented by an arbitrary singleton, triangle or any other function listed above. Note that the discrete universe of discourse is discretized into n' equal intervals. The degree of membership is also quantized into m' levels. This membership function is digitized, as shown, by the

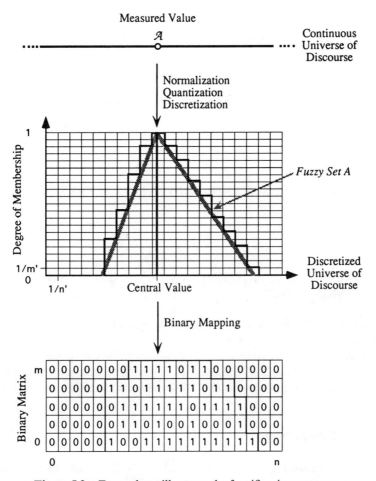

Figure 5.2 Example to illustrate the fuzzification process

step-shape function. As a result, the membership function is represented by a set of perpendicular segments.

Finally, the digitized representation is transformed into a Boolean matrix that has size $n*m$, where n denotes the number of m-bit binary vectors. Each vector is obtained from the digitized membership function in such a way that the m-bit number codes the degree of membership function on the appropriate position defined by the n' segment. In general, the binary mash may not be the same as the discrete mash (see Figure 5.2 for details). Note that if $2^m \neq m'$, the mapping must be nonlinear. Also, in practice n may not be equal to n', which creates the additional problem of horizontal nonlinear mapping from universe of discourse discretized into n' intervals into n vectors. The process of creating such mappings is heuristic and must be based on expert knowledge.

5.2.2 Digital Fuzzy Inferencing

In order to discuss fuzzy inferencing in detail, let us recall the fuzzy system characterized by the linguistic description in the form of fuzzy implication rules:

R_1: IF $A1$ IS A_1^1 AND IF $A2$ IS A_1^2 AND...AND IF AK IS A_1^K THEN $B1$ IS B_1^1 AND $B2$ IS B_1^2 AND...AND BL IS B_1^L,

ALSO

R_2: IF $A1$ IS A_2^1 AND IF $A2$ IS A_2^2 AND...AND IF AK IS A_2^K THEN $B1$ IS B_2^1 AND $B2$ IS B_2^2 AND...AND BL IS B_2^L,

ALSO

⋮

R_i: IF $A1$ IS A_i^1 AND IF $A2$ IS A_i^2 AND...AND IF AK IS A_I^K THEN $B1$ IS B_I^1 AND $B2$ IS B_I^2 AND...AND BL IS B_I^L,

ALSO

⋮

R_N: IF $A1$ IS A_N^1 AND IF $A2$ IS A_N^2 AND...AND IF AK IS A_N^K THEN $B1$ IS B_N^1 AND $B2$ IS B_N^2 AND...AND BL IS B_N^L,

For simplicity of discussion in this section, the system is restricted to a double-input single-output case.

The inference mechanisms employed in fuzzy logic controllers are generally based on a one-level generalized *Modus Ponens* scheme. As a result, there are four fuzzy reasoning methods to obtain the inference result from a system. These methods are briefly characterized below.

5.2.2.1 Mamdani's strategy

Mamdani's fuzzy reasoning method is based on MAX–MIN operator inferencing. Mamdani's fuzzy reasoning strategy is illustrated in Figure 5.3.

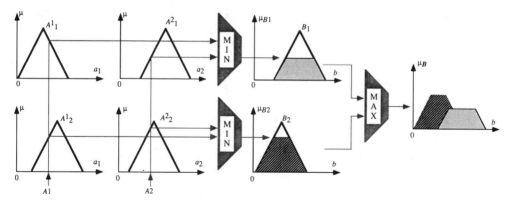

Figure 5.3 Example of Mamdani's reasoning strategy

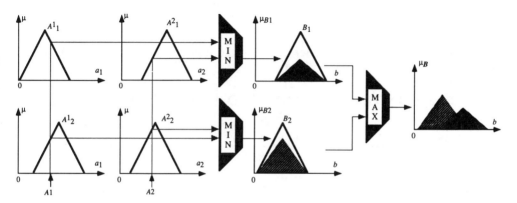

Figure 5.4 Example of Larsen's reasoning strategy

5.2.2.2 Larsen's strategy

Larsen's fuzzy reasoning method is based on PRODUCT operator inferencing. Larsen's fuzzy reasoning strategy is illustrated in Figure 5.4.

5.2.2.3 Tsukamoto's strategy

Tsukamoto's fuzzy reasoning method is based on a simplification of Mamdani's method, although all membership functions (antecedents and conclusions) are monotonic. Tsukamoto's fuzzy reasoning strategy is illustrated in Figure 5.5.

5.2.2.4 Takagi and Sugeno's strategy

Takagi and Sugeno's fuzzy reasoning method is based on a distinct model description. In this model the control variables are characterized by the functions of the process state

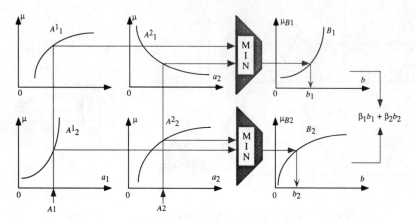

Figure 5.5 Example of Tsukamoto's reasoning strategy

variables. If we keep the previous terminology, the system characterized by Takagi and Sugeno's model is described by a set of equations:

R_1: IF $A1$ IS A_1^1 AND IF $A2$ IS A_1^2 AND...AND IF AK IS A_1^K THEN $B1 = f_1^1(A1,...,A_N^1,...,A2,...,A2_N,...,AK,...,A_N^K)$ AND $B2 = f_1^2(A1,...,A_N^1,...,A2,...,A2_N,...,AK,...,A_N^K)$ AND,..., AND $BL = f_1^L(A1,...,A_N^1,...,A2,...,A2_N,...,AK,...,A_N^K)$

ALSO

R_2: IF $A1$ IS A_2^1 AND IF $A2$ IS A_2^2 AND...AND IF AK IS A_2^K THEN $B1 = f_2^1(A1,...,A_N^1,...,A2,...,A2_N,...,AK,...,A_N^K)$ AND $B2 = f_2^2(A1,...,A_N^1,...,A2,...,A2_N,...,AK,...,A_N^K)$ AND,..., AND $BL = f_2^L(A1,...,A_N^1,...,A2,...,A2_N,...,AK,...,A_N^K)$

ALSO

$\vdots$

R_i: IF $A1$ IS A_i^1 AND IF $A2$ IS A_i^2 AND...AND IF AK IS A_i^K THEN $B1 = f_i^1(A1,...,A_N^1,...,A2,...,A2_N,...,AK,...,A_N^K)$ AND $B2 = f_i^2(A1,...,A_N^1,...,A2,...,A2_N,...,AK,...,A_N^K)$ AND,..., AND $BL = f_i^L(A1,...,A_N^1,...,A2,...,A2_N,...,AK,...,A_N^K)$

ALSO

$\vdots$

R_N: IF $A1$ IS A_N^1 AND IF $A2$ IS $A2_N$ AND...AND IF AK IS A_N^K THEN $B1 = f_N^1(A1,...,A_N^1,...,A2,...,A2_N,...,AK,...,A_N^K)$ AND $B2 = f_N^2(A1,...,A_N^1,...,A2,...,A2_N,...,AK,...,A_N^K)$ AND,..., AND $BL = f_N^L(A1,...,A_N^1,...,A2,...,A2_N,...,AK,...,A_N^K)$

The basic idea of Takagi and Sugeno's fuzzy reasoning model is illustrated in Figure 5.6.

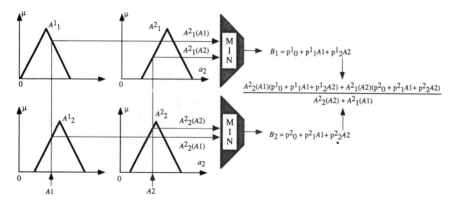

Figure 5.6 Example of Takagi and Sugeno's reasoning strategy

5.2.3 Digital Defuzzification

Generally, defuzzification describes the mapping from a space of fuzzy control action into a nonfuzzy control action. Defuzzification produces a nonfuzzy action that best represents the inferred fuzzy output. Sometimes, after the defuzzification, a denormalization procedure is required for practical applications.

In terms of the digital implementation of a defuzzification strategy, the usual approach involves arithmetic operations on a large number (depending on the granularity of the universe of discourse) of binary vectors. These operations include multiplications, summations and divisions. Defuzzification is usually one of the most time-consuming operation in fuzzy processing.

There are numerous defuzzification methods; however, only about five are practical. Although the choice is somehow subjective, these methods are briefly characterized in the following sections.

5.2.3.1 Centre-of-area (COA)

In general, centre-of-area (COA) is based on the computation of the position of a divisive axis between the left and right half area determined by the joint membership function of the fuzzy action. In most cases, the centre of the area is the same as the centre of gravity and, therefore, these names often denote the same method.

The simple example shown in Figure 5.7 illustrates COA's defuzzified value of a single-input single-output fuzzy system described by three rules of inferences.

5.2.3.2 Centre-of-gravity (COG)

Frequently used, the centre-of-gravity (COG) method computes the centroid of the area determined by the joint membership function of the fuzzy action (see also Figure 5.7).

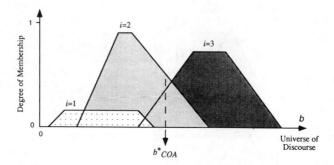

Figure 5.7 Result of defuzzification using the COA or COG methods

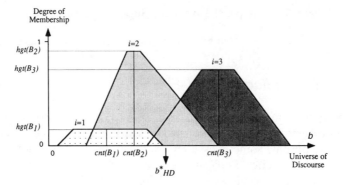

Figure 5.8 Result of defuzzification using the HD method

5.2.3.3 Height defuzzification (HD)

The height defuzzification (HD) method computes the weighted sum of the height values of all membership functions associated with conclusion terms.

Figure 5.8 illustrates the application of the HD method to find the defuzzified value of the single-input single-output fuzzy system described by three rules of inferences.

5.2.3.4 Centre-of-largest-area (COLA)

The centre-of-largest area (COLA) calculates the centre of the largest area determined by the joint membership function of the fuzzy action. This method is used in the case of multiple area fuzzy responses. After determining the largest area of the fuzzy conclusion, the strategy becomes analogous to COA.

The idea of the COLA method is illustrated in Figure 5.9.

5.2.3.5 Mean-of-maxima (MOM)

Not widely used, the mean-of-maxima (MOM) method calculates the arithmetic mean of all values with maximum membership. The idea of the MOM method is illustrated in Figure 5.10.

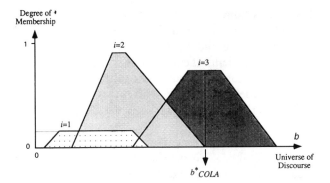

Figure 5.9 Result of defuzzification using COLA method

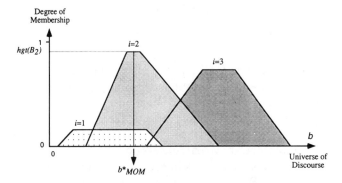

Figure 5.10 Result of defuzzification using MOM method

5.3 FUZZY LOGIC BASED CONTROLLERS

This section discusses hardware implementation issues of digital fuzzy logic controllers (FLCs). The section is organized in the following way. First, the set of major parameters characterizing a digital FLC is established. Then the mathematical background of a single-input single-output (SISO) FLC is discussed. The model for the double-input single-output (DISO) fuzzy logic controller is presented, followed by the generalized models for the multiple-input single-output (MISO) and the multiple-input multiple-output (MIMO) fuzzy logic controller.

5.3.1 Digital FLC Characteristics

In this section a set of major parameters to characterize the digital fuzzy logic controllers is established based on the information gathered from the publications referenced in this paper. We begin our discussion from the general viewpoint related to the application of FLC devices.

There are two basic parameters that determine the usefulness of a certain implementation of an FLC; these are the number of input variables and the number of output variables. In this case, the usefulness of an FLC can be viewed by means of the complexity of the process that can be controlled by an FLC under consideration.

Based on the information provided in Section 2, especially with respect to the fuzzification, inferencing and defuzzification, it has become obvious that the input trans-formation from the analogue measurement to the binary representation of fuzzy information is crucial for the FLC functionality. Also, the inverse transformation of fuzzy information into a crisp output determines the accuracy of the overall control action.

On the other hand, in terms of inferencing the basic characteristic of an FLC can be referred to as the 'storage ability'. This determines the FLC capacity to store linguistic rules as well as the capacity to store membership functions assigned to different linguistic variable values.

One of the major problems of FLC characteristics is related to their performance. In scientific and technical literature there can be found at least three performance measures. They are:

maximum frequency of the clock running the FLC device
number of fuzzy inferences per second, where the fuzzy inference is usually ambiguously defined, or even not defined; the fuzzy inference may be understood as a operation defined by a single rule, or an operation defined be a part of the rule
number of elementary fuzzy operations per second (e.g. MIN or MAX).

It has to be stressed that the above parameters do not provide a confident measure of the real speed of the FLC. Therefore, we propose to characterize the FLC be means of the input-to-output delay time (Θ_{IN-OUT}). Such a time is defined as a total delay time from the moment of providing the input variable to the FLC device to the moment of generating the crisp action at the output of this device. Such a time is, in our opinion, the most objective measure of the FLC performance because it does take into account the real outcome of the device. The parameters used so far may provide spectacular data; unfortunately these data lack performance value. Let us take as an example, the FLC device that can perform several million fuzzy logic inferences per second (FLIPS), but the control action could be available at the output after, let us say, milliseconds. In such a case the inference engine of the FLC is mostly occupied with a lot of number crunching that does not lead to the final result. As we will show below, such behaviour is determined by the internal FLC structure and can be avoid by appropriate structure modifications made at the design stage.

Let us specify the set of major FLC parameters that we have discussed in this section. They can be listed as follows:

number of input variables (inputs) (K)
number of output variables (outputs) (L)
number of linguistic rules in the knowledge base (N)
number of membership functions in the input universe of discourse (MB_{IN})
number of membership functions in the output universe of discourse (MB_{OUT})
number of binary vectors characterizing the membership function (n)
number of bits in a single binary vector (m)
input-to-output delay time (Θ_{IN-OUT}).

Let us algebraically analyze various configurations of FLC with regard to the set of specified parameters.

5.3.2 Single-Input Single-Output Fuzzy Logic Controllers

This section presents the analysis of the single-input single-output digital FLC. First, the classic method of building the FLC hardware is presented, based directly on the analytical equations. Second, the alternative method is discussed.

5.3.2.1 Classic method

Assume that the system can be characterized by a single input, represented by the variable A, and a single output represented by the variable B. A method of building a fuzzy model of a system consists of creation the set of linguistic statements, called fuzzy rules, where fuzzy subsets of input and output variables are used as antecedents and consequents.

If $A_1, A_2, \ldots, A_N$ are fuzzy subsets of A, and $B_1, B_2, \ldots, B_N$ are fuzzy subsets of B, then fuzzy relation R is defined by a set of fuzzy rules as follows:

$$
\begin{align*}
R_1: &\ \text{IF } A_1 \text{ THEN } B_1, \quad \text{ALSO} \\
R_2: &\ \text{IF } A_2 \text{ THEN } B_2, \quad \text{ALSO} \\
&\ \vdots \\
R_i: &\ \text{IF } A_i \text{ THEN } B_i, \quad \text{ALSO} \\
&\ \vdots \\
R_N: &\ \text{IF } A_N \text{ THEN } B_N.
\end{align*}
$$

More specifically $R_i = A \rightarrow B$, where $\rightarrow$ denotes the fuzzy implication. This implication is used to derive the variable B from the relation:

$$B = A \circ R \tag{5.1}$$

Therefore, the membership function of the relation R is defined as

$$\mu_{A \rightarrow B}(a, b) = f_{A \rightarrow B}(\mu_A(a), \mu_B(b)) \tag{5.2}$$

Fuzzy rules are combined by the connective 'ALSO' to yield the overall fuzzy relation R. This relation is constructed from all R_i rules, where $i = 1, \ldots, N$; and N is the total number of rules.

$$\mu_R(a, b) = f_{\text{ELSE}}(\mu_{R_i}(a, b)) = f_{\text{ELSE}}(F_{A \rightarrow B}(\mu_{A_i}(a), \mu_{B_i}(b))) \tag{5.3}$$

In order to derive the overall fuzzy relation, the connection 'ALSO' needs to be interpreted. The commonly used interpretation for 'ALSO' is the logical 'OR', which in this case may be translated as a union of fuzzy subsets:

$$R = \bigcup_{i=1}^{N} R_i = \text{MAX} f_{A \rightarrow B}(\mu_{A_i}(a), \mu_{B_i}(b)) \tag{5.4}$$

With a fuzzy observation $A1$ and the overall relation R, one can infer the resulting action $B1$ by applying the compositional rule of inference; that is

$$B1 = A1 \circ R \qquad (5.5)$$

The membership function of $B1$ can then calculated by the well-known MAX-MIN operation. Considering the ith rule R_i and the observation $A1$, the respective action $B1$ is given by

$$B1_i = A1 \circ R_i \qquad (5.6)$$

Therefore, the corresponding membership function is defined as follows:

$$\mu_{B1_i}(b) = \underset{a \in A}{\text{MAXMIN}}(\mu_{A1}(a), \mu_R(a,b))$$

$$= \text{MINMAX}\{\text{MIN}(\mu_{A1}(a), \mu_{A_i}(a)), \mu_{B_i}(b)\} \qquad (5.7)$$

$$= \text{MIN}(\Omega_i, \mu_{B_i}(b))$$

where

$$\Omega_i = \underset{a \in A}{\text{MAXMIN}}(\mu_{A1}(a), \mu_{A_i}(a)) \qquad (5.8)$$

Then the maximum of $B1_1, B1_2, \ldots, B1_N$ determines the final action $B1$, which is calculated as a union:

$$B1 = \underset{i=1}{\overset{N}{\cup}} B1_i \qquad (5.9)$$

5.3.2.2 Improved method

As mentioned above, there is an alternative method of performing fuzzy inference, as opposed to that presented in Section 5.3.2.1.

Based on equation (5.1), the alternative way of performing the fuzzy inference for single-input single-output system is to build the overall relation first, and then infer the output for a certain input. The fuzzy sub-relations $R_1, R_2, R_3, \ldots, R_N$ are created analogously by application of the same definition of fuzzy implication, where the MIN operator is used to compute the Cartesian product of two variables.

$$R_i = A_i \times B_i \qquad (5.10)$$

$$R_i(a, b) = \text{MIN}(A_i(a), B_i(b)) \qquad (5.11)$$
$$\forall (a, b) \in A \times B$$

where $i = 1, 2, \ldots, N$.

The final relation R is obtained as the union of $R_1, R_2, R_3, \ldots, R_N$, since the sentence connective 'ALSO' is defined as union. By rearranging equation (5.4), we obtain

$$R = \underset{i=1}{\overset{N}{\cup}} R_i = \text{MAX}[R_1(a,b), R_2(a,b), \ldots, R_N(a,b)] \qquad (5.12)$$
$$(a, b) \in A \times B$$

Now, to obtain a fuzzy answer $B1$ to a fuzzy input $A1$, it is enough to apply MAX-MIN composition to an overall rule R as follows:

$$B1 = \underset{a \in A}{\text{MAX}}[\text{MIN}(A1(a), R(a,b))] \tag{5.13}$$

$$B1 = \underset{a \in A}{\text{MAX}}[\text{MIN}(A1(a), \text{MAX}[R_1(a,b), R_2(a,b), \ldots, R_N(a,b)])] \tag{5.14}$$

$$B1 = \underset{a \in A}{\text{MAX}}(\text{MIN}[A1(a), \text{MAX}(\text{MIN}(A_1(a), B_1(b)), \text{MIN}(A_2(a), B_2(b)), \ldots$$
$$\ldots, \text{MIN}(A_N(a), B_N(b)))]) \tag{5.15}$$

5.3.3 Double-Input Single-Output Fuzzy Logic Controller

5.3.3.1 Classic method

In case of a double-input single-output controller, the classic method of inferring the response to the inputs consists of extending the model already introduced in Section 5.3.2.1.

If $A_1^1, A_2^1, \ldots, A_N^1$ are fuzzy subsets of A^1, and $A_1^2, A_2^2, \ldots, A_N^2$ are fuzzy subsets of A^2, and $B_1, B_2, \ldots, B_N$ are fuzzy subsets of B, then the fuzzy relation R is defined by a set of fuzzy rules as follows:

$$R_1: \text{IF } A_1^1 \text{ AND IF } A_1^2 \text{ THEN } B_1, \text{ ALSO}$$
$$R_2: \text{IF } A_2^1 \text{ AND IF } A_2^2 \text{ THEN } B_2, \text{ ALSO}$$
$$\vdots$$
$$R_i: \text{IF } A_i^1 \text{ AND IF } A_i^2 \text{ THEN } B, \text{ ALSO}$$
$$\vdots$$
$$R_N: \text{IF } A_N^1 \text{ AND IF } A_N^2 \text{ THEN } B_N.$$

Observe that in this case A^1 and A^2 represent input variables and B represents the output variable. More specifically, $R = (A^1, A^2) \rightarrow B$. The overall model of the double-input single-output, according to the definition of fuzzy implication, can be defined as

$$B = (A^1, A^2) \circ R \tag{5.16}$$

As a result, the membership function of the relation R is defined as

$$\mu_{(A^1,A^2) \rightarrow B}(a_1, a_2, b) = f_{(A^1,A^2) \rightarrow B}((\mu_{A^1}(a_1), \mu_{A^2}(a_2)), \mu_B(b)) \tag{5.17}$$

Fuzzy rules are combined by the connective 'ALSO' to yield the overall fuzzy relation R. This relation is constructed from all R_i rules, where $i = 1, \ldots, N$, and N is the total number of rules.

$$\mu_R(a_1, a_2, b) = f_{\text{ELSE}}(\mu_{R_i}(a_1, a_2, b)) = f_{\text{ELSE}}(f_{(A^1, A^2) \to B}((\mu_{A^1}(a_1), \mu_{A^2}(a_2)), \mu_B(b))) \quad (5.18)$$

In order to derive the overall fuzzy relation, the connective 'ALSO' needs to be interpreted. The commonly used interpretation for 'ALSO' is the logical 'OR', which in this case may be substituted by a union of fuzzy subsets:

$$R = \bigcup_{i=1}^{N} R_i = \text{MAX} f_{(A^1, A^2) \to B}((\mu_{A^1}(a_1), \mu_{A^2}(a_2)), \mu_B(b)) \quad (5.19)$$

With the fuzzy observations $A1$ and $A2$, and the overall relation R, one can infer the resulting fuzzy action $B1$ by applying the compositional rule of inference; that is

$$B1 = (A1, A2) \circ R = (A1, A2) \circ \bigcup_{i=1}^{N} R_i = \bigcup_{i=1}^{N} (A1, A2) \circ R_i \quad (5.20)$$

The membership function of $B1$ can then calculated by the well-known MAX-MIN operation. Considering the ith rule R_i and the observation $A1$, the respective action $B1$ is given by

$$B1_i = (A1, A2) \circ R_i \quad (5.21)$$

Therefore, the corresponding membership function is defined as follows:

$$\mu_{B1_i}(b) = \text{MAXMIN}(\mu_{A1}(a_1) \times \mu_{A2}(a_2), \mu_R(a_1, a_2, b))$$

$$a_1 \in A1, a_2 \in A2$$

$$= \text{MINMAX}\{\text{MIN}[\text{MIN}(\mu_{A1}(a_1), \mu_{A_i^1}(a_1)), \text{MIN}(\mu_{A2}(a_2), \mu_{A_i^2}(a_2))], \mu_{B_i}(b)\} \quad (5.22)$$

$$a_1 \in A1, a_2 \in A2$$

$$= \text{MIN}(\Omega_i, \mu_{B_i}(b))$$

$$b \in B$$

where, in this case, Ω_i is defined in a different way:

$$\Omega_i = \text{MIN}\{\text{MAXMIN}(\mu_{A1}(a_1), \mu_{A_i^1}(a_1)), \text{MAXMIN}(\mu_{A2}(a_2), \mu_{A_i^2}(a_2))\} \quad (5.23)$$

$$a_1 \in A1, a_2 \in A2$$

Then the maximum of $B1_1, B1_2, \ldots, B1_N$ determines the final action $B1$, which is calculated as a union:

$$B1 = \bigcup_{i=1}^{N} B1_i = \text{MAX}(B1_1, B1_2, \ldots, B1_N) \quad (5.24)$$

5.3.3.2 Alternative method

This method relies on the commutative characteristic of the MAX-MIN and UNION operators. The inputs are normalized into the same universe of discourse.

First, the individual fuzzy relations are built. A method of creating the fuzzy relation R_i, which represents the first fuzzy implication can be derived by the decomposition of the first rule into two parts:

$$R_i^1 = R_i^1 \times B_i \qquad (5.25)$$

and for the second variable

$$R_i^2 = A_i^2 \times B_i \qquad (5.26)$$

Note that R^1 and R^2 represent partial relation, or sub-relation, and therefore the overall relation $R = R^1 \cup R^2$.

The overall relation for the combination of the first variable A^1 and the output B is therefore given by

$$R^1 = R_1^1 \cup R_2^1 \cup, \ldots, \cup R_N^1 \qquad (5.27)$$

and for the combination of the first variable A^2 and the output B it is

$$R^2 = R_1^2 \cup R_2^2 \cup, \ldots, \cup R_N^2 \qquad (5.28)$$

By replacing the UNION operators with fuzzy MAX operators we obtain

$$R^1(a_1, b) = \text{MAX}\{R_1^1(a_1, b), R_2^1(a_1, b), \ldots, R_N^1(a_1, b)\} \qquad (5.29)$$

$$R^2(a_2, b) = \text{MAX}\{R_1^2(a_2, b), R_2^2(a_2, b), \ldots, R_N^2(a_2, b)\} \qquad (5.30)$$

As a result, two overall rules coexist in this model. As opposed to the classic method, let us use these partial relations to infer the fuzzy result.

Having determined two overall rules, the model's output $B1$ can be obtained using the MIN-superposition of two relations with respect to the input $A1$ and $A2$.

$$B1 = \text{MIN}\{[A1 \circ R^1(a_1, b)], [A2 \circ R^2(a_2, b)]\} \qquad (5.31)$$

$$b \in B$$

$$B1 = \text{MIN}\{\text{MAXMIN}[A1, R^1(a_1, b)], \text{MAXMIN}[A2, R^2(a_2, b)]\} \qquad (5.32)$$

$$b \in B$$

where $R^1(a_1, b)$ and $R^2(a_2, b)$ are given by equations (5.29) and (5.30), respectively.

5.3.4 Multiple-Input Single-Output Fuzzy Logic Controller

5.3.4.1 Classsic method

If $A_1^1, A_2^1, \ldots, A_N^1$ are fuzzy subsets of A^1, and $A_1^2, A_2^2, \ldots, A_N^2$ are fuzzy subsets of A^2, and..., and $A_1^K, A_2^K, \ldots, A_N^K$ are fuzzy subsets of A^K; $B_1, B_2, \ldots, B_N$ are fuzzy subsets of B then the fuzzy relation R is defined by a set of fuzzy rules as follows:

R_1: IF A_1^1 AND IF A_1^2 AND...AND IF A_1^K THEN B_1, ALSO
R_2: IF A_2^1 AND IF A_2^2 AND...AND IF A_2^K THEN B_2, ALSO
$\vdots$
R_i: IF A_i^1 AND IF A_i^2 AND...AND IF A_i^K THEN B_i, ALSO
$\vdots$
R_N: IF A_N^1 AND IF A_N^2 AND...AND IF A_N^K THEN B_N.

Observe that in this case the fuzzy relation is defined by $R = (A^1, A^2, \ldots, A^K) \to B$. If the output B needs to be inferred, then according to the definition of fuzzy implication:

$$B = (A^1, A^2, \ldots, A^K) \circ R \tag{5.33}$$

Also, using the procedure to build the overall fuzzy relation

$$R = \bigcup_{i=1}^{N} R_i \tag{5.34}$$

With the fuzzy observations $A1, A2, \ldots, AK$ and the overall relation R, one can infer the resulting action $B1$ by applying the compositional rule of inference; that is

$$B1 = (A1, A2, \ldots, AK) \circ R = (A1, A2, \ldots, AK) \circ \bigcup_{i=1}^{N} R_i = \bigcup_{i=1}^{N} (A1, A2, \ldots, AK) \circ R_i \tag{5.35}$$

The membership function of $B1$ can then calculated by the well-known MAX-MIN operation. Considering the ith rule R_i and the observation $A1$, the respective action $B1$ is given by

$$B1_i = (A1, A2), \ldots, AK) \circ R_i \tag{5.36}$$

So the membership function is given by

$$\mu_{B1_i}(b) = \text{MAXMIN}(\mu_{A1}(a_1) \times \mu_{A2}(a_2) \times \cdots \times \mu_{AK}(a_K), \mu_R(a_1, a_2, \ldots, a_K, b))$$
$$a_1 \in A1, a_2 \in A2, \ldots, A_K \in AK$$
$$= \text{MINMAX}(\text{MIN}(\text{MIN}(\mu_{A1}(a_1), \mu_{A_i^1}(a_1)), \text{MIN}(\mu_{A2}(a_2), \mu_{A_i^2}(a_2)) \ldots$$
$$a_1 \in A1, a_2 \in A2, \ldots, a_K \in AK \tag{5.37}$$
$$\ldots, \text{MIN}(\mu_{AK}(a_K), \mu_{A_i^K}(a_K)), \mu_{B_i}(b)))$$
$$= \text{MIN}(\Omega_i, \mu_{B_i}(b))$$
$$b \in B1$$

where Ω_i is defined by

$$\Omega_i = \text{MIN}(\text{MAXMIN}(\mu_{A1}(a_1), \mu_{A_i^1}(a_1)), \text{MAXMIN}(\mu_{A2}(a_2), \mu_{A_i^2}(a_2))\ldots \quad (5.38)$$

$$a_1 \in A1, a_2 \in A2, \ldots, a_K \in AK$$

$$\ldots, \text{MAXMIN}(\mu_{AK}(a_K), \mu_{A_i^K}(a_K))\}$$

Then the maximum of $B1_1, B1_2, \ldots, B1_N$ determines the final action $B1$, which is calculated as a union:

$$B1 = \bigcup_{i=1}^{N} B1_i = \text{MAX}(B1_1, B1_2, \ldots, B1_N) \quad (5.39)$$

5.3.4.2 Improved method

This method is based, by analogy with the previously used technique, on the rule decompositon into sub-relations.

One can assume that the first and subsequent rules can be decomposed into K separate sub-relations as follows:

$$R_i^k = A_i^k \times B_i^l \quad (5.40)$$

where i denotes the rule number $i = 1, 2, \ldots, N$; k denotes the input variable $k = 1, 2, \ldots, K$; and l denotes the output variable $k = 1, 2, \ldots, L$.

Therefore, if one wants to obtain overall the kth sub-rule, all contributions need to be united:

$$R^k = \bigcup_{i=1}^{N} R_i^k \quad (5.41)$$

or

$$R^k = \text{MAX}(R_1^k(a_k, b), R_2^k(a_k, b), \ldots, R_N^k(a_k, b)) \quad (5.42)$$

In this case, the model's output $B1$ related to a set of inputs $(A1, A2, \ldots, AK)$ can be obtained using the MIN-superposition of all K relations between the input Ak and the rule R^k.

$$B1 = \text{MIN}[A1 \circ R^1, A2 \circ R^2, \ldots, AK \circ R^K] \quad (5.43)$$

or

$$B1 = \text{MIN}\{\text{MAXMIN}[A1, R^1(a_1, b)], \ldots, \text{MAXMIN}[AK, R^K(a_K, b)]\} \quad (5.44)$$

5.3.5 Multiple-Input Multiple-Output Fuzzy Logic Controller

5.3.5.1 Classic method

If $A_1^1, A_2^1, \ldots, A_N^1$ are fuzzy subsets of A^1, and $A_1^2, A_2^2, \ldots, A_N^2$ are fuzzy subsets of A^2, and…, and $A_1^K, A_2^K, \ldots, A_N^K$ are fuzzy subsets of A^K, and $B_1^1, B_2^1, \ldots, B_N^1$ are fuzzy subsets of B^1, and

$B_1^2, B_2^2, \ldots, B_N^2$ are fuzzy subsets of B^2, and..., and $B_1^K, B_2^K, \ldots, B_N^K$ are fuzzy subsets of B^K then the fuzzy relation R is defined by a set of fuzzy rules as follows:

R_1: IF A_1^1 AND IF A_1^2 AND...AND IF A_1^K THEN B_1^1 AND B_1^2 AND...AND B_1^L,

ALSO

R_2: IF A_2^1 AND IF A_2^2 AND...AND IF A_2^K THEN B_2^1 AND B_2^2 AND...AND B_2^L,

ALSO

$\vdots$

R_i: IF A_i^1 AND IF A_i^2 AND...AND IF A_i^K THEN B_i^1 AND B_i^2 AND...AND B_i^L,

ALSO

$\vdots$

R_N: IF A_N^1 AND IF $A2_N$ AND...AND IF A_N^K THEN B_N^1 AND B_N^2 AND...AND B_N^L.

Observe that in this case $A^1, A^2, \ldots, A^K$ represent input variables and $B^1, B^2, \ldots, B^L$ represent the output variables. More specifically

$$R = (A^1, A^2, \ldots, A^K) \rightarrow (B^1, B^2, \ldots, B^L) \tag{5.45}$$

If output $B^1, B^2, \ldots, B^L$ need to be inferred, then according to the definition of fuzzy implication

$$(B^1, B^2, \ldots, B^L) = (A^1, A^2, \ldots, A^K) \circ R \tag{5.46}$$

With fuzzy observations $A1$ and $A2, \ldots, AK$ and the overall relation R, one can infer the resulting action $B1$ by applying the compositional rule of inference; that is

$$B1 = (A1, A2, \ldots, AK) \circ R = (A1, A2, \ldots, AK) \circ \bigcup_{i=1}^{N} R_i = \bigcup_{i=1}^{N} (A1, A2, \ldots, AK) \circ R_i \tag{5.47}$$

The membership function of $B1$ can then calculated by the well-known MAX-MIN operation. By considering the ith rule R_i and the observation $A1$, the respective action $B1$ is given by

$$B1_i = (A1, A2, \ldots, AK) \circ R_i \tag{5.48}$$

where $l = 1, 2, \ldots, L$.

Therefore, the corresponding membership function is defined as follows:

$$\mu_{B1_i}(b) = \text{MAXMIN}(\mu_{A1}(a_1) \times \mu_{A2}(a_2) \times \cdots \times \mu_{AK}(a_K), \mu_R(a_1, a_2, \ldots, b_1, \ldots, b_L))$$
$$a_1 \in A1, a_2 \in A2, \ldots, A_K \in AK$$

$$= \text{MINMAX}\{\text{MIN}(\text{MIN}(\mu_{A1}(a_1), \mu_{A_i^1}(a_1)), \text{MIN}(\mu_{A2}(a_2), \mu_{A_i^2}(a_2))\ldots$$

$$a_1 \in A1, a_2 \in A2, \ldots, a_K \in AK \qquad (5.49)$$

$$\ldots, \text{MIN}(\mu_K(a_K), \mu_{A_i^K}(a_K)), \mu_R(a_1, a_2, \ldots, b_1, \ldots, b_L))\}$$

$$= \text{MIN}(\Omega_i^l, \mu_{A_i^l}(b_l))$$

$$b_l \in Bl$$

where, in this case, Ω_i is defined in a different way:

$$\Omega_i^l = \text{MIN}\{\text{MAXMIN}(\mu_{A1}(a_1), \mu_{A_{1i}}(a_1)), \text{MAXMIN}(\mu_{A2}(a_2), \mu_{A_{2i}}(a_2)), \ldots$$

$$a_1 \in A1, a_2 \in A2, \ldots, a_K \in AK \qquad (5.50)$$

$$\ldots, \text{MAXMIN}(\mu_K(a_K), \mu_{A_i^K}(a_K))\}$$

Then the maximum of $B1_1, B1_2, \ldots, B1_N$ determines the final action $B1$ which can be calculated as a union:

$$B1 = \bigcup_{i=1}^{N} B1_i \qquad (5.51)$$

This can be performed for: $l = 1, 2, \ldots, L$ output variables.

5.3.5.2 Alternative method

By decomposing the relations defined in a single rule, the set of sub-relations can be obtained for all rules.

For the first rule and for the first input variable we can build the set of following sub-sub-relations:

$$R_1^{11} = A_1^1 \times B_1^1 \qquad (5.52)$$

$$R_1^{12} = A_1^1 \times B_1^2 \qquad (5.53)$$

$$\vdots$$

$$R_1^{1l} = A_1^1 \times B_1^l \qquad (5.54)$$

$$\vdots$$

$$R_1^{1L} = A_1^1 \times B_1^L \qquad (5.55)$$

Similarly, for the first rule and for the kth input variable we can build the set of following sub-sub-relations:

$$R_1^{k1} = A_1^k \times B_1^1 \qquad (5.56)$$

$$R_1^{k2} = A_1^k \times B_1^2 \qquad (5.57)$$

$$\vdots$$

$$R_1^{kl} = A_1^k \times B_1^l \qquad (5.58)$$
$$\vdots$$
$$R_1^{kL} = A_1^k \times B_1^L \qquad (5.59)$$

Following the similar technique, for the first rule and the last (Kth) input variable we have

$$R_1^{K1} = A_1^K \times B_1^1 \qquad (5.60)$$
$$R_1^{K2} = A_1^K \times B_1^2 \qquad (5.61)$$
$$\vdots$$
$$R_1^{Kl} = A_1^K \times B_1^l \qquad (5.62)$$
$$\vdots$$
$$R_1^{KL} = A_1^K \times B_1^L \qquad (5.63)$$

As a result, we obtained $K*L$ sub-sub-relations for the first rule. With the same strategy we can define sub-sub-relations for each of N rules.

Completing the description we can define the following set of sub-sub-relations for Nth rule.

For the Nth rule and for the first input variable we can build the set of following sub-sub-relations:

$$R_N^{11} = A_N^1 \times B_N^1 \qquad (5.64)$$
$$R_N^{12} = A_N^1 \times B_N^2 \qquad (5.65)$$
$$\vdots$$
$$R_N^{1l} = A_N^1 \times B_N^l \qquad (5.66)$$
$$\vdots$$
$$R_N^{1L} = A_N^1 \times B_N^L \qquad (5.67)$$

Similarly, for the Nth rule and for the kth input variable we can build the set of following sub-sub-relations:

$$R_N^{k1} = A_N^k \times B_N^1 \qquad (5.68)$$
$$R_N^{k2} = A_N^k \times B_N^2 \qquad (5.69)$$
$$\vdots$$
$$R_N^{kl} = A_N^k \times B_N^l \qquad (5.70)$$
$$\vdots$$
$$R_N^{kL} = A_N^k \times B_N^L \qquad (5.71)$$

Following the similar technique, for the Nth rule and the last (Kth) input variable we have

$$R_N^{K1} = A_N^K \times B_N^1 \qquad (5.72)$$

$$R_N^{K2} = A_N^K \times B_N^2 \qquad (5.73)$$

$$\vdots$$

$$R_N^{Kl} = A_N^K \times B_N^l \qquad (5.74)$$

$$\vdots$$

$$R_N^{KL} = A_N^K \times B_N^L \qquad (5.75)$$

Now our system is described by a set of $N*K*L$ sub-sub-relations. Let us group these relations in a way that will be suitable for further inferencing. Grouping sub-sub-relations with respect to the index indicating the number of input and output variable (kl) throughout all N rules leads to the following sub-relations:

$$R^{11} = \bigcup_{i=1}^{N} R_i^{11} \qquad (5.76)$$

$$\vdots$$

$$R^{1L} = \bigcup_{i=1}^{N} R_i^{1L}$$

$$\vdots$$

$$R^{K1} = \bigcup_{i=1}^{N} R_i^{K1} \qquad (5.77)$$

$$\vdots$$

$$R^{KL} = \bigcup_{i=1}^{N} R_i^{KL} \qquad (5.78)$$

Now we have $K*L$ sub-relations that can be used to infer the fuzzy results in the case that the input set of variables is $A1$ and $A2,\ldots, AK$. The model's outputs can be obtained using MIN-superposition of all L relations:

$$B1 = \text{MIN}[(A1 \circ R^{11}), (A2 \circ R^{21}), \ldots, (AK \circ R^{K1})] \qquad (5.79)$$

$$BL = \text{MIN}[(A1 \circ R^{1L}), (A2 \circ R^{2L}), \ldots, (AK \circ R^{KL})] \qquad (5.80)$$

In this way, we can infer all L fuzzy outputs by the MIN-superposition of all sub-rules incorporating a respective output variable. More specifically one can rewrite equations (5.81) and (5.82) into

$$B1 = \text{MIN}\{\text{MAXMIN}[A1, R^{11}(a_1, B_1)], \ldots, \text{MAXMIN}[AK, R^{K1}(a_K, b_1)]\} \qquad (5.81)$$

$$BL = \text{MIN}\{\text{MAXMIN}[A1, R^{1L}(a_1, b_L)], \ldots, \text{MAXMIN}[AK, R^{KL}(a_K, b_L)]\} \qquad (5.82)$$

5.4 HARDWARE IMPLEMENTATION: COMPARATIVE STUDY

This section discusses some implementation issues related to the FLC and is organized as follows. First, the mapping of each FLC model into hardware is presented, followed by the analysis of the cost function and the overall performance that can be reached. Second, the hardware implementations issues are discussed in terms of hardware cost, performance, memory size and clock frequency, discussed in detail. Finally, general conclusions are provided.

5.4.1 Hardware mapping of FLC models

This section presents the key elements of the analytical FLC models mapped into a hardware-realizable scheme. We present the hardware schemes for FLC configurations discussed in the earlier sections of this paper.

SISO classic implementation   Taking straightforwardly the FLC model presented in Section 5.3.2.1, we can map the analytical model into a hardware model. One of the possible solutions is illustrated in Figure 5.11.

As can be noted from Figure 5.11, the FLC is mapped into the fuzzifier, inference engine and defuzzifier. For the sake of uniformity we assume that fuzzifier and defuzzifier modules are exactly the same for all implementations. Therefore, we will further concentrate on inference module details.

In case of SISO FLC, the inference module is constructed of N MIN units and N MAX units performing the antecedent part of rules. MIN unit must perform the minimum operation on n m-bit vectors, as assumed above. Then the MAX unit determines the maximum value represented by one of these n m-bit vectors. To do the minimum, the antecedent membership functions (representing the values of input linguistic variables) must be stored and ready to enter the MIN units. These functions, as another n m-bit vector, are stored in the rule base memory. Proceeding with fuzzy inference, the consequent membership functions (representing the values of output linguistic variables) are entered into N MIN units. These functions are stored the rule base memory. Consequent MIN units perform the same type operation as antecedent MIN units. The results of consequent MIN units are united by means of a multi-input MAX unit. Such a unit is usually implemented as a MAX unit binary tree, which in this case contains $N-1$ MAX units. Finally the unified defuzzifier generates a crisp output of the FLC.

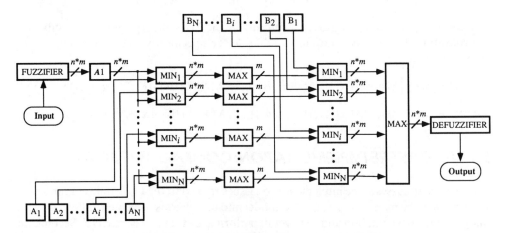

Figure 5.11   Diagram of the SISO classic controller

Let us now derive the cost function for the SISO FLC hardware. By investigating the scheme of Figure 5.11 we have

$$\Psi_{1N1(C)} = \text{FUZZ} + N(\text{MAX} + 2\text{MIN}) + (N-1)\text{MAX} + 2nm\text{MEM} + \text{DEFUZZ} \quad (5.83)$$

where Ψ_{1N1} denotes the hardware cost function for a single-input single-output FLC featuring N rules of inference in its rule base; FUZZ denotes the whole fuzzifier module, DEFUZZ denotes the defuzzifier module, and MEM represents the elementary memory cell.

Based on the same diagram, the overall input-to-output delay time ($\Theta_{\text{IN-OUT}}$) can be estimated as

$$\Theta_{\text{IN-OUT}} = \tau_{\text{FUZZ}} + 2\tau_{\text{MIN}} + (n-1)\tau_{\text{MAX}} + (N-1)\tau_{\text{MAX}} + \tau_{\text{DEFUZZ}} \quad (5.84)$$

where τ_{FUZZ} denotes the total delay time of fuzzifier, τ_{MIN} is the delay time introduced by a single MIN unit, τ_{MAX} is the delay time introduced by the MAX unit, and τ_{DEFUZZ} is the delay time introduced by a defuzzification unit.

SISO alternative implementation By analyzing the FLC model presented in Section 5.3.2.2, we can map it into a hardware model which is illustrated in Figure 5.12.

As can be noted from Figure 5.12, the FLC is mapped into the fuzzifier, inference engine and defuzzifier as usual. However, we have to point to the few distinct differences. First of all, in this case learning the overall rule can be done off-line in the part of the scheme contained in the rectangle. That is a dramatic difference compared with the classic implementations. Second, the multiple-MIN unit performing the minimum operation on the overall rule and the incoming fuzzy vector can be viewed as $(n-1)$ ordinary MIN units working concurrently. Also, the final MAX unit can be decomposed in the same way.

If we closely look at the rule learning module, we note that the process can be implemented with two MIN units and a single MAX unit plus register REG storing the

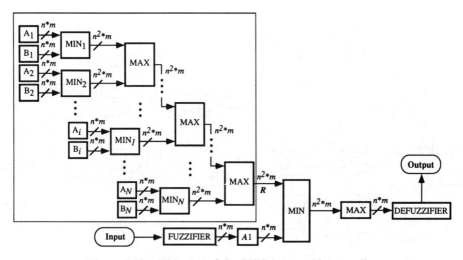

Figure 5.12 Diagram of the SISO improved controller

166 DESIGN CONSIDERATIONS OF DIGITAL FUZZY LOGIC CONTROLLERS

temporary MAX result. Because the learning process can be performed off-line, the extensive (parallel) hardware is not necessary.

As a result, we can describe the hardware cost fot the SISO improved method by

$$\Psi_{1N1(I)} = \text{FUZZ} + \text{MIN}(n^2) + \text{MAX} + n^2 m\text{MEM} + \text{DEFUZZ} + \text{REG} \qquad (5.85)$$

As far as the delay time is concerned, the estimation can be described by the following equation:

$$\Theta_{\text{IN-OUT}} = \tau_{\text{FUZZ}} + \tau_{\text{MIN}} + (n-1)\tau_{\text{MAX}} + \tau_{\text{DEFUZZ}} \qquad (5.86)$$

Note that this equation does not take into account the time needed for learning the overall rule; however, it does take the delay needed to perform multiple maximum operations before defuzzification.

DISO classic implementation The double-input single-output (DISO) FLC model presented in Section 5.3.3.1, can be mapped into a hardware model which is illustrated in Figure 5.13.

The DISO FLC model can be implemented in a fashion similar to that applied to create the SISO classic scheme. Analogously, the DISO FLC is mapped into the fuzzifier, inference engine and defuzzifier. This time, however, due to the second variable the scheme must be enhanced by a second fuzzifier, another block of memory unit for the second input variable, and a set of N MIN units.

Because of the close analogy to the SISO model we can directly derive an expression for the hardware cost as follows:

$$\Psi_{2N1(C)} = 2\text{FUZZ} + 2N(\text{MAX} + n\text{MIN}) + n(N-1)\text{MAX}$$
$$+ N\text{MIN} + nN\text{MIN} + 3nm\text{MEM} + \text{DEFUZZ} \qquad (5.87)$$

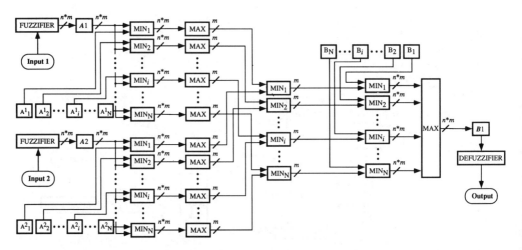

Figure 5.13 Diagram of the DISO classic controller

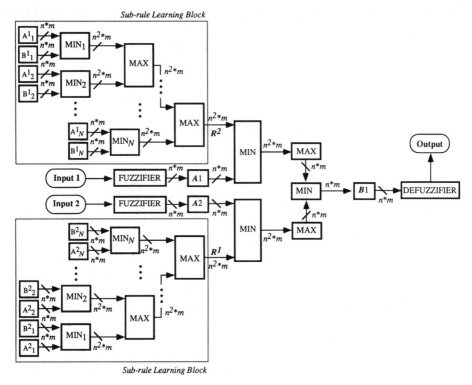

Figure 5.14 Diagram of the DISO improved controller

As far as the delay time is concerned, the estimation can again be derived from the SISO model by introducing additional delay on the MIN unit. As a result the following equation can be obtained:

$$\Theta_{\text{IN-OUT}} = \tau_{\text{FUZZ}} + 3\tau_{\text{MIN}} + (n-1)\tau_{\text{MAX}} + (N-1)\tau_{\text{MAX}} + \tau_{\text{DEFUZZ}} \tag{5.88}$$

DISO alternative implementation Analyzing the FLC model presented in Section 5.3.3.2, we can map it into a hardware scheme which is illustrated in figure 5.14.

As can be noted from the scheme of Figure 5.14, the FLC is mapped into the typical functional blocks. The only difference is that in the DISO case there are two separate input paths (fuzzifiers and sub-rule memories).

As a result, the hardware cost for the DISO improved controller can be estimated by

$$\Psi_{2N1(I)} = 2\text{FUZZ} + 2\text{MIN}(n^2) + 2n\text{MAX} + 2n^2m\text{MEM} + n\text{MIN} + \text{DEFUZZ} + 2\text{REG} \tag{5.89}$$

Now we should point out one important assumption that was made while constructing the cost function. We assumed that the multiple MIN or MAX units are realized in a parallel way. In other words, for an n m-bit vector, there are available separate n MIN units to perform the operation concurrently. Obviously, it may be implemented serially, but the overall performance would be extremely slow, and therefore we do not consider such an alternative.

As far as the delay time is concerned, the estimation can be described by the following equation:

$$\Theta_{\text{IN-OUT}} = \tau_{\text{FUZZ}} + 2\tau_{\text{MIN}} + (n-1)\tau_{\text{MAX}} + \tau_{\text{DEFUZZ}} \tag{5.90}$$

Again, the above result is obtained under the assumption that the learning process is performed off-line.

MIMO classic implementation The multiple-input multiple-output (MIMO) FLC model presented in Section 5.3.5.1, can be mapped into a hardware model which is illustrated in Figure 5.15. We can skip the detailed analysis of the MISO FLC model, as a special case of the MIMO model.

The MIMO FLC model can be implemented with the approach that we have used so far. The MIMO FLC is mapped into the fuzzifier, inference engine and defuzzifier. This time, however, due to the presence of K input variables, the scheme must be enhanced by a set of fuzzifiers, and memory units separate for each input as well as output variables. Also, N multi-input MIN units must be added between antecedent and consequent parts of the scheme.

By taking a MIMO scheme into consideration one can derive a cost function as

$$\Psi_{KNL(C)} = K\text{FUZZ} + KN(\text{MAX} + n\text{MIN}) + L(N-1)n\text{MAX} + N(K-1)\text{MIN}$$
$$+ LNn\text{MIN} + KLnm\text{MEM} + L\text{DEFUZZ} \tag{5.91}$$

So far as the delay time is concerned, the esstimation can be derived from the scheme presented in Figure 5.15, by introducing additional delay on the multi-input MIN unit. As a result, the following equation can be obtained:

$$\Theta_{\text{IN-OUT}} = \tau_{\text{FUZZ}} + 2\tau_{\text{MIN}} + (n-1)\tau_{\text{MAX}} + (n-1)\tau_{\text{MIN}} + (N-1)\tau_{\text{MAX}} + \tau_{\text{DEFUZZ}} \tag{5.92}$$

MIMO alternative implementation Finally, by analyzing the FLC model presented in Section 5.3.5.2, we can map it into a hardware scheme which is illustrated in Figure 5.16.

As can be noted from the scheme of Figure 5.16, the FLC is mapped into the typical functional blocks. However, in this case, the difference is that there need to be K separate memory units for each of the L output variables.

As a result, the hardware cost for the MIMO improved controller can be estimated by

$$\Psi_{KNL(I)} = K\text{FUZZ} + KL\text{MIN}(n^2) + KLn\text{MAX} + KLn^2m\text{MEM} + Kn\text{MAX}$$
$$+ n\text{MIN} + L\text{DEFUZZ} + KL\text{REG} \tag{5.93}$$

So far as the delay time is concerned, the estimation can be described by the following equation:

$$\Theta_{\text{IN-OUT}} = \tau_{\text{FUZZ}} + \tau_{\text{MIN}} + (n-1)\tau_{\text{MAX}} + (K-1)\tau_{\text{MIN}} + \tau_{\text{DEFUZZ}} \tag{5.94}$$

HARDWARE IMPLEMENTATION: COMPARATIVE STUDY   169

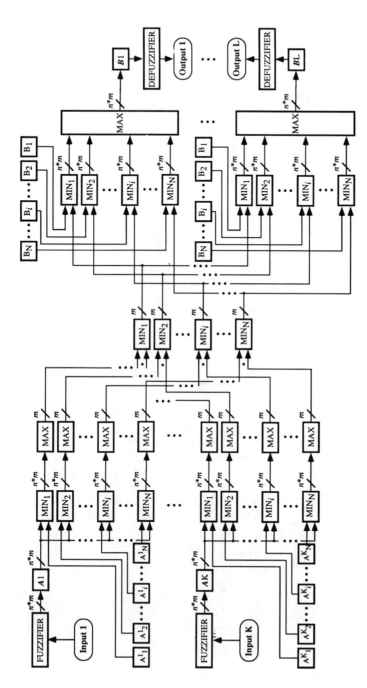

Figure 5.15 Diagram of the MIMO classic controller

170 DESIGN CONSIDERATIONS OF DIGITAL FUZZY LOGIC CONTROLLERS

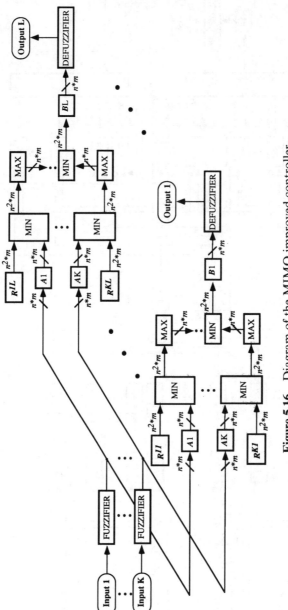

Figure 5.16 Diagram of the MIMO improved controller

Again, the above result is obtained under the assumption that the learning process is performed off-line.

5.4.2 Hardware implementation issues

VLSI implementation of FLC raises several practical difficulties that can be clearly visible by checking cost functions and overall delays for various configurations. Therefore, let us use equations (5.92) and (5.93) as the general formula for the classic FLC model, as well as equations (5.94) and (5.95) as the general formula for the alternative FLC model.

We compare various implementations of the presented FLC models based on the criteria derived from the set of characteristic parameters. We use the formula parameterized by

number of input variables/inputs (K)
number of output variables/outputs (L)
number of linguistic rules in the knowledge base (N)
number of binary vectors characterizing the membership function (n)
number of bits in a single binary vector (m)

Hardware cost Comparing the classic and alternative approaches of FLC implementations in terms of hardware cost we can conclude the following.

Assuming the same cost for fuzzifier and defuzzifier modules, the hardware cost of the classic FLC depends linearly on the number of rules. Due to its characteristic feature, the improved FLC hardware cost does not depend on the number of rules.

The hardware cost for both implementations depends on the number of input and the number of output variables.

The hardware cost for both implementations depends on the number of binary vectors characterizing the membership function of the input and the output variables.

The hardware cost for both implementations does not depend on the number of membership functions in the input and the output universe of discourse.

Performance We will compare the performance of the two FLC implementations by means of the introduced overall delay from input to output. By analyzing formula given by equations (5.93) and (5.95) and assuming the same performances of the typical functional units, one can conclude the following.

The performance of the classic FLC does depend on the number of input variables, the number of rules, and the number of the number binary vectors characterizing the membership function.

The performance of the improved FLC depends on the number of the number binary vectors characterizing the membership function, and the number of input variables (through the multi-input MIN unit). However, it does not depend on the number of rules N.

Maximum clock frequency The maximum clock frequency for the FLC scheme is determined by the slowest unit. The slowest unit in the controller is probably the defuzzifier. As we described at the beginning of this chapter, the defuzzification involves a substantial number of additions and divisions, and that is the major problem when considering the real hardware implementation. Obviously, the pipeline technique helps to improve the speed, but the defuzzifier will still eventually determine the maximum clock frequency for the FLC device.

Rule memory Comparing the classic and alternative approaches in terms of the rule base memory size, one can conclude the following. The memory size for both implementations depends on the number of binary vectors characterizing the membership function, the number of bits in a single vector, the number of input variables, and the number of output variables. However, it needs to be stressed that for improved FLC the memory size depends quadratically on the number of binary vectors characterizing the membership function (n^2), whereas for classic FLC this relation is linear.

5.4.3 Summary

In summary, we provide some general rules for the choice of a specific approach to the hardware implementations based on the issues discussed above.

Depending on the specific area of application, one can choose the best FLC hardware implementation by considering the following recommendations.

For real-time control applications, where the performance plays the most important role and high hardware cost may be sacrificed, the best solution is the alternative approach. This approach provides a performance that is not depend on the number of inputs nor the number of outputs; however the hardware cost depends on these two parameters.

For moderate performance applications where the minimum hardware cost needs to be maintained, the classic approach seems to be the best. If one can implement the proposed in literature improvements [2] to minimize the hardware cost and maximize the performance, the resulting FLC hardware may be quite attractive.

For the applications to simple control systems (e.g. SISO or DISO), the alternative approach is better, because it offers a better performance while keeping hardware costs that are comparable to the classic one. In such cases, the improved solution offers even more flexibility, providing a possibility for on-line rule learning, for the cost of additional hardware.

5.5 FINAL REMARKS

As the recent trend shows, there is a substantial transition of digital fuzzy technology from research and development laboratories to mass production. This transition has occurred during the last few years, and is clearly a result of increasing market demand for fuzzy logic based products. The market includes not only high-tech electronic equipment, but also high-volume consumer products and industrial applications.

The transition towards industrial applications is, however, relatively slow. Motivated by an economic factor, the fuzzy technology as applied, even digital, does not offer spectacular advantages over the classic solutions. In many cases, fuzzy logic digital systems appear to be more burdensome for many industrial control applications. The tremendous advantages of the digital fuzzy logic systems that include flexibility, programmability and compatibility with existing digital hardware have to be oriented towards more sophisticated systems, where the classic approaches simply fail or are more expensive and time-consuming. Many recent studies show that the fuzzy controllers cannot successfully compete with the classic and well-established and verified PD or PID controllers in the domain of simple and well-known applications. Those include, above all, most linear SISO systems.

Moreover, stand-alone fuzzy technology cannot offer the characteristics (e.g. robustness, learning ability, knowledge acquisition) that can be obtained from the merge of techniques that include neural networks, chaotic computation or genetic algorithms. In the case of emerging technologiess, the new class of compatibility problems need to be addressed even if elements of the merged system are implemented with the same digital technique.

As a result, within a few years one can expect the development of a large number of systems based on the merging of fuzzy technology with other new technologies to provide more competitive solutions for difficult or even (to-date) insolvable technical problems.

REFERENCES

1. Chiueh, T. (1991) Optimization of fuzzy logic implementation. *Proceedings of 21st International Symposium of Multiple Valued Logic*, Victoria, Canada, pp. 348–255.
2. Chiueh, T. (1992) Optimization of fuzzy logic inference architecture. *IEEE Computer*, 25, 67–71.
3. Corder, R. (1989) A high-speed fuzzy processor. *Proceedings of 3rd IFSA Congress*, Seattle, pp. 379–381.
4. Corder, R. (1989) Architecture for custom VLSI processor based embedded fuzzy expert systems. *Proceedngs of 3rd IFSA Congress*, pp. 382–389.
5. Dettloff, W., Watanabe, H., and Yount, K. (1989) VLSI fuzzy logic inference engine for real-time process control. *Proceedings of the IEEE Custom Integrated Circuits Conference*, San Diego, pp. 12.4.1–12.4.5.
6. Driankov, D., Hellendorn, H., and Reinfrank, M. (1993) *An Introduction to Fuzzy Control* (Springer-Verlage, New York).
7. Eichweld, H., Lohner, M., and Muller, M. (1992) Architecture of a CMOS fuzzy logic controller with optimized memory organization and operator design. *Proceedings of 1st International Conference on Fuzzy Systems*, San Diego, pp. 1317–1323.
8. Grantner, J., and Patyra, M. J. (1993) Impplementation of fuzzy logic finite state machines. *5th International Fuzzy Systems Association World Congress*, Seoul, pp. 781–784.
9. Grantner, J., Patyra, M. J., and Stachowicz, M. (1992) Intelligent fuzzy controller for event-driven real time systems. *NAFIPS Congress'92*, Puerto Vallarta, Maxico, pp. 312–321.
10. Grantner, J., Patyra, M. J., and Stachowicz, M. S. (1993) Architecture for event-driven intelligent fuzzy controller. *IEEE FUZZ'93*, San Francisco, pp. 273–278.
11. Howard, N., Taylor, R., and Allison, N. (1992) The design and implementation of a massively-parallel fuzzy architecture. *Proceedings of 1st International Conference on Fuzzy Systems*, San Diego, pp. 545–552.
12. Ikeda, H., Kisu, N., Hiramoto, Y., and Nakamura, S. (1992) A fuzzy inference coprocessor using a flexible active-driver architecture. *Proceedings of 1st International Conference on Fuzzy Systems*, San Diego, pp. 537–544.

13 Ikeda, H., Hiramoto, Y., Kisu, N., and Nakamura, S. (1991) A fuzzy processor for a sophisticated automatic transmission control. *Proceedings of the 4th IFSA World Congress*, Brussels, pp. 53–56.
14 Kang, H., and Vachtsevanos, G. (1992) Fuzzy hupercubes: a possibilistic inferencing paradigm. *Proceedings of 1st International Conference on Fuzzy Systems*, San Diego, pp. 553–559.
15 Katsumata, A., Tokunaga, H., and Yasunobu, S. (1991) Fuzzy set processor (FSP) for fuzzy information processing. *Proceedings of International Fuzzy Engineering Symposium*, Yokohama, pp. 399–406.
16 Kosko, B. (1994) *Fuzzy Thinking* (Prentice Hall, Englewood Cliffs, NJ).
17 Lee, C. C. (1990) Fuzzy logic in control systems: fuzzy logic controller—Part I. *IEEE Transactions on Systems, Man and Cybernetics*, **20**, 404–418.
18 Lee, C. C. (1990) Fuzzy logic in control systems: fuzzy logic controller—Part II. *IEEE Transactions on Systems, Man and Cybernetics*, **20**, 419–435.
19 Manzoul, M. and Tayal, S. (1990) Systolic VLSI array for multi-variable fuzzy cntrol systems. *Cybernetics and Systems: An International Journal*, **21**, 27–42.
20 Manzoul, M., and Serrate, H. (1988) Fuzzy systolic arrays. *Proceedings of 19th International Symposium on Multiple-Valued Logic*, Palma de Mallorca, Spain.
21 Manzoul, M., and Jayabharathi, D. (1992) Fuzzy controller on FPGA chip. *Proceedings of 1st International Conference on Fuzzy Systemss*, San Diego, pp. 1309–1316.
22 Marinos, P. (1966) Fuzzy logic. Bell Telephone Labs., Technical Memorandum 66-3341-1.
23 Marinos, P. (1969) Fuzzy logic and its application to switching systems. *IEEE Transactions on Computers*, **18**, 343–348.
24 Nakamura, K., Sakashita, N., Nitta, Y., Shimomura, K., Ohno, T., Eguchi, K., and Tokuda, T. (1993) A 12b resolution 200kFLIPS fuzzy inference processor. *IEEE International Solid State Circuit Conference Digest of Technical Papers*, Buena Vista, pp. 182–183.
25 Ostrowski, D., Cheung, P., and Roubaud, K. (1993) An outline of the intuitive design of fuzzy logic and its efficient implementation. *Proceedings of International Conference on Fuzzy Systems*, San Francisco, pp. 184–189.
26 Patyra, M. J., and Grantner, J. (1993) Hardware implementation issues of multivariable fuzzy control systems. *Proceedings of the 3rd IEEE International Conference on Industrial Fuzzy Control and Intelligent Systems*, Houston, pp. 179–184.
27 Sasaki, M., Ueno, F., Inoue, T. (1993) 7.5MFLIPS fuzzy microprocessor using SIMD and logic in memory structure. *Proceedings of International Conference on Fuzzy Systems*, San Francisco, pp. 527–534.
28 Srini, V. (1975) Realization of fuzzy forms. *IEEE Transactions on Computers*, **24**, 941–943.
29 Stachowicz, M. S., Grantner, J., and Kinney, L. L. (1992) Pipeline architecture boots performance of fuzzy logic controller. *IFSICC'92 International Fuzzy Systems and Intelligent Control Conference*, Louisville, pp. 190–198.
30 Symon, J., and Watanabe, H. (1990) Fuzzy logic inference engine board system. *Proceedings of the International Conference on Fuzzy Logic and Neural Networks*, Iizuka, Japan, pp. 161–164.
31 Togai, M., and Chiu, S. (1987) A fuzzy Logic Chip and a fuzzy inference accelerator for real-time approximate reasoning. *Proceedings of 17th International Symposium of Multiple Valued Logic*, pp. 25–29.
32 Togai, M., and Watanabe, H. (1985) A VLSI implementation of a fuzzy-inference engine: toward an expert system on a chip. *Proceedings of the Second Conference on Artificial Intelligence Applications*, Miami Beach, pp. 193–197.

33 Togai, M., and Watanabe, H. (1986) Expert system on a chip: an engine for real-time approximate reasoning. *IEEE EXPERT*, **1**, 55–62.
34 Togai, M., and Watanabe, H. (1986) A VLSI implementation of a fuzzy-inference engine: toward an expert system on a chip. *Information Sciences*, **38**, 147–163.
35 Watanabe, H. (1991) Some consideration on design of fuzzy information processors—from a computer architectural point of view. *Proceedings of International Fuzzy Engineering Symposium*, Yokohama, pp. 387–398.
36 Watanabe, H. (1992) RISC approach to design of fuzzy processor architecture. *Proceedings of International Conference on Fuzzy Systems*, San Diego, pp. 431–440.
37 Watanabe, H., and Chen, D. (1993) Evaluation of fuzzy instructions in a RISC processor. *Proceedings of International Conference on Fuzzy Systems*, San Francisco, pp. 521–526.
38 Watanabe, H., and Dettloft, W. (1988) Fuzzy logic inference processor for real time control: a second generation full custom design. *Proceedings of the Annual Asilomar Conference on Signals, Systems and Computers*, pp. 729–735.
39 Watanabe, H., Dettloff, W., Symon, J., and Yount, K. (1991) VLSI fuzzy chip and inference accelerator board system. *Proceedings of 21st International Symposium of Multiple Valued Logic*, Victoria, Canada, pp. 120–127.
40 Watanabe, H., Dettloff, W., and Youn, K. (1989) VLSI chip for fuzzy logic inference. *Proceedings of 3rd IFSA World Congress*, Seattle, pp. 292–295.
41 Watanabe, H., Dettloff, W., and Yount, K. (1990) A VLSI fuzzy logic controller with reconfigurable, cascadable architecture. *IEEE JSSC*, **25**, 376–381.
42 Watanabe, T., Matsumoto, M., and Enokida, M. (1989) Synthesis of synchronous fuzzy sequential circuits. *Proceedings of 3rd IFSA World Congress*, Seattle, pp. 288–291.

6
Parallel Algorithm for Fuzzy Logic Controller

J. L. Grantner
Department of Electrical and Computer Engineering, Western Michigan University, USA

6.1 INTRODUCTION

Fuzzy control has emerged as one of the most active and successful areas for research in the applications of fuzzy set theory, in particular, in the realm of industrial processes. The computational requirements of many real-time process control, autonomous navigation and sensor data processing algorithms based on fuzzy logic can only be met by the assistance of dedicated hardware. There has been an increasing need for high-speed fuzzy controllers which can efficiently be implemented by VLSI integrated circuit technology.

In this chapter a parallel algorithm and its conceptual hardware realization are proposed to assist the acceleration of both the model building and the inference computations for a fuzzy controller. First, mathematical models will be presented to build up linguistic models and perform inference computations with SISO, MISO and MIMO systems. By analyzing the suggested mathematical models, one can easily see that a single algorithm can be developed that will be parametrized according to the operation to be performed: either model building or inference. Next, conceptual hardware realization of SISO, MISO and MIMO fuzzy controllers will be discussed. To achieve a high processing rate, a pipelined architecture will be suggested. The maximum sustainable processing rate will be formulated to predict the performance of the hardware accelerator. Simulation results have shown that using two-valued logic the proposed fuzzy controller can efficiently be implemented by VLSI technology [5].

6.2 MATHEMATICAL MODELS FOR FUZZY MODEL BUILDING AND INFERENCE COMPUTATIONS

6.2.1 Single-Input Single-Output System

On the basis of a verbal description which is called a linguistic model introduced by Zadeh [1], an overall fuzzy relation R for a single-input single-output (SISO) system is created by the formula below, where → is the symbol of the operation or operations by which fuzzy implications are defined, and * stands for an operation which interprets the sentence connective 'also'. XI and ZI stand for fuzzy (linguistic) inputs and outputs, respectively.

$$R = \mathop{*}_{I=1}^{N} (XI \rightarrow ZI)$$

Let the verbal description of the process behaviour contain N relations:

$$R1: \text{IF } X \text{ is } (X1) \text{ THEN } Z \text{ is } (Z1)$$
$$\text{ALSO} \qquad (6.1)$$
$$\vdots$$
$$RN: \text{IF } X \text{ is } (XN) \text{ THEN } Z \text{ is } (ZN)$$

Fuzzy implication is interpreted as intersection. Discrete fuzzy sets are assumed, inputs are restricted to fuzzy numbers and fuzzy singletons [2]. Fuzzy relations $R1, R2, \ldots, RN$ are created analogously by application of the same definition of fuzzy implication:

$$\begin{aligned} R1 &= X1 \times Z1 \\ &\vdots \\ RN &= XN \times ZN \end{aligned} \qquad (6.2)$$

In enhanced form

$$\forall (u, w) \in U \times W \ R1(u, w) = \min(X1(u), Z1(w))$$
$$\cdots \qquad (6.3)$$
$$\forall (u, w) \in U \times W \ RN(u, w) = \min(XN(u), ZN(w))$$

where U and W stand for the universe of discourse for fuzzy inputs and outputs, respectively.

The overall relation R (being the process model) is obtained as the union of $R1, R2, \ldots, RN$ since the sentence connective 'also' is defined as union [2].

$$R = R1 \cup R2 \cup \cdots \cup RN \qquad (6.4)$$

$$\forall (u, w) \in U \times W, R(u, w) = \max(R1(u, w), \ldots, RN(u, w)) \qquad (6.5)$$

Min and max operations are chosen from the sets of triangular norms and conorms for intersection and union, respectively [2, 3].

The compositional rule of inference for approximate reasoning is suggested by Zadeh [3]. Max–min composition [2] is chosen to infer fuzzy conclusion Z to a fuzzy observation X:

$$Z = X \circ R \qquad (6.6)$$

$$\forall (u, w) \in U \times W, Z(w) = \max_{u \in U} [\min(X(u), R(u, w))] \qquad (6.7)$$

6.2.2 Multiple-Input Single-Output System

To describe the multiple-input single-output (MISO) system, let the system's performance be given by N relations and let the system have M fuzzy inputs $X^{(1)}, X^{(2)}, \ldots, X^{(M)}$ and a single fuzzy output Z:

$R1$: IF $X^{(1)}$ is $(X1^{(1)})$ AND $X^{(2)}$ is $(X1^{(2)})\ldots$ AND $X^{(M)}$ is $(X1^{(M)})$ THEN Z is $(Z1)$
ALSO (6.8)
$\vdots$
RN: IF $X^{(1)}$ is $(XN^{(1)})$ AND $X^{(2)}$ is $(XN^{(2)})\ldots$ AND $X^{(M)}$ is $(XN^{(M)})$ THEN Z is (ZN)

where $X1^{(1)}, X1^{(2)}, \ldots, X1^{(M)}$, and $Z1$ stand for the values of fuzzy inputs $X^{(1)}, X^{(2)}, \ldots, X^{(M)}$ and fuzzy output Z, respectively, when rule $R1$ is created, $XN^{(1)}, XN^{(2)}, \ldots, XN^{(M)}$ input values, and ZN output value, respectively, are used to create rule RN. $X^{(1)}, X^{(2)}, \ldots, X^{(M)}$ are normalized to the same universe of discourse. The first and following rules are decomposed into M separate sub-rules [5, 6] and are given in equations (6.9) and (6.10), respectively.

$$R1^{(1)} = X1^{(1)} \times Z1$$
$$\vdots \qquad (6.9)$$
$$R1^{(M)} = X1^{(M)} \times Z1$$

$$\vdots$$

$$RN^{(1)} = XN^{(1)} \times ZN$$
$$\vdots \qquad (6.10)$$
$$RN^{(M)} = XN^{(M)} \times ZN$$

In the enhanced form, equations (6.9) and (6.10) can be rewritten as

$$R1^{(1)}(u,w) = \min(X1^{(1)}(u), Z1(w))$$
$$\vdots \qquad (6.11)$$
$$R1^{(M)}(u,w) = \min(X1^{(M)}(u), Z1(w))$$

$$\vdots$$

$$RN^{(1)}(u,w) = \min(XN^{(1)}(u), ZN(w))$$
$$\vdots \qquad (6.12)$$
$$RN^{(M)}(u,w) = \min(XN^{(M)}, ZN(w))$$

The M overall sub-rules are then given by

$$R^{(1)} = R1^{(1)} \cup R2^{(1)} \cup \cdots \cup RN^{(1)}$$
$$\cdots \qquad (6.13)$$
$$R^{(M)} = R1^{(M)} \cup R2^{(M)} \cup \cdots \cup RN^{(M)}$$

The enhanced form of equation (6.13) is as follows:

$$R^{(1)}(u,w) = \max\{R1^{(1)}(u,w), R2^{(1)}(u,w), \ldots, RN^{(1)}(u,w)\}$$
$$\vdots \quad (6.14)$$
$$R^{(M)}(u,w) = \max\{R1^{(M)}(u,w), R2^{(M)}(u,w), \ldots, RN^{(M)}(u,w)\}$$

In this case, the fuzzy output of the MISO model is obtained using the min-superposition of all the M outputs inferred from the sub-models:

$$Z = \min\{(X^{(1)} \circ R^{(1)}), \ldots, (X^{(M)} \circ R^{(M)})\} \quad (6.15)$$

It should be noted that the projection applied in equation (6.15) to obtain one-dimensional fuzzy output will result in some loss of accuracy. In many cases that loss can be tolerated [7].

In the extended form, equation (6.15) is rewritten as

$$Z(w) = \min\left\{\left[\max_{u}(\min(X^{(1)}(u), R^{(1)}(u,w)))\right], \ldots, \left[\max_{u}(\min(X^{(M)}(u), R^{(M)}(u,w)))\right]\right\} \quad (6.16)$$

6.2.3 Multiple-Input Multiple-Output System

Let the linguistic model of the process contain N relations, and fuzzy sets describing the particular states that occur in the verbal description of inputs $X^{(1)}, X^{(2)}, \ldots,$ and $X^{(M)}$, and outputs $Z^{(1)}, Z^{(2)}, \ldots,$ and $Z^{(K)}$ be given by:

R1: IF $X^{(1)}$ is $(X1^{(1)})$ AND $X^{(2)}$ is $(X1^{(2)})$...AND $X^{(M)}$ is $(X1^{(M)})$,
    THEN $Z^{(1)}$ is $(Z1^{(1)})$ AND $Z^{(2)}$ is $(Z1^{(2)})$...AND $Z^{(K)}$ is $(Z1^{(K)})$
    ALSO
    $\vdots$ (6.17)
RN: IF $X^{(1)}$ is $(XN^{(1)})$ AND $X^{(2)}$ is $(XN^{(2)})$...AND $X^{(M)}$ is $(XN^{(M)})$,
    THEN $Z^{(1)}$ is $(ZN^{(1)})$ AND $Z^{(2)}$ is $(ZN^{(2)})$...AND $Z^{(K)}$ is $(ZN^{(K)})$

In effect, the rule base of a MIMO system is composed of a set of MISO sub-rule bases, therefore the general rule structure of a MIMO system can be represented as a collection of MISO systems [2], [8].

On the grounds of equations (6.9)–(6.16), the first and following rules of equation (6.17) are decomposed into M separate sub-rules for each output $Z^{(i)}$ ($i = 1, \ldots, K$) [5] as follows, for the first rule:

$$R1^{(11)} = X1^{(1)} \times Z1^{(1)}$$
$$\vdots \quad (6.18)$$
$$R1^{(M1)} = X1^{(M)} \times Z1^{(1)}$$
$$\vdots$$
$$RN^{(1K)} = X1^{(1)} \times Z1^{(K)}$$
$$\vdots \quad (6.19)$$
$$R1^{(MK)} = X1^{(M)} \times Z1^{(K)}$$

and for the Nth rule

$$RN^{(11)} = XN^{(1)} \times ZN^{(1)}$$
$$\vdots$$
$$RN^{(M1)} = XN^{(M)} \times ZN^{(1)}$$
(6.20)

$$\vdots$$

$$RN^{(1K)} = XN^{(1)} \times ZN^{(K)}$$
$$\vdots$$
$$RN^{(MK)} = XN^{(M)} \times ZN^{(K)}$$
(6.21)

where $R1^{(11)}, \ldots, R1^{(M1)}$ stand for sub-rules created for MIMO rule $R1$ in respect to output $Z^{(1)}, \ldots,$ and $R1^{(1K)}, \ldots, R1^{(MK)}$ stand for sub-rules created for MIMO rule $R1$ in respect to output $Z^{(K)}$, respectively. A similar notational scheme applies to sub-rules $RN^{(11)}, \ldots, RN^{(M1)}, \ldots, RN^{(1K)}, \ldots, RN^{(MK)}$ in respect to outputs $Z^{(1)}, \ldots, Z^{(K)}$, respectively.

The enhanced forms of equations (6.18)–(6.21) are omitted for the sake of simplicity (refer to equations (6.11)–(6.12)). Therefore, the set of overall sub-rules can be expressed by

$$R^{(11)} = R1^{(11)} \cup R2^{(11)} \cup \cdots \cup RN^{(11)}$$
$$\vdots$$
$$R^{(M1)} = R1^{(M1)} \cup R2^{(M1)} \cup \cdots \cup RN^{(M1)}$$
(6.22)

$$\vdots$$

$$R^{(1K)} = R1^{(1K)} \cup R2^{(1K)} \cup \cdots \cup RN^{(1K)}$$
$$\vdots$$
$$R^{(MK)} = R1^{(MK)} \cup R2^{(MK)} \cup \cdots \cup RN^{(MK)}$$
(6.23)

The enhanced forms of equations (6.22)–(6.23) are omitted for the sake of simplicity (refer to equation (6.14)). With this MIMO model, each fuzzy output is obtained using the min-superposition of all the M outputs inferred from the corresponding sub-models:

$$Z^{(1)} = \min\{(X^{(1)} \circ R^{(11)}), \ldots, (X^{(M)} \circ R^{(M1)})\}$$
$$Z^{(K)} = \min\{(X^{(1)} \circ R^{(1K)}), \ldots, (X^{(M)} \circ R^{(MK)})\}$$
(6.24)

The extended form of equation (6.24) is omitted for the sake of simplicity (refer to equation (6.16)).

6.3 PARALLEL ALGORITHM

The approach in this chapter is to construct a compact aggregated rule base, rather than storing each rule individually. The benefits of this method are: better sensitivity in approximate reasoning [9] and low memory requirement. By analyzing the mathematical models for SISO, MISO and MIMO systems (equations (6.3), (6.5), (6.7), (6.12), (6.14)–(6.16), and (6.22)–(6.24)), it can be seen that max operation follows min operation either to add a new rule to the current knowledge base or to perform an inference computation. Thus a single algorithm [14] can be developed such that its inputs and outputs will be parametrized according to the operation to be carried out: either model building or inference computation. The core sequence will be followed by another min operation for inference in MISO and MIMO systems.

The algorithm for building up overall rule R made of N rules is given below using two concurrent language statements parbegin and parend:

$$I = 1; \quad (6.27)$$
repeat
  while $I = N + 1$ do skip;
  $i = 1$;
  repeat
    while $i = p + 1$ do skip;
    parbegin
      $t_{Ii}(u, w) = \min(XI(u_i), ZI(w_1)), \ldots,$
      $\ldots, \min(XI(u_i), ZI(w_p));$
    parend
    parbegin
      $r_{Ii}(u, w) = \max(t_{Ii}(u_i, w_1), r_{(I-1)i}(u_i, w_1)), \ldots,$
      $\ldots, \max(t_{Ii}(u_i, w_p), r_{(I-1)i}(u_i, w_p));$
    parend
    $i = i + 1;$
  until false;
  $I = I + 1;$
until false;

where $XI(u_i)$ and $ZI(w_j)$ are elements of the fuzzy inputs and outputs that are fed into the algorithm for adding rule I to the rule base, respectively ($i = 1, \ldots, p; j = 1, \ldots, p; I = 1, \ldots, N$), r_{Ii} is a vector of p elements

$$r_{Ii}: r_{Ii}^1, r_{Ii}^2, \ldots, r_{Ii}^p$$

and t_{Ii} is a temporary vector (not stored) of p elements

$$t_{Ii}: t_{Ii}^1, t_{Ii}^2, \ldots, t_{Ii}^p$$

The accumulated knowledge base after the Ith cycle is represented by the array of p^2 elements

$$R(I): \begin{bmatrix} r_{I1}^1 & r_{I1}^2 & \cdots & r_{I1}^p \\ r_{I2}^1 & r_{I2}^2 & \cdots & r_{I2}^p \\ \vdots & \vdots & \ddots & \vdots \\ r_{Ip}^1 & r_{Ip}^2 & \cdots & r_{Ip}^p \end{bmatrix} \text{ or } \begin{bmatrix} r_{I1} \\ r_{I2} \\ \vdots \\ r_{Ip} \end{bmatrix} \quad (6.28)$$

and all elements of $R(0)$ are initially reset to zero (non membership) value.

The algorithm follows equations (6.3) and (6.5), for SISO systems, and equations (6.11), (6.12), (6.14), and (6.18)–(6.23), for MISO and MIMO systems, respectively. It deals with elements of just one vector (one row of the RI matrix) in parallel. Since there is no data dependency between rows of the R matrix, the degree of parallelism can be increased and multiple rows can be processed simultaneously.

It follows from the recursive structure of the algorithm that the process of constructing a rule base made of N rules does not require more memory than that of a single rule. Digitized membership functions are assumed and, hence, the memory requirement can be given by the formula

$$M = \log_2 n \cdot k_1 \cdot k_2 \quad \text{[bit]} \tag{6.29}$$

With the algorithm (6.27), $p = k_1 = k_2$ is assumed. The values for p and n will be discussed in Section 6.4.1. The total memory requirement is constant if n and p are also constant, no matter of how many rules are used to construct the linguistic model, and the degree of parallelism employed. This is a very attractive feature of the algorithm in terms of VLSI implementation.

Due to the recursive nature of the algorithm (6.27), after a new rule has been added to the rule base, the condition

$$\forall (u, w) \in U \times W \quad R(u, w) = 1 \tag{6.30}$$

can be evaluated by checking the vector r_{Ii} in each cycle to see all of its elements are of full (1) membership grade. The algorithm for setting a two-valued error flag EF (embedded into the algorithm (6.27)) is as follows:

initially, EF_{00} is set to a '1' (error condition)

then EF_{Ii} is updated in each i cycle

$$EF_{Ii} = EF_{I(i-1)} \wedge [r_{Ii}^1 \text{ is } (1)] \wedge \cdots \wedge [r_{Ii}^p \text{ is } (1)] \tag{6.31}$$

where $\wedge$ stands for the Boolean AND operation and the logic value of the conditional statement

$$[r_{Ii}^k \text{ is } (1)] \quad (k = 1, \ldots, p)$$

is a '1' iff r_{Ii}^k is of (1) full membership.

Thus at the completion of adding a new rule to overall rule R, it will be known if the model has just failed. This is an advantageous feature for those real-time applications where the linguistic model of the process is created on-line. With the algorithm for inference computations, the external cycle for I is not needed, therefore it will be omitted. The input parameters are: X (fuzzy input) and R (overall rule). The algorithm infers fuzzy output Z (refer to equation (6.17)) as follows:

$$
\begin{aligned}
&i = 1; \\
&\text{repeat} \\
&\quad \text{while } i = p + 1 \text{ do skip;} \\
&\quad \text{parbegin} \\
&\qquad t_i(u, w) = \min(X(u_i), r_i(u_i, w_1)), \ldots, \\
&\qquad \ldots, \min(X(u_i), r_i(u_i, w_p)); \\
&\quad \text{parend} \\
&\quad \text{parbegin} \\
&\qquad Z(w) = \max(t_i(u_i, w_1), Z(w_1)), \ldots, \\
&\qquad \ldots, \max(t_i(u_i, w_p), Z(w_p)); \\
&\quad \text{parend} \\
&\quad i = i + 1; \\
&\text{until false;}
\end{aligned}
\quad (6.32)
$$

where $X(u_i)$ and $Z(w_j)$ are elements of the fuzzy input and output vectors, respectively, $(i = 1, \ldots, p; j = 1, \ldots, p)$, t_i is a temporary vector (not stored) of p elements, and r_i is a vector of p elements (row i of overall rule matrix R, refer to (6.27)). The vector $Z(w)$ holds the inferred conclusion after the pth cycle and is represented by a vector of p elements:

$$Z(w): Z(w_1), Z(w_2), \ldots, Z(w_p)$$

and all elements of $Z(w)$ are intially reset to zero (non-membership) value.

Comparison of algorithms (6.27) and (6.32) shows that the inner cycles are essentially the same and, hence, just one basic cycle is to be implemented. One input set to the min and max operators, respectively, is parametrized and the outputs of the max operators are directed to either the R matrix or the Z vector.

It should be noted that the parallel algorithm shown in (6.27) and (6.32) is a generic one because the min and max operators can be substituted by any pair of t-norms and t-conorms, provided that they possess the commutative and associative property (e.g. algebraic product and max operators can be used to implement Larsen's R_p inference method [2]).

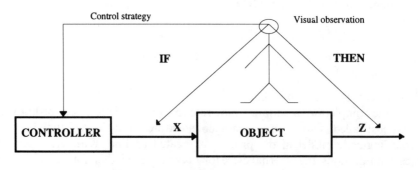

Figure 6.1 Design of linguistic model

The way how algorithms (6.27) and (6.32) work in the case of a SISO system will be illustrated by a simple example. The graphical representation of fuzzy sets and relations [10] will be used.

Let the linguistic model of the process be created by using visual observations of a skilled operator, as shown in Figure 6.1.

Let the performance of the process be given by just four statements:

Rule 1: If X is negative small   then Z is positive small
Rule 2: If X is positive small   then Z is positive medium
Rule 3: If X is positive medium   then Z is positive big
Rule 4: If X is positive big   then Z is negative small

Formally, the process will be described by four fuzzy IF–THEN rules as follows:

$R1$: IF X is $X1$ THEN Z is $Z1$
$R2$: IF X is $X2$ THEN Z is $Z2$
$R3$: IF X is $X3$ THEN Z is $Z3$
$R4$: IF X is $X4$ THEN Z is $Z4$

Graphical illustration of fuzzy sets $X1,\ldots,X4$, and $Z1,\ldots,Z4$ is given in Figure 6.2.

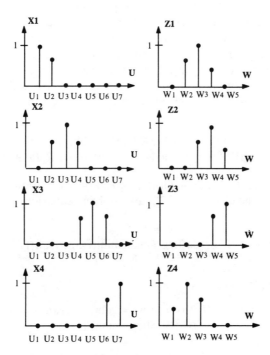

Figure 6.2 Graphical representation of fuzzy sets $X1,\ldots,X4, Z1,\ldots,Z4$

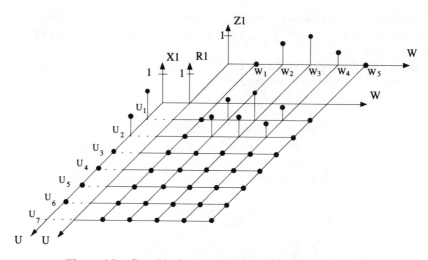

Figure 6.3 Graphical representation of fuzzy relation $R1$

By applying equations (6.2) and (6.3), fuzzy relation (rule) $R1$ is given by

$$R1 = X1 \times Z1$$
$$\forall (u, w) \in U \times W \; R1(u, w) = \min(X1(u), Z1(w))$$

The process of building up rule $R1$ is illustrated in Figure 6.3.

By using the algorithm (6.27), first u_1 of $X1$ is paired with all w ($w_1, \ldots, w_5$) elements of $Z1$ simultaneously and the minimum of each pair is taken. Then these minimum values are paired with elements of Row 1 of R (initially reset to zero) and the maxima are taken to produce the new contents or Row 1. After step 7, $R1$ has been added to the rule base which, at this point in time, consists of just $R1$, since $R1$ was the first rule in the sequence of building up R. The remaining relations $R2$, $R3$ and $R4$ are added to the rule base analogously. The final relation R is given below (refer to equations (6.4) and (6.5)) and is shown in Figure 6.4.

$$R = R1 \cup R2 \cup R3 \cup R4$$
$$\forall (u, w) \in U \times W \; R(u, w) = \max(R1(u, w), R2(u, w), R3(u, w), R4(u, w))$$

Next it will be shown how fuzzy conclusion Z to fuzzy observation X is inferred from the linguistic model using the algorithm (6.32). By applying equations (6.6) and (6.7)

$$Z = X \circ R$$
$$\forall (u, w) \in U \times W \; Z(w) = \max_{u \in U}[\min(X(u), R(u, w))]$$

they can be rewritten as

$$\forall (u, w) \in U \times W \; \tilde{R}(u, w) = \min(X(u), R(u, w))$$
$$\forall w \in W \; Z(w) = \max_{u \in U} \tilde{R}(u, w)$$

The process of fuzzy inference is illustrated in Figure 6.5.

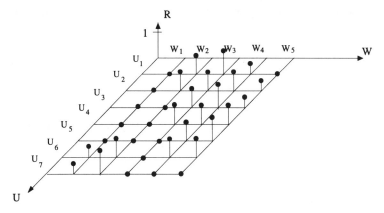

Figure 6.4 Graphical representation of overall rule R

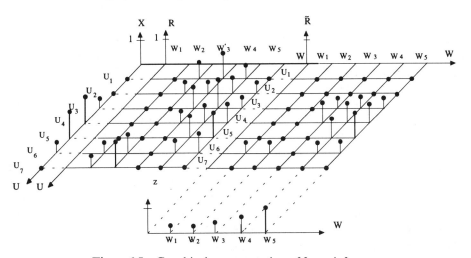

Figure 6.5 Graphical representation of fuzzy inference

Using algorithm (6.32), in step 1, u_1 is paired with the $w_1,\ldots,w_5$ elements of Row 1 of R, respectively, and the minima are taken. At this point, the first row of the auxiliary data array is created. These elements will be paired with elements of Z (the vector Z was reset to zero prior to step 1) and the maxima are taken which, in turn, will be loaded into Z. The auxiliary data array is not stored. After step 7, the fuzzy conclusion is obtained in Z.

6.4 CONCEPTUAL HARDWARE IMPLEMENTATION

6.4.1 SISO System

The conceptual hardware implementation of the combined fuzzy model/inference unit (often referred to as the fuzzy engine) will be presented here in a form which is obtained by the direct mapping of the algorithms (6.27) and (6.32) into two-valued Boolean logic. No higher degree of parallelism over that shown in Section 6.3 will be considered in this section.

First, a decision is to be made on the format of the digitized membership function. It will be assumed that the degree of membership function is a discrete valued function with a five-element domain set (it can be extended up to eight using a three-bit binary code) [11]. The universe of discourse of a fuzzy subset is limited to a finite set of 25 elements. A graphic interpretation [10] of fuzzy sets $X^{(1)}$, $X^{(2)}$ and Z is given in Figure 6.6.

In case of two-valued logic, three bits are used to represent each element of the set. Hence, 75 bits are used to digitize the membership function. The values of the parameters to obtain the memory requirement of the knowledge base and the number of operators are: $n = 5(8)$, $k_1 = k_2 = 25$.

The functional block diagram of the fuzzy model/fuzzy inference unit for SISO systems is given in Figure 6.7. This unit is considered the key segment of the pipelined fuzzy logic controller (FLC) hardware accelerator that will be outlined in Section 6.4.3. The direct data path from the host interface to the rule memory is not shown for the sake of simplicity.

The process of building up the knowledge base can be described as follows: after registers XI and ZI have been loaded, adding a new rule to the rule set (overall rule R) takes k_1 (currently 25) clock periods, where k_1 is the dimension of the fuzzy set XI. The MUX2 multiplexer at the input of the minimum unit selects the ZI register. During the first clock period, u_1 (of XI) is paired with all w elements of ZI and these pairs are fed to the inputs of the minimum unit. If the current rule is the first one in the sequence of constructing the linguistic model of a process, throughout k_1 cycles 0 (non-membership) elements will be paired with the outputs of the minimum unit and be fed to the inputs of the maximum unit. Throughout the construction of the first rule, MUX3 will direct the zero vector to the inputs of the maximum unit. The whole word of maximum values is stored at the first location of the R rule memory, during the first clock period. During the jth clock period, u_j is compared to all w elements of ZI simultaneously, and then the vector of maximum values is stored in the jth location of R.

If the current rule is not the first one in the sequence, the MUX3 multiplexer at the input of the maximum unit selects the ith row of R ($1 \leq i \leq k_1$) during the ith clock period, and the contents of this row in R will be updated from the outputs of the maximum unit.

Hence, constructing a knowledge base of N rules takes $N \times k_1$ clock periods in case of the architecture shown in Figure 6.7. The clock periods needed to load registers XI and ZI are ignored at this point.

The memory requirement of the model is just $M = 3 \times 25 \times 25 = 1875$ bits which is rather low (less than 1/4 of a kilobyte) no matter what the number of rules used to build up the model.

Computing the fuzzy inference also takes k_1 clock periods. In this case, the MUX2 multiplexer at the input of the minimum unit pairs the u_i element of XI with all r^j-elements of the ith row in R. If $i = 1$ (first clock period), then the MUX3 multiplexer at the input of the maximum unit selects 0 as the other operand for each element at the output of the minimum unit. Alternatively, the Z register should be cleared to 0 (non-membership) value prior to the inference cycle. The outputs of the maximum unit are fed to the inputs of the Z register. During clock periods 2 to k_1, the outputs of the Z register are fed back to the inputs of the maximum unit through the MUX3 multiplexer. The contents of the R rule memory remain unchanged during the fuzzy inference process. After the last clock period,

CONCEPTUAL HARDWARE IMPLEMENTATION 189

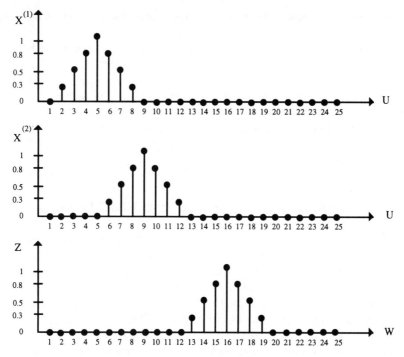

Figure 6.6 Graphic interpretation of fuzzy sets $X^{(1)}$, $X^{(2)}$ and Z

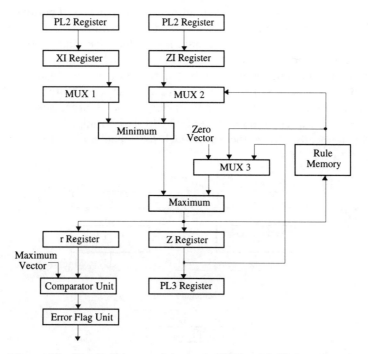

Figure 6.7 Block diagram of the basic SISO model/inference unit

the register Z holds the result of the $X \circ R$ max–min composition operation in the digitized fuzzy data format.

To detect whether the error condition (6.30) has been met, an error flag (EF) is added to the fuzzy engine. If the status of the error flag is a logical '1' after a new rule has been added to the rule base, then all elements of R are of 1 (full membership) grade. The error flag can be used to generate an interrupt request to the host machine. The system can then recover from this erronous state by either downloading a 'safe' linguistic model to the R memory or starting over the model building process with a new rule set.

Another SISO model/inference unit with a higher degree of parallelism will be illustrated in Section 6.5.

6.4.2 Hardware Architectures for MISO and MIMO Systems

A MISO linguistic model is decomposed into a set of SISO sub-models (M sub-models for M inputs), as shown in Section 6.2 (equations (6.9)–(6.14)). The fuzzy conclusion is obtained by using the min-superposition of all the outputs inferred from the sub-models (equations (6.15) and (6.16)). Hence, a MISO FLC can be constructed from a set of simultaneously working SISO FLCs. For inference computations, a binary tree of minimum units is connected to the outputs of the SISO units. The block diagram of the model/inference section of the MISO FLC is given in Figure 6.8.

The most challenging issues are related to the hardware implementation of MIMO fuzzy controllers. Of the few published solutions, a systolic array was suggested by Manzoul and Tayal [12]. Taylor *et al.* proposed the fuzzy automata machine (FAMe) [13] which is a massively parallel cellular automata based on reconfigurable logic devices (RLDs). The RLDs are programmed for fuzzy logic operations. Although the FAMe approach is attractive, the final performance is less than one might expect because the cellular array is not optimal for the problem. On the grounds of equations (6.22)–(6.24), in order to carry out simultaneous evaluation of all the components of each parallel output, $M \times K$ processing elements (PEs) are needed for an M-input/K-output MIMO FLC. In the case of a smaller number of inputs and/or outputs, some processing elements would not be used. Each PE works as a separate SISO FLC [6], equipped with its own rule base memory which locally stores the sub-model $R^{(ij)}$. The architecture for the whole

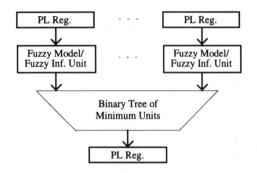

Figure 6.8 Hardware architecture of MISO FLC

MIMO FLC, called the directed data stream (DDS) architecture, was proposed by Patyra [5].

6.4.3 Fuzzy Controller Hardware Accelerator

The architecture of a fuzzy logic controller hardware accelerator that performs fuzzy model building, inference and defuzzification computations will be presented in this section [15]. This FLC hardware accelerator does not include a fuzzifier unit, and the remaining units are integrated in a common pipelined structure. The accelerator consists of three basic units: the host interface, the combined fuzzy model/inference unit and the defuzzifier unit. To achieve a high processing rate for real-time applications, the units are connected in a three-level functional pipeline as it is illustrated in Figure 6.9.

The pipelined architecture allows the simultaneous operation of the three functional units. The speed characteristics will be formulated in Section 6.5. The interface unit is customized to the host machine which can be either a PC, a workstation or any digital equipment. The host machine supplies fuzzy input data to the hardware accelerator in digitized membership format. For stand-alone applications (i.e. intelligent sensor or process controller) the interface unit is replaced by some process interface (i.e. A/D and D/A converters) and a fuzzifier unit.

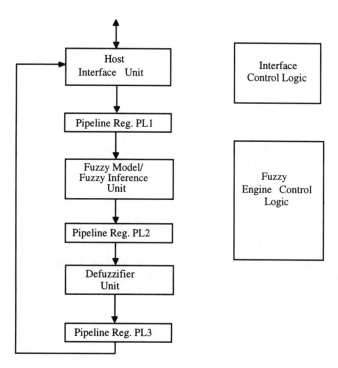

Figure 6.9 Block diagram of FLC hardware accelerator

6.5 PERFORMANCE CHARACTERISTICS

As illustrated in Figure 6.9, the three function units of the hardware accelerator form a pipeline. The subtasks are as follows:

    T1: I/O data transfer from/to the host machine
    T2: adding a new rule to the knowledge base or performing a fuzzy inference computation
    T3: computing a defuzzification strategy

In the case of model building, pipeline step T3 is omitted. Once the pipeline is filled, the hardware accelerator produces new fuzzy and/or crisp output data at a rate which is determined by the longest time needed to complete any pipeline subtask. For the further discussion, subtask T2 will be considered the critical one.

6.5.1 Maximum Sustainable Processing Rate

On the basis of algorithms (6.27) and (6.32), elements of one row of the R matrix are processed in parallel, so it will take n clock periods to complete this subtask. However, there is no data dependency among rows of the R matrix due to the commutative and distributive properties of the min and max operations, hence, the degree of parallelism can be increased to the level where all rows are processed in parallel. Hence, the maximum sustainable rate of fuzzy logic operations per second is given by the formula

$$\frac{1}{NT} = \frac{1}{([n \div k] + p)T} \qquad (6.33)$$

where n stands for the number of elements in the universe of discourse for fuzzy subset X (inputs), k rows of the R matrix (overall rule) are processed in parallel, p is a constant value ($p = 0$, if $k = 1$, and $p = 1$ or 2 if $k > 1$) and T stands for the clock period time. For example, by quadrupling the internal units of the basic fuzzy model/inference unit, the time required to complete this pipeline subtask either for model building or an inference computation can be reduced to

$$[25 \div 4] + 1(2) = 8(9)$$

clock periods. The value of p, that is, the propagational delay through the two-level Max-tree in terms of clock periods, and the clock rate can be determined by gate-level simulation in case of VLSI implementation. The block diagram of a quadruple configuration of the model/inference (M/I) unit for a SISO FLC is shown in Figure 6.10.

It should be noted that the total R memory requirement of the configuration shown in Figure 6.10 remains unchanged in respect to the basic SISO engine given in Figure 6.7. One quarter of the rows of matrix R are stored in rule memories R_a, R_b, R_c and R_d, respectively.

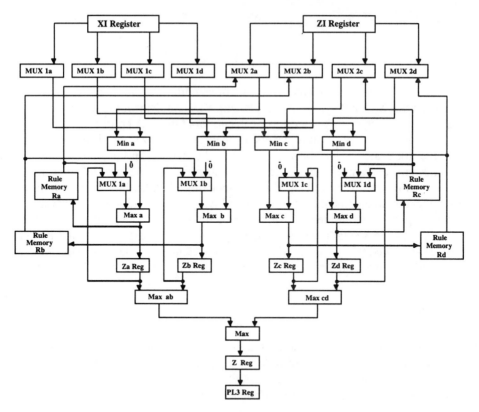

Figure 6.10 Four-way parallel configuration of SISO M/I engine

6.5.2 Improvements

To improve the processing rate of the M/I engine even further, fuzzy logic optimization methods [16] are to be combined with parallel processing of the rows of the R rule memory. The basic thought is that either for model building or inference computations, only those elements of fuzzy set X (input) need to be considered whose degrees of membership are greater than 0. Due to the properties of the min and max operations used for both model building and inference computations, the elements of zero degree of membership (non-membership) will neither contribute any value to the knowledge base nor play any role in inferring a fuzzy consequence. If the first and last elements of the support of the membership function for X are marked by hardware pointers (this information can be provided by either the host machine or the fuzzifier unit with a stand-alone configuration), then just the corresponding rows of R will be processed. Hence, in (6.33), n will be reduced to just the number of elements making up the support, rather than the total number of elements in the universe of discourse for X. Assuming that the support of X does not exceed eight elements and a eight-way parallel M/I inference engine ($n=k$) is available, the ideal peak processing rate that is given by the formula

$$\frac{1}{NT} = \frac{1}{(1+p)T} \qquad (6.34)$$

can be achieved. The biggest challenge is then how to implement that many wide parallel data paths in the M/I unit in case of a single-chip VLSI implementation. Focusing on just the support of X helps to alleviate this problem, too. Only a support-wide parallel data path is, needed, along with multiplexers, instead of a full-size one. There is a penalty, though the clock rate needs to be lowered due to the increased propagation delay via the additional data path multiplexers for X. To remedy this problem, the clock rate can further be improved by using behavioural pipelining [17], i.e. breaking up long delay paths in the functional pipeline stages and employing as many additional hardware units working in parallel as needed. Assuming a comparatively conservative clock rate at 30 MHz, and letting p equal 1, the M/I unit can perform at approximately 16 MFLIPS rate.

It should be noted that to sustain such a high processing rate, the whole pipeline should be balanced, otherwise another unit's lower performance will become a bottleneck.

6.6 CONCLUSIONS

Mathematical models of fuzzy model building and inference computations for SISO, MISO and MIMO fuzzy logic controllers were formulated. By analyzing these models, a new parallel algorithm and its conceptual hardware implementation were presented. The characteristic features of the algorithm are as follows:

a high degree of parallelism to perform fuzzy logic operations
a constant low memory requirement to store the complete knowledge base, no matter what the mumber of rules used to construct it
an error flag for the prompt detection if the linguistic model has failed by adding a new rule to the knowledge base
the same hardware architecture can be used for both model building and inference computations
the algorithm can be used in SISO, MISO and MIMO systems.

A formula was given to evaluate the performance of the hardware accelerator. By combining parallel processing with fuzzy logic optimization methods, it was also shown that significant further improvement can be achieved in the FLC's performance.

REFERENCES

1 Zadeh, L. A. (1975) The concept of a linguistic variable and its application to approximate reasoning—I, II and III. *Information Sciences*, **8**, 199–249; **8**, 301–357; **9**, 43–80.
2 Lee, C. C. (1990) Fuzzy logic in control systems: fuzzy logic controller—Part I, Part II. *IEEE Transactions on Systems, Man and Cybernetics*, **20**, 404–418; 419–435.
3 Zadeh, L. A. (1973) Outline of a new approach to the analysis of complex systems and decision process. *IEEE Transactions on Systems Man and Cybernetics*, **3**, 28–44.

4. Yager, R. R. (1991) Connectives and quantifiers in fuzzy sets. *Fuzzy Sets and Systems*, **40**, 39–75.
5. Patyra, M., and Grantner, J. (1993) Hardware implementation issues of multivariable fuzzy control systems. Invited paper to the *Third International Conference on Industrial Fuzzy Control and Intelligent Systems (IFES'93)*, College Station, Texas, USA, pp. 179–184.
6. Grantner, J., and Patyra, M. (1993) VLSI implementation of fuzzy logic finite state machines. *IFSA'93 World Congress*, Seoul, Korea, pp. 781–784.
7. Gupta, M. M., Kiszka, J. B., and Trojan, G. M. (1986) Multivariable structure of fuzzy control systems. *IEEE Transactions on Systems, Man and Cybernetics*, **16**, 638–656.
8. Kosko, B. (1992) *Neural Networks and Fuzzy Systems* (Prentice Hall).
9. Koczy, L. T., and Hirota, K. (1992) A fast algorithm for fuzzy inference by compact rules. In L. A. Zadeh and J. Kacprzyk (Eds) *Fuzzy Logic for the Management of Uncertainty* (John Wiley).
10. Stachowicz, M., and Kochanska, M. (1982) Graphic interpretation of fuzzy sets and fuzzy relations. *Second World Conference on Mathematics at the Service of Man*, Universidad Politechnica de Las Palmas, Las Palmas, Spain, pp. 620–629.
11. Stachowicz, M. S., Grantner, J., and Kinney, L. L. (1991) Two-valued logic for linguistic data acquisition. *Workshop of the North American Fuzzy Information Processing Society (NAFIPS)*, University of Missouri-Columbia, pp. 168–172.
12. Manzoul, M. A., and Tayal, S. (1990) Systolic VLSI array for multi-variable fuzzy control systems. *International Journal of Cybernetics and Systems*, **21**, 27–42.
13. Howard, N., Taylor, R., and Allison, N. (1992) The design and implementation of a massively-parallel-fuzzy architecture. *IEEE-FUZZ'92 Conference*, San Diego, pp. 545–552.
14. Grantner, J. (1994) Design of event-driven real-time linguistic models based on fuzzy logic finite state machines for high-speed intelligent fuzzy logic controllers. Dissertation for the Degree Candidate of Technical Science, Hungarian Academy of Sciences, Budapest.
15. Stachowicz, M. S., Grantner, J., and Kinney, L. L. (1992) Pipelined processing of linguistic data published in *AMSE Review* with the following reference: *Advances in Modeling and Analysis*, B, 1992, **23**, 1–4.
16. Chiueh, T.-C. (1991) Optimization of fuzzy logic implementation. *21st International Symposium on Multiple Valued Logic*, Victoria, Canada, pp. 348–355.
17. Arato, P., Beres, I., Rucinski, A., Davis, R., and Torbert, R. (1993) The SAM high-level datapath synthesis method for pipelined custom application specific integrated circuits. *CCC'93 Conference*, Budapest, Hungary, pp. 149–158.

7
Fuzzy Flip-flop

Kazuhiro Ozawa
Faculty of Economics, Hosei University, Japan

Kaoru Hirota
Department of Systems Science, Interdisciplinary Graduate School of Science and Technology, Tokyo Institute of Technolgy, Japan

Laszlo T. Koczy
Department of Communication Electronics, Technical University of Budapest, Hungary

7.1 INTRODUCTION

A great deal of research had been directed towards the realization of a 'fuzzy computer'. A few types of fuzzy processors which can perform various fuzzy operations, e.g. fuzzy negation, min, max and fuzzy inference, have been proposed and realized by Yamakawa [1,2] and Togai [3]. These fuzzy inference chips have opened new opportunities for artificial intelligence. However, all of them were based on single-step fuzzy inference. In oder to realize multi-stage fuzzy inference, fuzzy memory modules are indispensable. In the case of an 'ordinary' computer, a binary flip-flop circuit, which can memorize single bits of information has been widely used as a fundamental element of memory modules.

In this chapter we propose and define a fuzzy flip-flop which is an extended form of an ordinary, binary flip-flop, specifically a J-K flip-flop. A truth table for a J-K flip-flop is fuzzified, where binary NOT, AND and OR operations are extended to fuzzy negation, t-norm, and s-norm, respectively. Two types of fundamental, characteristic equations of fuzzy flip-flop are introduced. They are reset type and set type equations, both of which are fuzzy extensions of a characteristic equation of a J-K flip-flop. Results of their characteristics are demonstrated graphically, especially in the case in which fuzzy negation, t-norm and s-norm relate to: complementation, min and max operations; complementation, algebraic product, and algebraic sum operations, respectively. Both reset and set type equations are unified in the case of complementation, min and max operations, and complementation, algebraic product and sum operations, and fundamental characteristic equations of both type of fuzzy flip-flops are introduced. The possibility of hardware implementation of fuzzy flip-flop circuits is also proposed using fuzzy gate circuits; moreover, hardware circuits in voltage mode and discrete mode of both types are presented and tested with illustrations. Through such experiments, a possibility of VLSI implementation is confirmed for the purpose of realizing fuzzy memory modules in the framework of a multi-stage fuzzy inference computer. Finally, an idea of fuzzy shift register which can store and shift the membership function information defined on a finite support set is presented, and a VLSI design of the fuzzy register as a fuzzy memory module is presented using a VHDL compiler and simulator; optimization of VLSI of the fuzzy register circuit is analyzed.

7.2 OUTLINE OF BINARY FLIP-FLOP AND FUNDAMENTAL FUZZY OPERATIONS

7.2.1 A Binary Logic J-K Flip-Flop

A flip-flop circuit, especially a J-K flip-flop that can memorize a single bit of information, has been of great use in memory modules of computer hardware. The next state $Q(t+1)$ of a J-K flip-flop is characterized as a function of both the present state $Q(t)$ and the present two inputs $J(t)$ and $K(t)$ (cf. a truth table of a J-K flip-flop in Table 7.1). A simplified notation J, K and Q is sometimes used instead of $J(t)$, $K(t)$ and $Q(t)$, respectively, in what follows. The minterm expression of $Q(t+1)$ is expressed as

$$Q(t+1) = \bar{J}\bar{K}Q + J\bar{K}\bar{Q} + J\bar{K}Q + JK\bar{Q} \tag{7.1}$$

and it is simplified as

$$Q(t+1) = J\bar{Q} + \bar{K}Q \tag{7.2}$$

which is well known as a characteristic equation of a J-K flip-flop. On the other hand, another mutually equivalent maxterm expression can be given by

$$Q(t+1) = (J+K+Q)(J+\bar{K}+Q)(J+\bar{K}+\bar{Q})(\bar{J}+\bar{K}+\bar{Q}) \tag{7.3}$$

Table 7.1 Truth-table of J-K flip-flop

J	K	$Q(t)$	$Q(t+1)$
0	0	0	0
0	0	1	1
0	1	0	0
0	1	1	0
1	0	0	1
1	0	1	1
1	1	0	1
1	1	1	0

OUTLINE OF BINARY FLIP-FLOP AND FUNDAMENTAL FUZZY OPERATIONS

which can be simplified as

$$Q(t+1) = (J+Q)(\bar{K}+\bar{Q}) \tag{7.4}$$

The equivalence of (7.2) and (7.4) can be easily confirmed using such well-known properties as double negation, de Morgan's law, commutative law, associative law, distributive law, absorption law and complemented law, among which it should be noted that the last law, which is not valid in fuzzy logic, is used.

$$\begin{aligned}
J\bar{Q} + \bar{K}Q &= \overline{\overline{J\bar{Q} + \bar{K}Q}} \\
&= \overline{\overline{J\bar{Q}} \cdot \overline{\bar{K}Q}} \\
&= \overline{(\bar{J} + \bar{\bar{Q}}) \cdot (\bar{\bar{K}} + \bar{Q})} \\
&= \overline{(\bar{J} + Q) \cdot (K + \bar{Q})} \\
&= \overline{\bar{J}K + \bar{J}\bar{Q} + KQ + Q\bar{Q}} \\
&= \overline{\bar{J}K + \bar{J}\bar{Q} + KQ} \tag{7.5} \\
&= \overline{\bar{J}K(Q + \bar{Q}) + \bar{J}\bar{Q} + KQ} \\
&= \overline{\bar{J}Q + \bar{J}\bar{Q}K + KQ + KQ\bar{J}} \\
&= \overline{\bar{J}\bar{Q} + KQ} \\
&= \overline{\bar{J}\bar{Q}} \cdot \overline{KQ} \\
&= (J + Q) \cdot (\bar{K} + \bar{Q})
\end{aligned}$$

7.2.2 Definition of Fuzzy Negation, t-norm and s-norm

Every two-valued Boolean function can be expressed by finite combinations of NOT, AND, and OR operations. Together, these equations constitute a complete system of Boolean algebra. In the case of fuzzy logic, they are extended to include the concepts of fuzzy negation, *t*-norm and *s*-norm, respectively. They are summarized as follows.

7.2.2.1 Fuzzy negation

$$\cdot^{\circledast} : [0,1] \to [0,1] \tag{7.6}$$

$$N1: 0^{\circledast} = 1 \tag{7.7}$$

$$N2: (A^{\circledast})^{\circledast} = A \tag{7.8}$$

$$N3: A < B \Rightarrow A^{\circledast} > B^{\circledast} \tag{7.9}$$

Example:

$$\text{complementation} \quad A^{\circledR} = 1 - A \quad (7.10)$$

7.2.2.2 t-norm

$$\circled{T}: [0,1] \times [0,1] \to [0,1] \quad (7.11)$$

$$\text{T1: } A\circled{T}1 = A, A\circled{T}0 = 0 \quad (7.12)$$

$$\text{T2: } A < B \Rightarrow A\circled{T}C < B\circled{T}C \quad (7.13)$$

$$\text{T3: } A\circled{T}B = B\circled{T}A \quad (7.14)$$

$$\text{T4: } A\circled{T}(B\circled{T}C) = (A\circled{T}B)\circled{T}C \quad (7.15)$$

Example:

$$\text{logical product } A \wedge B = \min\{A, B\} \quad (7.16)$$

$$\text{algebraic product } A \cdot B = AB \quad (7.17)$$

$$\text{bounded product } A \cdot B = 0 \vee (A + B - 1) \quad (7.18)$$

$$\text{drastic product } A \wedge B = \begin{cases} A & \cdots & B = 1 \\ B & \cdots & A = 1 \\ 0 & \cdots & A, B < 1 \end{cases} \quad (7.19)$$

7.2.2.3 s-norm (t-conorm)

$$\circled{S}: [0,1] \times [0,1] \to [0,1] \quad (7.20)$$

$$\text{S1: } A\circled{S}1 = 1, A\circled{S}0 = A \quad (7.21)$$

$$\text{S2: } A < B \Rightarrow A\circled{S}C < B\circled{S}C \quad (7.22)$$

$$\text{S3: } A\circled{S}B = B\circled{S}A \quad (7.23)$$

$$\text{S4: } A\circled{S}(B\circled{S}C) = (A\circled{S}B)\circled{S}C \quad (7.24)$$

Example:

$$\text{logical sum} \quad A \vee B = \max\{A, B\} \quad (7.25)$$

$$\text{algebraic sum} \quad A \dotplus B = A + B - AB \quad (7.26)$$

$$\text{bounded sum} \quad A \oplus B = 1 \wedge (A + B) \tag{7.27}$$

$$\text{drastic sum} \quad A \mathbin{\dot{\vee}} B = \begin{cases} A & \cdots & B = 0 \\ B & \cdots & A = 0 \\ 1 & \cdots & A, B > 0 \end{cases} \tag{7.28}$$

The dual relations t-norm and s-norm with respect to fuzzy negation, which are generalized forms of de Morgan's law, are usually composed as

$$(A^{\circledR} \circledt B^{\circledR})^{\circledR} = A \circleds B \tag{7.29}$$

or equivalently

$$(A^{\circledR} \circleds B^{\circledR})^{\circledR} = A \circledt B \tag{7.30}$$

(Each of the logical operations (7.16) and (7.25), algebraic operations (7.17) and (7.26), bounded operations (7.18) and (7.27), and drastic operations (7.19) and (7.28) together with complementation (7.10) satisfy (7.29) and (7.30).) Moreover, the following distributive law is sometimes assumed:

$$A \circledt (B \circleds C) = (A \circledt B) \circleds (A \circledt C) \tag{7.31}$$

or

$$A \circleds (B \circledt C) = (A \circleds B) \circledt (A \circleds C) \tag{7.32}$$

(In the case of logical product and logical sum, these equations hold true.) From a practical point of view, complementation, min and max are usually used as standard operations of fuzzy negation, t-norm, and s-norm, respectively. This relationship exists because: the system (composed of complementation, min and max) easily allows for an understanding of the physical meaning; and the system constitutes a complete pseudo-Boolean algebra. So most of the properties in binary logic hold true except for the complemented law.

7.3 DEFINITION OF FUZZY FLIP-FLOP

In this section, the definition of a fuzzy flip-flop is presented. Using fuzzy negation, t-norm and s-norm, we can extend (7.2) and obtain (7.33)

$$Q_R(t+1) = (J \circledt Q^{\circledR}) \circleds (K^{\circledR} \circledt Q) \tag{7.33}$$

In the same way, (7.4) is extended to obtain (7.34)

$$Q_S(t+1) = (J \circleds Q) \circledt (K^{\circledR} \circleds Q^{\circledR}) \tag{7.34}$$

As mentioned in (7.5) it is clear that equation (7.2) is equal to equation (7.4). In fuzzy logic, however, equation (7.33) dose not always equal equation (7.34) because the complemented law and the distributive law do not hold true. However the following proposition can be satisfied.

Theorem 1 If the following relation concerning fuzzy negation, t-norm and s-norm is valid

$$A \oplus (B \otimes C) \geq (A \oplus B) \otimes (A \oplus C) \tag{7.35}$$

then

$$(J \otimes Q) \oplus (K^\circledR \otimes Q^\circledR) \geq (J \oplus Q^\circledR) \otimes (K^\circledR \oplus Q) \tag{7.36}$$

i.e. $Q_S(t+1) \geq Q_R(t+1)$ \hfill (7.37)

will be obtained.

Proof

$$(J \otimes Q) \oplus (K^\circledR \otimes Q^\circledR)$$
$$\geq \{J \oplus (K^\circledR \otimes Q^\circledR)\} \otimes \{Q \oplus (K^\circledR \otimes Q^\circledR)\}$$
$$\geq (J \oplus K^\circledR) \otimes (J \oplus Q^\circledR) \otimes (Q \oplus K^\circledR) \otimes (Q \oplus Q^\circledR)$$
$$= \{(J \oplus Q^\circledR) \otimes (K^\circledR \oplus Q)\} \otimes \{(J \oplus K^\circledR) \otimes (Q \oplus Q^\circledR)\}$$
$$\geq \{(J \oplus Q^\circledR) \otimes (K^\circledR \oplus Q)\} \otimes 0$$
$$= (J \oplus Q^\circledR) \otimes (K^\circledR \oplus Q) \tag{7.38}$$

Table 7.2 shows the values of equation (7.33) and equation (7.34) when J and K are restricted to two values $\{0, 1\}$. If $J = 0$, $K = 1$ (i.e. if the reset input is situated in the case of the J–K flip-flop), then the next state (7.33) is equal to 0 (i.e. reset), but that of (7.34) is $Q \oplus Q^\circledR (\geq 0)$. On the other hand, if $J = 1$, $K = 0$ (i.e. if set input is situated in the case of the J–K flip-flop), then the next state (7.34) equals 1, but that of (7.33) is $Q^\circledR \otimes Q (\leq 1)$. So we will define a reset type fuzzy flip-flop by a characteristic equation (7.33), and a set type fuzzy flip-flop by equation (7.34). (Here, it should be noted again that $Q \oplus Q^\circledR$ and $Q^\circledR \otimes Q$ are not always equal to 0 and 1, respectively, because of the lack of complemented law in fuzzy logic.)

Table 7.2 Values of (7.33) and (7.34) ($J, K \in \{0, 1\}$)

J	K	set type (3-2)	reset type (3-1)	F^2
0	0	Q	Q	Q
0	1	$Q \oplus Q^\circledR$	0	0
1	0	1	$Q^\circledR \otimes Q$	1
1	1	$Q^\circledR$	$Q^\circledR$	$\bar{Q}$

7.4 FUZZY FLIP-FLOP USING COMPLEMENTATION, MIN AND MAX OPERATIONS

In order to construct a fuzzy flip-flop circuit, we will discuss the fundamental system which is composed of complementation for fuzzy negation, min for t-norm, and max for s-norm. In this system, equation (7.33) (reset type fuzzy flip-flop) and equation (7.34) (set type fuzzy flip-flop) are expressed as

$$Q_R(t+1) = \{J \wedge (1-Q)\} \vee \{(1-K) \wedge Q\} \tag{7.39}$$

$$Q_S(t+1) = (J \vee Q) \wedge \{(1-K) \vee (1-Q)\} \tag{7.40}$$

respectively. A fuzzy flip-flop characterized by (7.39) is called a min max reset type fuzzy flip-flop, and that of (7.40) a min max set type fuzzy flip-flop.

Figure 7.1 shows the next states of the $Q_R(t+1)$ of the reset type fuzzy flip-flop in the case of $Q(t) = 0, 0.25, 0.5, 0.75$ and 1.0, respectively.

Similarly, the next states $Q_S(t+1)$ of the set type fuzzy flip-flop are shown in Figure 7.2. In the case of $Q(t) = 0$ and $Q(t) = 1$, the values of the next state $Q(t+1)$ in both types are equal. If $0 < Q(t) < 1$, then $Q(t+1)$ of the set type is of greater or equal value than that of the reset type. This fact will be easily confirmed by Theorem 1. Because the complementation, min and max operations satisfy the distributive law, equation (7.37) holds true.

However, it should be noted that the values $Q(t+1)$ of both types are continuously connected on the line segment $J = K$. This can be analytically proven as follows.

In the case $J = K < Q$ (7.34) is expressed as

$$\{J \wedge (1-Q)\} \vee \{(1-J) \wedge Q\} \tag{7.41}$$

So

$$Q > J \geq J \wedge (1-Q), (1-J) > (1-Q) \geq J \wedge (1-Q) \tag{7.42}$$

should hold, and

$$(1-J) \wedge Q > J \wedge (1-Q) \tag{7.44}$$

is obtained. Thus, equation (7.39) is equal to

$$(1-J) \wedge Q \tag{7.55}$$

On the other hand, equation (7.40) is clearly equal to

$$Q \wedge (1-J) \tag{7.56}$$

thus both equation (7.39) and (7.40) are the same.

In the case $J = K = Q$, clearly (7.39) and (7.40) are equal to the same value as

$$J \wedge (1-J) \tag{7.57}$$

In the case $J = K > Q$, (7.39) is expressed as

$$\{J \wedge (1-Q)\} \vee \{(1-J) \wedge Q\} \tag{7.58}$$

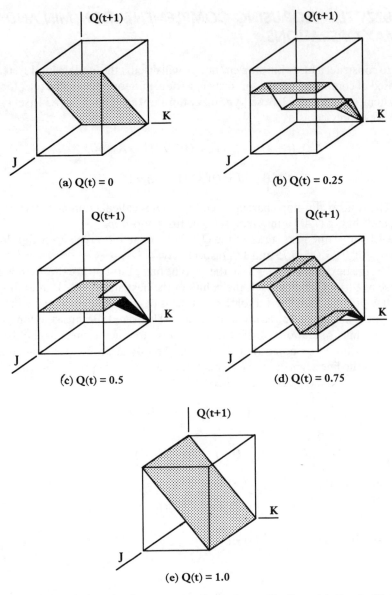

Figure 7.1 Characteristics of min max reset type fuzzy flip-flop: (a) $Q = 0$, (b) $Q = 0.25$, (c) $Q = 0.5$, (d) $Q = 0.75$, (e) $Q = 1.0$

Here

$$J > Q \geq (1 - J) \wedge Q \tag{7.59}$$

$$(1 - Q) > (1 - J) \geq (1 - J) \wedge Q \tag{7.60}$$

and

$$J \wedge (1 - Q) > (1 - J) \wedge Q \tag{7.61}$$

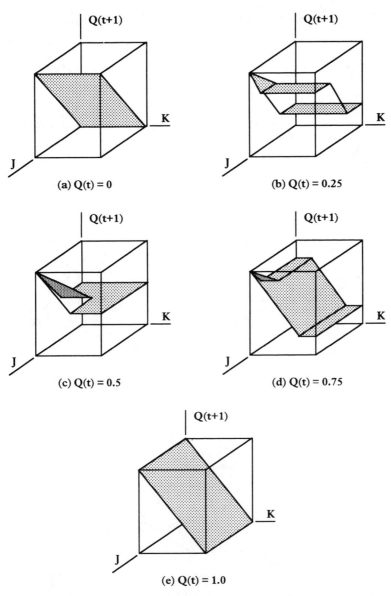

Figure 7.2 Characteristics of min max set type fuzzy flip flop: (a) $Q=0$, (b) $Q=0.25$, (c) $Q=0.5$, (d) $Q=0.75$, (e) $Q=1.0$

thus (7.39) is equal to

$$J \wedge (1-Q) \tag{7.62}$$

On the other hand, (7.40) is clearly equal to

$$J \wedge (1-Q) \tag{7.63}$$

So they are the same.

Therefore, in order to extend the binary J-K flip-flop to a fuzzy flip flop smoothly, we

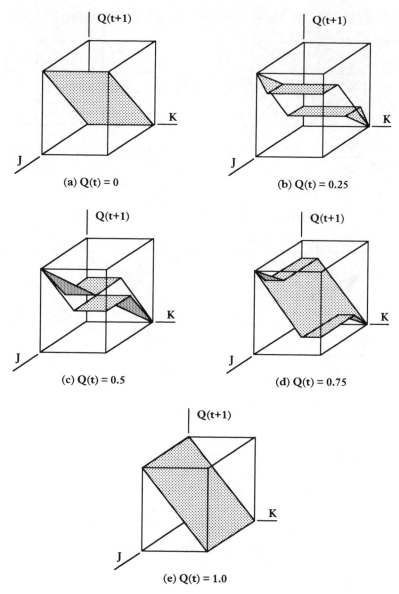

Figure 7.3 Characteristics of min max fuzzy flip-flop: (a) $Q = 0$, (b) $Q = 0.25$, (c) $Q = 0.5$, (d) $Q = 0.75$, (e) $Q = 1.0$

must introduce the newly defined function

$$Q(t+1) = \begin{cases} (J \vee Q) \wedge \{(1-K) \vee (1-Q)\} & (J \geq K) \\ \{J \wedge (1-Q)\} \vee \{(1-K) \wedge Q\} & (J \leq K) \end{cases} \quad (7.64)$$

Equation (7.64) is defined as the fundamental equation of the min max type fuzzy flip-flop. Figure 7.3 shows values of equation (7.64) in the case of $Q(t) = 0, 0.25, 0.5, 0.75$ and 1.0.

7.5 FUZZY FLIP-FLOP USING COMPLEMENTATION, ALGEBRAIC PRODUCT AND ALGEBRAIC SUM

In the case of complementation for fuzzy negation, and algebraic product and algebraic sum for t-norm and s-norm, respectively, the characteristic equations (7.33) and (7.34) of the fuzzy flip-flop can be expressed as

$$Q_R(t+1) = \{J \cdot (1-Q)\} \dot{+} \{(1-K) \cdot Q\} \tag{7.65}$$

$$Q_S(t+1) = (J \dot{+} Q) \cdot \{(1-K) \dot{+} (1-Q)\} \tag{7.66}$$

The equations (7.65) and (7.66) can be transformed into more simplified forms by using the definition of algebraic product and algebraic sum:

$$Q_R(t+1) = J + Q - 2JQ - KQ + JQ^2 + JQK - JQK^2 \tag{7.67}$$

$$Q_S(t+1) = J + Q - JQ - JKQ - KQ^2 + JKQ^2 \tag{7.68}$$

These characteristics are demonstrated graphically in Figures 7.4 and 7.5, respectively. In these cases, the $Q(t)$ are fixed as parameters, and thus the next states $Q(t+1)$'s are expressed as second-order polynomials of J and K, and they are expressed as quadrics.
The difference between $Q_R(t+1)$ and $Q_S(t+1)$ is calculated as

$$Q_S(t+1) - Q_R(t+1) = \{J(1-K) + K(1-J)\}Q(1-Q) \geq 0 \tag{7.69}$$

so it is shown that

$$Q_R(t+1) \leq Q_S(t+1) \tag{7.70}$$

7.6 FUNDAMENTALS OF IMPLEMENTATION OF THE MIN MAX FUZZY FLIP-FLOP

The fuzzy flip-flop circuits can be constructed based on their characteristic equations (7.33) and (7.34). Figure 7.6 shows these circuits, i.e. reset type and set type of fuzzy flip-flop, using fuzzy negation, t-norm and s-norm gates.
From the viewpoint of practical applications, complementation, min and max are the most important fuzzy operations; so fuzzy negation, t-norm and s-norm are regarded as complementation, min and max, respectively, in the following. Real implementation of the complementation circuit has been proposed in [4, 5], and min and max circuits have also been proposed in [2]. Figure 7.7 shows gate circuits which have been constructed using transistors and resistors. In these circuits, fuzzy values [0, 1] are indicated in voltage levels [0 V, 5 V]. Then implementation of the fuzzy flip-flop circuit characterized by (7.64) can be realized using the complementation, min and max system; however, a comparator and electric relay circuits are necessary for this hardware implementation. Of course, this requirement is undesirable for real construction of fuzzy memory modules. Accordingly,

208 FUZZY FLIP-FLOP

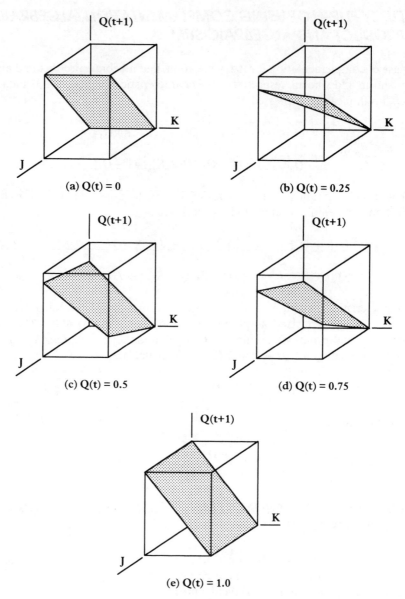

Figure 7.4 Characteristics of reset type fuzzy flip-flop using algebraic product and algebraic sum: (a) $Q = 0$, (b) $Q = 0.25$, (c) $Q = 0.5$, (d) $Q = 0.75$, (e) $Q = 1.0$

we must investigate the following equation:

$$Q(t+1) = (J \circledS K^{\circledR}) \circledT (J \circledS Q) \circledT (K^{\circledR} \circledS Q^{\circledR}) \tag{7.71}$$

This equation (7.71) has characteristics of both reset type and set type; significantly, it should be noted that (7.33), (7.34) and (7.71) are the same in two-valued logic. This

FUNDAMENTALS OF IMPLEMENTATION OF THE MIN MAX FUZZY FLIP-FLOP

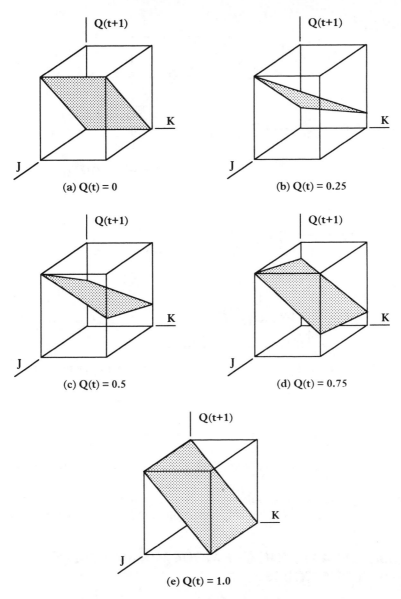

Figure 7.5 Characteristics of set type fuzzy flip-flop using algebraic product and algebraic sum: (a) $Q = 0$, (b) $Q = 0.25$, (c) $Q = 0.5$, (d) $Q = 0.75$, (e) $Q = 1.0$

expression (7.71) can be obtained by modifying the equation (7.5) in the case of the binary J-K flip-flop, i.e.

$$Q(t+1) = \overline{JK + \overline{J}Q} + KQ$$
$$= (J + \bar{K})(J + Q)(\bar{K} + \bar{Q}) \qquad (7.72)$$

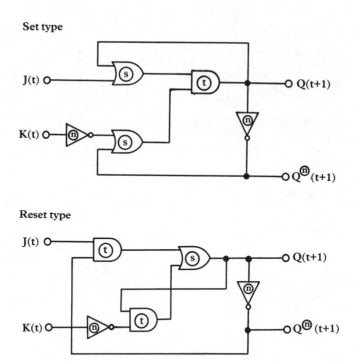

Figure 7.6 Reset type and set type fuzzy flip-flop circuit using fuzzy negation, t-norm and s-norm

In the complementation, min and max system, equation (7.71) is expressed as

$$Q(t+1) = \{J \vee (1-K)\} \wedge (J \vee Q) \wedge \{(1-K) \vee (1-Q)\} \tag{7.73}$$

One can easily confirm that equation (7.73) is equivalent to equation (7.64). So equation (7.73) is also considered the fundamental equation of the min max type fuzzy flip-flop.

7.7 DISCRETE AND VOLTAGE MODE MIN MAX FUZZY FLIP-FLOP CIRCUITS

The min max type fuzzy flip-flop circuit in discrete mode can be constructed, based on the characteristic equation (7.73). In this case the fuzzy information [0, 1] is digitized by $\{0000, 0001, \cdots, 1111\}$; the clocked min max type fuzzy flip-flop circuit can be constructed using digitized fuzzy gates circuits which are shown in Figure 7.8. The complementation circuits is constructed by four inverter gates. The min and max gate circuits are constructed based on the four bits magnitude comparator and several AND and OR gate circuits (binary gate circuits). Here all of them are C-MOS IC. Figure 7.9a shows the circuit diagram, and Figure 7.9b shows the waveform chart of this circuit. Figure 7.10 shows mounted board of this circuit. In this circuit the fuzzy information is expressed in parallel.

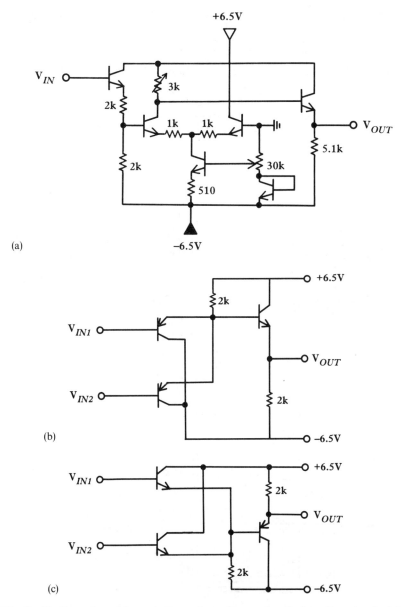

Figure 7.7 Realization of complementation, min and max circuits in voltage levels: (a) complementation circuit, (b) min circuit, (c) max circuit

Based on the equation (7.73), a voltage mode fuzzy flip-flop circuit using the complementation gate, min gate and max gate can be constructed (cf. Figure 7.11). The inputs $J(t)$ and $K(t)$ are driven by a synchronized clock pulse in the two forefront minimum circuits. The output $Q(t)$ is memorized by the two sample-and-hold circuits. These circuits run as shown in the waveform timechart (Figure 7.11b). Figure 7.12 shows the detailed circuit

212 FUZZY FLIP-FLOP

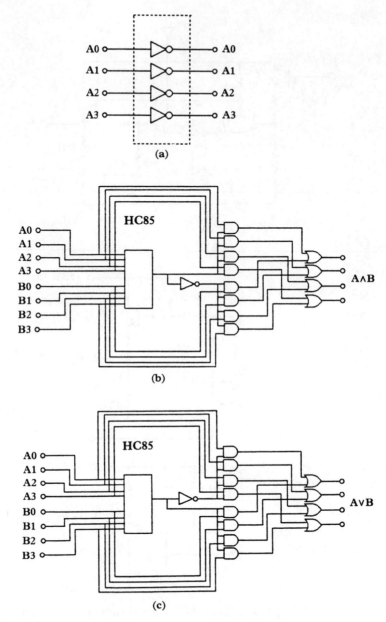

Figure 7.8 Realization of complementation, min and max circuits in discrete mode: (*a*) complementation circuit, (*b*) min circuit, (*c*) max circuit

diagram and Figure 7.13 shows the mounted board of this fuzzy flip-flop circuit. Its static characteristics have been confirmed experimentally, i.e. the same figures as Figure 7.3 have been obtained by substituting [0, 1] for [0 V, 5 V]. Figure 7.14 shows dynamic characteristics, where the voltage of the clock pulse is 5.3 V, and the clock cycle time is

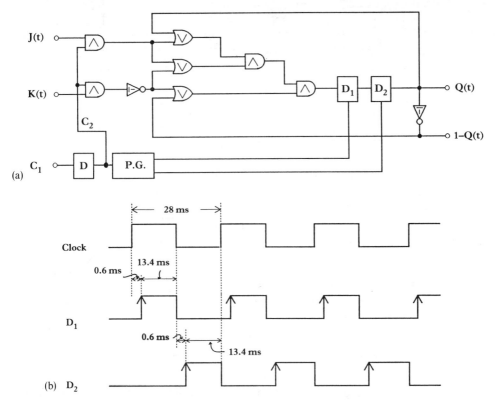

Figure 7.9 Min max type fuzzy flip-flop circuit in discrete mode: (*a*) circuit diagram, (*b*) waveform chart

Figure 7.10 Mounted board of min max type fuzzy flip-flop in discrete mode

8.2 ms. For the purpose of real VLSI implementation of fuzzy memory modules of a min max type fuzzy flip-flop, it is at least necessary to discuss several characteristics of this printed board circuit. They are the access time, temperature characteristics, electricity consumption and durability against noise.

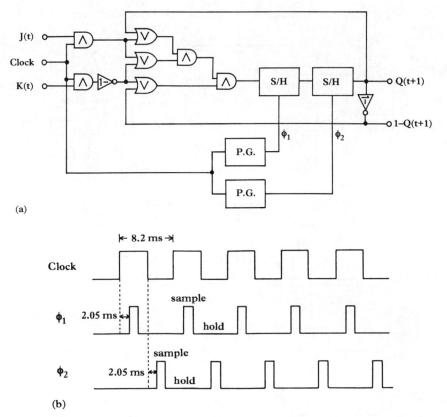

Figure 7.11 Min max type fuzzy flip-flop circuit in voltage mode: (*a*) circuit diagram, (*b*) waveform chart

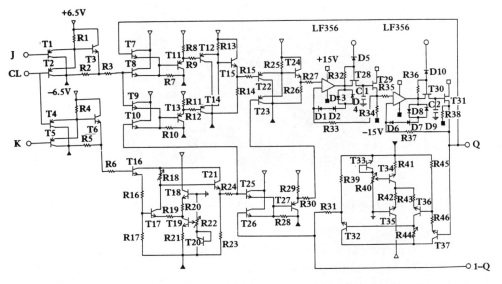

Figure 7.12 Detailed circuit diagram of min max type fuzzy flip-flop in voltage mode

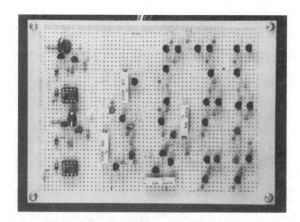

Figure 7.13 Mounted board of min max type fuzzy flip-flop in voltage mode

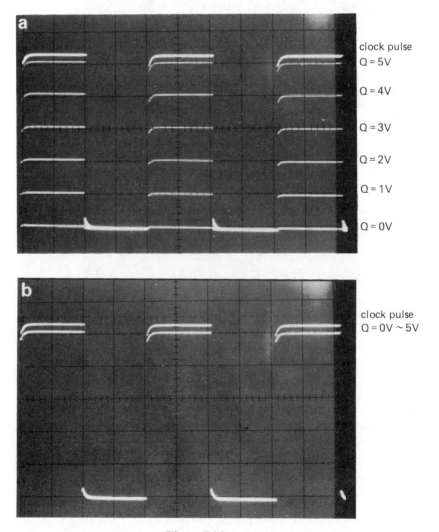

Figure 7.14

216 FUZZY FLIP-FLOP

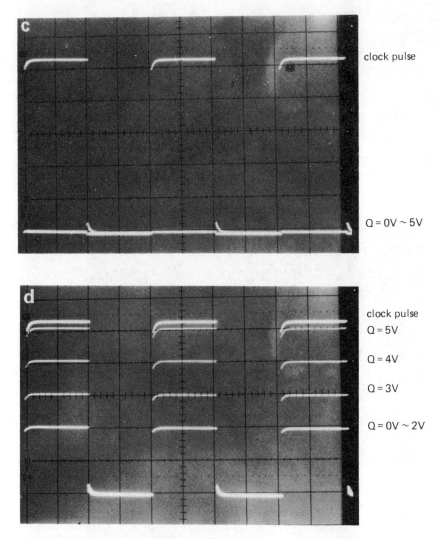

Figure 7.14 Dynamic characteristics of min max type fuzzy flip-flop: (a) $J = 0$, $K = 0$, (b) $J = 5\,\text{V}$, $K = 0$, (c) $J = 0$, $K = 5\,\text{V}$, (d) $J = 2\,\text{V}$, $K = 0$

Although the cycle time is 8.2 ms in our experiment (this is because we have no good clock pulse oscillator), the access time may be improved to the order of several hundred seconds, because the complementation circuit, min circuit, max circuit and sample and hold circuit work in less than 100 ns, 10 ns and 100 ns, respectively [5, 2, 6]. Concerning the temperature characteristic, it is confirmed that this printed board circuit works inside the temperature range 0 and 50° C. Although the electricity consumption is 1.3 W in this board, it can be improved by using C-MOS technology. It has also been confirmed that this board circuit works well in a noisy environment, e.g. near a personal computer and a fluorescent lamp.

7.8 FUNDAMENTALS OF IMPLEMENTATION OF THE ALGEBRAIC FUZZY FLIP-FLOP

In order to construct the simple hardware circuit of the algebraic fuzzy flip-flop, we would like to introduce the unified equation of reset type and set type. From equations (7.67) and (7.68) the following algebraic equation will be obtained:

$$Q(t+1) = (a_0 + a_1 J + a_2 K + a_3 JK) + (b_0 + b_1 J + b_2 K + b_3 JK)Q$$
$$+ (c_0 + c_1 J + c_2 K + c_3 JK)Q^2 \qquad (7.74)$$

Coefficients $a_0, a_1, \ldots, c_3$ can be assigned by taking the binary J-K flip-flop conditions into consideration.

In the case of memory state (i.e. $J = 0, K = 0$)

$$a_0 + b_0 Q + c_0 Q^2 = Q$$
$$a_0 = 0, \quad b_0 = 1, \quad c_0 = 0$$

In the case of reset state (i.e. $J = 0, K = 1$)

$$a_2 + (1 + b_2)Q + c_2 Q^2 = 0$$
$$a_2 = 0, \quad b_2 = -1, \quad c_2 = 0$$

In the case of set state (i.e. $J = 1, K = 0$)

$$a_1 + (1 + b_1)Q + c_1 Q^2 = 1$$
$$a_1 = 1, \quad b_1 = -1, \quad c_1 = 0$$

In the case of memory inversion state (i.e. $J = 1, K = 1$)

$$(1 + a_3) + (1 - 1 - 1 + b_3)Q$$
$$(1 + c_3)Q + c_3 Q^2 = 1 - Q$$
$$a_3 = 0, \quad b_3 = 0, \quad c_3 = 0$$

Therefore the following chracteristic equation of a algebraic fuzzy flip-flop will be obtained:

$$Q(t+1) = J + Q - JQ - KQ \qquad (7.75)$$

This equation is considered the fundamental equation of the algebraic fuzzy flip-flop. Based on the definition of the algebraic fuzzy flip-flop (equation (7.75)), the characteristics can be calculated as Figure 7.15, using computer simulation experiments. In this calcula-

218 FUZZY FLIP-FLOP

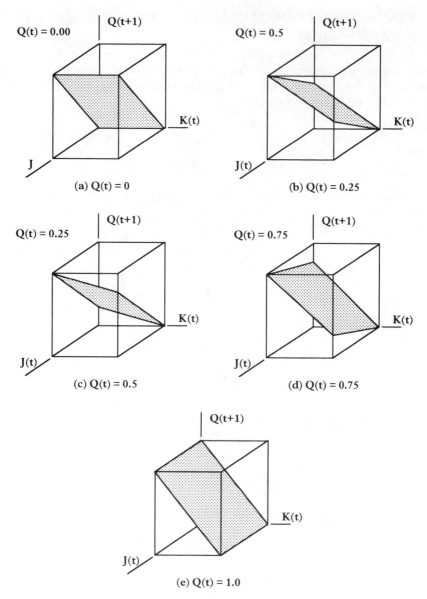

Figure 7.15 Characteristics of the algebraic fuzzy flip-flop (equation (7.75))

tion, the outputs $Q(t)$ are fixed as 0, 2.25, 0.5, 0.75 and 1.0, respectively. The algebraic fuzzy flip-flop which is defined here include the fundamental characteristics of the binary J-K flip-flop (see above). The characteristics are nicely symmetric and very simple. For example, when $J = 0.5$, $K = 0.5$, the next states $Q(t+1) = 0.5$ in all cases. This means that when J and K are maximally uncertain, i.e. $J = K = 0.5$, the next state is maximally uncertain, i.e. $Q(t+1) = 0.5$.

7.9 DISCRETE AND VOLTAGE MODE ALGEBRAIC FUZZY FLIP-FLOP CIRCUITS

Based on the characteristic equation (7.77), the algebraic fuzzy flip-flop circuit in discrete mode can be constructed. However, in order to construct the circuit in discrete mode, two multipliers and three full adders are required. Of course this requirement is undesirable for real construction of simple fuzzy memory modules. Thus the characteristic equation is modified as follows:

$$Q(t+1) = J(1-Q) + Q(1-K) \qquad (7.76)$$

This expression is desirable for the discrete mode circuit. If the fuzzy information $[0, 1]$ is digitized by $\{0000, 0001, \ldots, 1111\}$, the algebraic fuzzy flip-flop circuit can be constructed using two four bits $\times$ 4 bits binary multipliers and a four bits binary full adder (TTL IC). Figure 7.16 shows the circuit diagram of it. Figure 7.17 shows the mounted board of the discrete mode algebraic fuzzy flip-flop circuit. In this circuit the fuzzy information, expressed by four bits, is processed in parallel. It should be noted that the four bits truncation operation is performed after the multiplication operation, because four bits digitized multiplication generates eight bits information. So the results (eight bits information) is approximate to four bits information.

Secondly the voltage mode algebraic fuzzy flip-flop circuit is implemented. The fuzzy information (given by $[0, 1]$) is expressed by continuous voltage mode $[0\,V, 5\,V]$. In order to construct the simple circuit, the equation (7.75) is modified as follows:

$$Q(t+1) = J + Q - (J+K)Q \qquad (7.77)$$

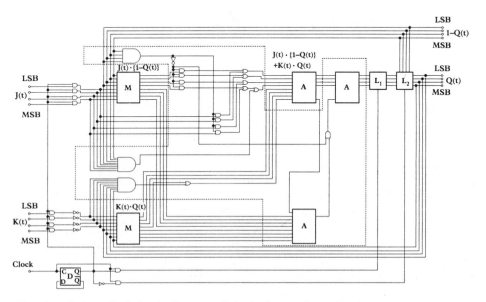

Figure 7.16 Detailed circuit diagram of algebraic type fuzzy flip-flop in discrete mode

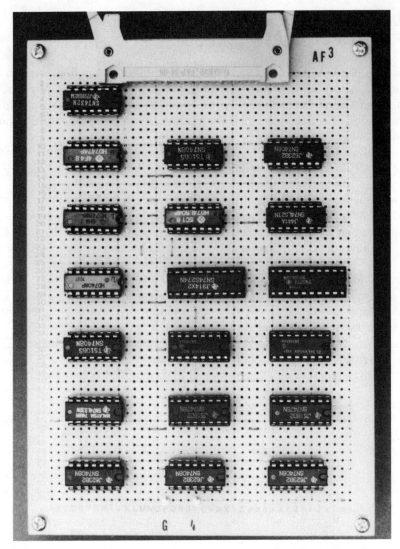

Figure 7.17 Mounted board of min max type fuzzy flip-flop in discrete mode

This expression is the empirical equation for a continuous mode algebraic fuzzy flip-flop circuit. Therefore we can constructed the circuit using a current mode multiplier and three voltage mode adders. Figure 7.18a shows the circuit diagram of the continuous mode algebraic fuzzy flip-flop circuit and Figure 7.19 shows the circuit board. After the multiple operation, the current value is transformed to the voltage value in [0 V, 5 V]. The inputs J and K are driven by a synchronized clock pulse in the two forefront minimum circuits. The output Q is memorized by the two sample and hold circuits in order to use the information in the next state. These two sample and hold circuits run as shown in the waveform timechart (Figure 7.18b). Five power sources are necessary in this circuit (5 V, 6.5 V, 15 V).

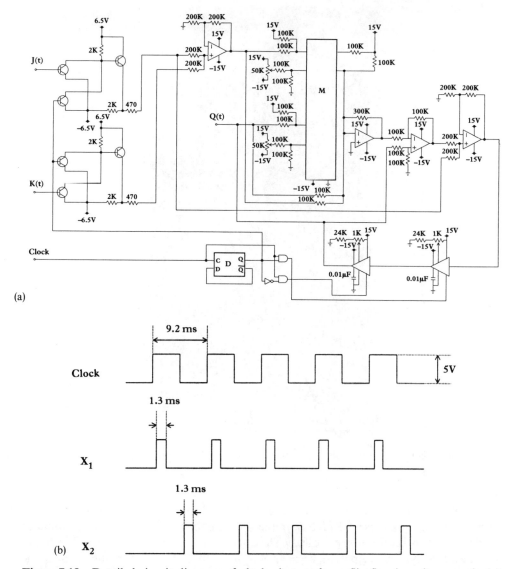

Figure 7.18 Detailed circuit diagram of algebraic type fuzzy flip-flop in voltage mode: (*a*) circuit diagram, (*b*) waveform timechart

7.10 COMPARISON OF THE PERFORMANCE OF MIN MAX TYPE VERSUS ALGEBRAIC TYPE FUZZY FLIP-FLOP CIRCUIT

We have already proposed and defined the concept of the fuzzy flip-flop, and the characteristics have also been shown using computer simulation experiment. Based on the definition and the simulation, min max fuzzy flip-flop circuits and algebraic fuzzy flip-flop circuits, in discrete mode and continuous mode have been constructed. In this section the performances of the previous four circuits will be discussed (cf. Table 7.3). In total

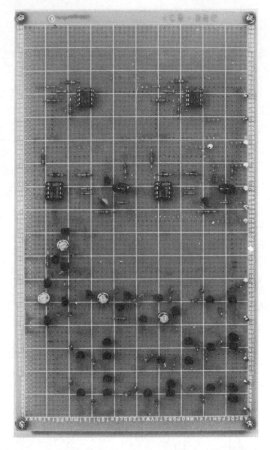

Figure 7.19 Mounted board of min max type fuzzy flip-flop in voltage mode

propagation delay, discrete mode algebraic fuzzy flip-flop is 274 ns, the most high-speed, because of using high-speed TTL ICs. In the total number of transistors, the continuous mode min max fuzzy flip-flop is the least. The discrete mode min max fuzzy flip-flop circuit has the lowest power dissipation, because of using C-MOS ICs. In continuous mode the fuzzy information [0, 1] is expressed in voltage as [0 V, 5 V], whereas in the discrete mode the fuzzy value is digitized by $\{0000, 0001, 1111\}$. Generally, discrete mode circuits, using digital IC, are superior to noise immunity; furthermore the number of power source is only one. This is the merit of fuzzy memory modules.

7.11 FUZZY REGISTER CIRCUIT

In this chapter a fuzzy register circuit is proposed as a fuzzy memory module, and its VLSI implementation is presented using VHDL (very high speed integrated circuit hardware description language) compiler and logic simulator. The detailed characteristics of the circuit are shown, i.e. the area of the circuit and the delay time.

Table 7.3 Comparison of the performance of min max type versus algebraic type fuzzy flip-flop circuits

	Algebraic F^3		Min max F^3	
	Discrete	Continuous	Discrete	Continuous
Total propagation delay	274 nsec	< 100 μsec	900 nsec	330 nsec
Total number of transistors	580 + α	< 300	2850	100
Power dissipation	2.75 W	1.2 W (Max)	1.48 mW	1.3 W
Possibility of VLSI	possible	impossible	possible	impossible
Used IC	TTL IC	linear IC (RC4200, LF356, LF398)	C-MOS IC	non
Expression of fuzzy value	4 bit parallel	voltage [0 V, 5 V]	4 bit parallel	voltage [0 V, 5 V]
Fundamental cells of the circuit	4 × 4 multipliers 4 bit full adders	analog multipliers, analog adders	1 −, min, max gates	1 −, min, max gates
Noise immunity	good	fair	good	fair
The number of power sources	1	3 pairs (+ 6.5 V, + 5 V, + 15 V)	1	2 pairs (+ 6.5 V, + 15 V)

The fuzzy flip-flop has been defined and introduced as an extended form of that of binary J-K flip-flop using the concepts of fuzzy negation, *t*-norm and *s*-norm operation. Definitions of two basic types of fuzzy flip-flops (min max type and algebraic type) are shown in the following equations: min max type

$$Q(t) = \{J \vee (1 - K)\} \wedge (J \vee Q) \wedge \{(1 - K) \vee (1 - Q)\} \quad (7.78)$$

Algebraic type

$$Q(t) = J + Q - JQ - KQ \quad (7.79)$$

where the variable $t - 1$ is omitted in the right-hand side of equations (7.78) and (7.79). The detailed characteristics of these fuzzy flip-flops have been analyzed in Sections 7.6 and 7.8, respectively. Both of these fuzzy flip-flops include four basic characteristics which are the memory of fuzzy information ($J = K = 0$), the output of the inverse fuzzy information ($J = K = 1$), the set ($J = 1, K = 0$), and the reset ($J = 0, K = 1$) fuzzy information. The fuzzy register can be designed using previous two types of fuzzy flip-flops.

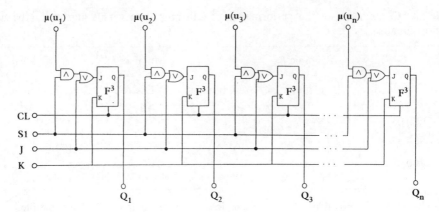

Figure 7.20. Block diagram of the fuzzy register circuit

Table 7.4 Control pulse of fuzzy register circuit

	hold	reset	set	inversion	data in
CL	↑	↑	↑	↑	↑
S1	0	0	0	0	1
J	0000	0000	1111	1111	0000
R	0000	1111	0000	1111	0000

Figure 7.20 and Table 7.4 show the fuzzy register circuit diagram and its control signal states, respectively, where F^3 shows the above-mentioned fuzzy flip-flop circuit. Gate symbols, where ∧ and ∨ are filled in the military standard symbol AND and OR, are min and max gates, respectively. The input/output of any fuzzy information come in and out from the fuzzy register through the terminal $\mu(u_i)$ and $Q_i(t)$, respectively. J, K and S1 (which is the control pulse of the circuit) are supposed to be supplied as shown in Table 7.4. CL means the clock pulse of the fuzzy register circuit. In this case the fuzzy information is digitized by four bits.

7.12 VLSI DESIGN OF THE FUZZY REGISTER CIRCUIT

7.12.1 VLSI Design of the Min Max Type Fuzzy Flip-Flop Circuit

In this section as an example a VLSI design of min max type fuzzy flip-flop as the lower cell of the fuzzy register is presented. The VLSI source of the min max type fuzzy flip-flop is designed using VHDL, and optimization of the circuit uses the VHDL compiler and simulator (by Synopsys Inc.). Figure 7.21 shows the VHDL source code of the min max type fuzzy flip-flop. The fuzzy information, each of which is expressed by four bits, are processed in parallel. It is defined as UNSIGNED (three down to zero) in the entity block

```vhdl
use WORKARITHMETIC.all;
library SYNOPSYS;
use SYNOPSYS.TYPES.all;
entity MINMAX is
  port(CLK: in BIT;
    J, K: in UNSIGNED(3 downto 0);
    Q, QN: buffer UNSIGNED (3 downto 0));
end MINMAX1;
architecture BEHAVIOR of MINMAX is
begin
  process
  variable U1, U2, U3, U4, U5, U6: UNSIGNED (3 downto 0);
  begin
    wait until CLK' event and CLK = '1';
    U4:= "1111" – K;
    if (J > U4) then
      U1:= J;
    else
      U1:= "1111" – K;
    end if;
    if (J > Q) then
      U2:= J;
    else
      U2:= Q;
    end if;
    U5:= "1111" – K;
    U6:= "1111" – Q;
    if (U5 > U6) then
      U3:= "1111" – K;
    else
      U3:= "1111" – Q;
    end if;
    if (U1 < U2) then
      if (U1 < U3) then
        Q ⇐ U1;
      else
        Q ⇐ U3;
      end if;
    else
      if (U2 < U3) then
        Q ⇐ U2;
      else
        Q ⇐ U3;
      end if;
    end if;
    QN ⇐ "1111" – Q;
  end process;
end BEHAVIOR;
```

Figure 7.21 VHDL source list of min-max type fuzzy flip-flop

226 FUZZY FLIP-FLOP

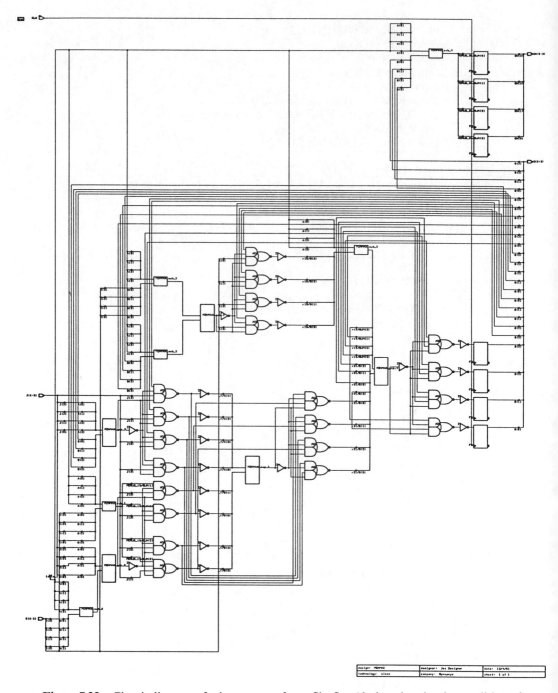

Figure 7.22 Circuit diagram of min max type fuzzy flip-flop (designed under the condition of the smallest circuit area)

in the list. The architecture block in the list describes the operation of a min max type fuzzy flip-flop. All of these functions in the process block are synchronously processed with a clock pulse (CLK). Figure 7.22 shows the circuit diagram, designed by a VHDL compiler, of the min max type fuzzy flip-flop.

This diagram is optimized under the condition of the smallest area of the VLSI. The main element of the circuit is a four bit magnitude comparator; the others are a D-type flip-flop and basic gate circuits. The detailed circuit diagram of the four bit magnitude comparator is shown in Figure 7.23. Table 7.5 shows the area information of the circuit, and Table 7.6 shows the delay information. The area of the circuit is 223.0, which is the relative value when the area of the AND gate is equal to 1.0. The time unit of the delay information is ns.

On the other hand, under the condition of the fastest delay time, the min max type fuzzy flip-flop circuit is designed in Figure 7.24. In this case the main units of the circuit are the multiplexor and four bit magnitude comparator. Table 7.7 shows the area information of the circuit, and Table 7.8 shows the delay time information. The total area is 294.0. The area is increased 1.32 times over that of the previous smallest condition. However delay times of the rise and fall edge are improved as 1.96 ns and 1.68 ns, respectively.

Finally, a part of simulation results of min max type fuzzy flip-flop circuit is shown in Table 7.9. The simulation is performed following the TEST BENCH program described by VHDL (Figure 7.2) using the VHDL simulator. In the TEST BENCH program, $J = 0000$ and $K = 1111$ are first fed into the fuzzy flip-flop, i.e. to reset the fuzzy flip-flop, then the analysis will start. The clock cycle is 20 ns. In Table 7.9, when the input/output is changed, EVENT is shown. Close agreement between simulated and calculated values was obtained in all cases.

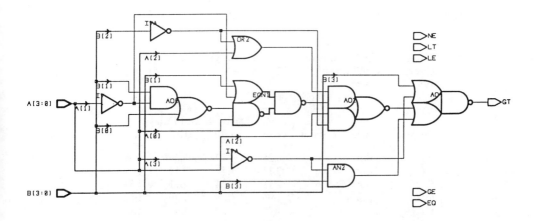

Figure 7.23 Comparator circuit in Figure 7.22

Table 7.5 Area information for Figure 7.22
```
****************************
Report : reference
Design : MINMAX
Version: v2.0b
Date    : Sat Jan 11 13:47:57 1992
****************************
```

Attributes:
  b – black box (unknown)
  bo – allows boundary optimization
  d – dont_touch
  h – hierarchical
  n – noncombinational
  r – removable
  s – synthetic module
  u – contains unmapped logic

Reference	Library	Unit Area	Count	Total Area	Attributes
A02	class	2.00	20	40.00	
FD1	class	7.00	8	56.00	n
IV	class	1.00	19	19.00	
IVA	class	1.00	2	2.00	
MINMAX_cmp_0		16.00	1	16.00	s, h
MINMAX_cmp_1		17.00	1	17.00	s, h
MINMAX_cmp_2		17.00	1	17.00	s, h
MINMAX_cmp_3		16.00	1	16.00	s, h
MINMAX_cmp_4		16.00	1	16.00	s, h
MINMAX_sub_0		4.00	1	4.00	s, h
MINMAX_sub_1		4.00	1	4.00	s, h
MINMAX_sub_2		4.00	1	4.00	s, h
MINMAX_sub_3		4.00	1	4.00	s, h
MINMAX_sub_4		4.00	1	4.00	s, h
MINMAX_sub_5		4.00	1	4.00	s, h

Total 15 references   223.00

Table 7.6 Delay time information for Figure 7.22
Operating Conditions:
Wire Loading Model:
Wire Loading Model Mode: top

Point	Type	Fanout	Max Delay rise	Max Delay fall	Min Delay rise	Min Delay fall
Q[3]	out	6	2.26	1.79	2.26	1.79
Q[1]	out	6	2.11	1.74	2.11	1.74
Q[2]	out	6	2.11	1.74	2.11	1.74
Q[0]	out	6	1.96	1.68	1.96	1.68
QN[0]	out	1	1.09	1.37	1.09	1.37
QN[1]	out	1	1.09	1.37	1.09	1.37
QN[2]	out	1	1.09	1.37	1.09	1.37
QN[3]	out	1	1.09	1.37	1.09	1.37

1
design_analyzer >

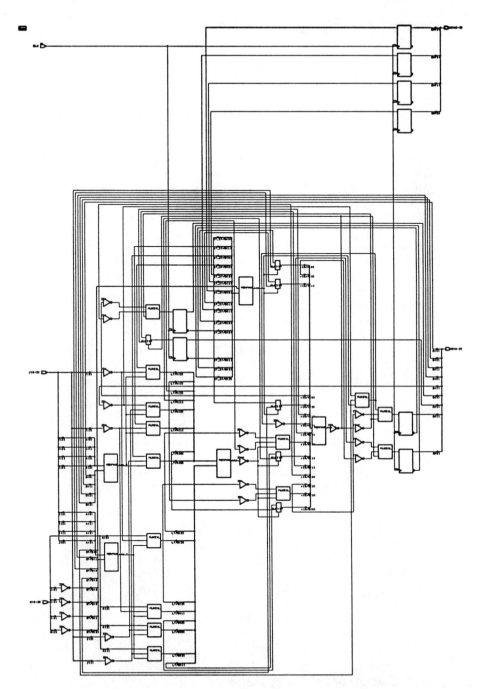

Figure 7.24 Circuit diagram of min max type fuzzy flip-flop (designed under the condition of the fastest delay time area)

Table 7.7 Area information for Figure 7.24

Reference	Library	Unit Area	Count	Total Area	Attributes
A02	class	2.00	1	2.00	
B5I	class	2.00	1	2.00	
FD1	class	7.00	8	56.00	n
IV	class	1.00	2	2.00	
IVA	class	1.00	15	15.00	
IVDAP	class	2.00	1	2.00	
IVP	class	1.00	2	2.00	
MINMAX_cmp_0		20.00	1	20.00	s, h
MINMAX_cmp_1		33.00	1	33.00	s, h
MINMAX_cmp_2		23.00	1	23.00	s, h
MINMAX_cmp_3		22.00	1	22.00	s, h
MINMAX_cmp_4		28.00	1	28.00	s, h
MINMAX_sub_0		4.00	1	4.00	s, h
MINMAX_sub_1		4.00	1	4.00	s, h
MINMAX_sub_2		4.00	1	4.00	s, h
MINMAX_sub_3		4.00	1	4.00	s, h
MINMAX_sub_4		4.00	1	4.00	s, h
MINMAX_sub_5		4.00	1	4.00	s, h
MUX21H	class	4.00	2	8.00	
MUX21HP	class	5.00	2	10.00	
MUX21L	class	3.00	15	45.00	
Total 21 references				294.00	

Table 7.8 Delay time information for Figure 7.24

Point	Type	Fanout	Max Delay rise	Max Delay fall	Min Delay rise	Min Delay fall
Q[1]	out	4	1.96	1.68	1.96	1.68
Q[0]	out	4	1.82	1.63	1.82	1.63
Q[3]	out	4	1.67	1.58	1.67	1.58
Q[2]	out	4	1.67	1.58	1.67	1.58
QN[3]	out	1	1.09	1.37	1.09	1.37
QN[2]	out	1	1.09	1.37	1.09	1.37
QN[1]	out	1	1.09	1.37	1.09	1.37
QN[0]	out	1	1.09	1.37	1.09	1.37

7.12.2 VLSI Design of the Fuzzy Register

In this section, VLSI of the fuzzy register circuit is designed. Based on the VHDL source of the fuzzy register circuits, VLSI of the fuzzy register circuit is designed as shown in Figure 7.26. Here the fuzzy register is designed using four fuzzy flip-flop circuits in order to simplify the circuit diagram. Here membership information, inputs J and K and outputs

Table 7.9 Part of the simulation results of min max type fuzzy flip-flop

10 NS		
	M:	EVENT/SYSTEM/CLOCK (value = '1')
	M3:	EVENT/SYSTEM/Q (value = X"0")
20 NS		
	M1:	EVENT/SYSTEM/J (value = X"1")
	M:	EVENT/SYSTEM/CLOCK (value = '0')
30 NS		
	M:	EVENT/SYSTEM/CLOCK (value = '1')
	M3:	EVENT/SYSTEM/Q (value = X"1")
40 NS		
	M1:	EVENT/SYSTEM/J (value = X"2")
	M:	EVENT/SYSTEM/CLOCK (value = '0')
50 NS		
	M:	EVENT/SYSTEM/CLOCK (value = '1')
	M3:	EVENT/SYSTEM/Q (value = X"2")
60 NS		
	M1:	EVENT/SYSTEM/J (value = X"3")
	M:	EVENT/SYSTEM/CLOCK (value = '0')
70 NS		
	M:	EVENT/SYSTEM/CLOCK (value = '1')
	M3:	EVENT/SYSTEM/Q (value = X"3")
	M3:	EVENT/SYSTEM/Q (value = X"7")
2000 NS		
	M1:	EVENT/SYSTEM/J (value = X"A")
	M:	EVENT/SYSTEM/CLOCK (value = '0')
2010 NS		
	M:	EVENT/SYSTEM/CLOCK (value = '1')
	M3:	EVENT/SYSTEM/Q (value = X"8")
2020 NS		
	M1:	EVENT/SYSTEM/J (value = X"B")
	M:	EVENT/SYSTEM/CLOCK (value = '0')
2030 NS		
	M:	EVENT/SYSTEM/CLOCK (value = '1')
	M3:	EVENT/SYSTEM/Q (value = X"7")
2040 NS		
	M1:	EVENT/SYSTEM/J (value = X"C")
	M:	EVENT/SYSTEM/CLOCK (value = '0')
2050 NS		
	M:	EVENT/SYSTEM/CLOCK (value = '1')
	M3:	EVENT/SYSTEM/Q (value = X"8")

Q are digitized by $\{0000, 0001, \ldots, 1111\}$, respectively. MEN1, MEN2, MEN3 and MEN4 are inputs of the membership value to the circuit and CL is the clock pulse input. Table 7.10 shows the area information of the fuzzy register circuit. In this case the total number of fuzzy flip-flops used is 11 to take into consideration the practical use. The total

```
use WORK.ARITHMETIC.all;
library SYNOPSYS;
use SYNOPSYS.TYPES.all;
entity testbench is
   port (CLK: out BIT;
      J, K: out UNSIGNED (3 downto 0)
   );
end testbench;
architecture first of testbench is
begin
stimulus: process
   variable count1: UNSIGNED (3 downto 0):= "0000";
   variable count2: UNSIGNED (3 downto 0):= "1111";
begin
   CLK <= '0';
   J <= count1;
   K <= count2;
   wait for 10 ns;
   CLK <= '1';
   wait for 10 ns;
   CLK <= '0';
   count1 := count1 + "0001";
   if (count1 = UNSIGNED' ("1111")) then
      if (count2 = UNSIGNED' ("0000")) then
      wait;
      end if;
      count1 := "0000";
      count2 := count2 - "0001";
   end if;
end process stimulus;
end first;
```

Figure 7.25 Test bench program for the simulation of the min max type fuzzy flip-flop

Table 7.10 Area information of fuzzy register circuit using 11 min max type fuzzy flip-flops

Attributes:
   b – black box (unknown)
  bo – allows boundary optimization
   d – dont_touch
   h – hierarchical
   n – noncombinational
   r – removable
   s – synthetic module
   u – contains unmapped logic

Reference	Library	Unit Area	Count	Total Area	Attributes
IVA	class	1.00	1	1.00	
MINMAX1		223.00	11	2453.00	h, n
ND2	class	1.00	92	92.00	
Total 3 references				2546.00	

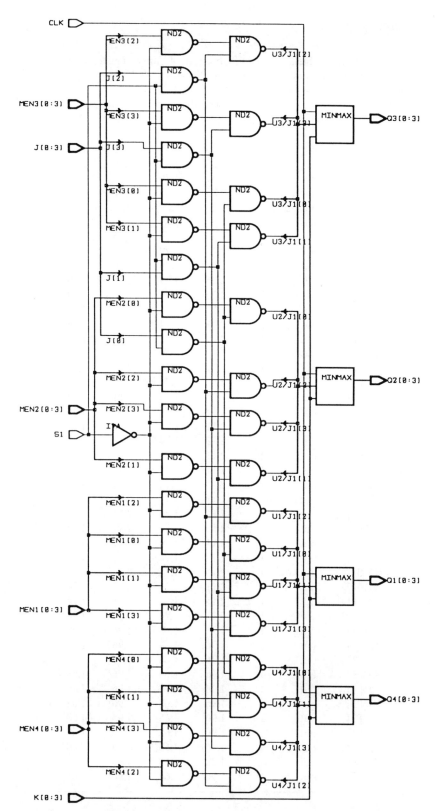

Figure 7.26 Fuzzy register circuit using min max type fuzzy flip-flop

Table 7.11 Delay time information of fuzzy register circuit using 11 min max type fuzzy flip-flops

Point	Type	Fanout	Max Delay		Min Delay	
			rise	fall	rise	fall
Q1[0]	out	1	2.26	1.79	2.26	1.79
Q2[0]	out	1	2.26	1.79	2.26	1.79
Q3[0]	out	1	2.26	1.79	2.26	1.79
Q4[0]	out	1	2.26	1.79	2.26	1.79
Q5[0]	out	1	2.26	1.79	2.26	1.79
Q6[0]	out	1	2.26	1.79	2.26	1.79
Q7[0]	out	1	2.26	1.79	2.26	1.79
Q8[0]	out	1	2.26	1.79	2.26	1.79
Q9[0]	out	1	2.26	1.79	2.26	1.79
Q10[0]	out	1	2.26	1.79	2.26	1.79
Q11[0]	out	1	2.26	1.79	2.26	1.79
Q1[1]	out	1	2.11	1.74	2.11	1.74
Q1[2]	out	1	2.11	1.74	2.11	1.74
Q2[1]	out	1	2.11	1.74	2.11	1.74
Q2[2]	out	1	2.11	1.74	2.11	1.74
Q3[1]	out	1	2.11	1.74	2.11	1.74
Q3[2]	out	1	2.11	1.74	2.11	1.74
Q4[1]	out	1	2.11	1.74	2.11	1.74
Q4[2]	out	1	2.11	1.74	2.11	1.74
Q5[1]	out	1	2.11	1.74	2.11	1.74
Q5[2]	out	1	2.11	1.74	2.11	1.74
Q6[1]	out	1	2.11	1.74	2.11	1.74
Q6[2]	out	1	2.11	1.74	2.11	1.74
Q7[1]	out	1	2.11	1.74	2.11	1.74
Q7[2]	out	1	2.11	1.74	2.11	1.74
Q8[1]	out	1	2.11	1.74	2.11	1.74
Q8[2]	out	1	2.11	1.74	2.11	1.74
Q9[1]	out	1	2.11	1.74	2.11	1.74
Q9[2]	out	1	2.11	1.74	2.11	1.74
Q10[1]	out	1	2.11	1.74	2.11	1.74
Q10[2]	out	1	2.11	1.74	2.11	1.74
Q11[1]	out	1	2.11	1.74	2.11	1.74
Q11[2]	out	1	2.11	1.74	2.11	1.74
Q1[3]	out	1	1.96	1.68	1.96	1.68
Q2[3]	out	1	1.96	1.68	1.96	1.68
Q3[3]	out	1	1.96	1.68	1.96	1.68
Q4[3]	out	1	1.96	1.68	1.96	1.68
Q5[3]	out	1	1.96	1.68	1.96	1.68
Q6[3]	out	1	1.96	1.68	1.96	1.68
Q7[3]	out	1	1.96	1.68	1.96	1.68
Q8[3]	out	1	1.96	1.68	1.96	1.68
Q9[3]	out	1	1.96	1.68	1.96	1.68
Q10[3]	out	1	1.96	1.68	1.96	1.68
Q11[3]	out	1	1.96	1.68	1.96	1.68

area is 2546.0. The fuzzy flip-flop (min max type) as the fundamental unit of fuzzy register accounts for 96.3 percent of all area of the circuit. The remaining 0.7 percent is the fundamental binary gate circuits for the control pulse selection. Table 7.11 shows the delay time information of the present fuzzy register circuit. The maximum delay times are 2.26 ns and 1.79 ns at rise and fall, respectively. In this section we used the min max type fuzzy flip-flop as the fundamental unit of the fuzzy register; as a matter of course it can be realized using other types of fuzzy flip-flop.

7.13 CONCLUSION

The fuzzy flip-flop that can memorize fuzzy information was proposed and defined. Fuzzy negation, t-norm and s-norm (t-conorm) operations were used to establish two types of characteristic equation, a reset and set type fuzzy flip-flop. Computer simulations of each type were illustrated to show the peculiar characteristics of fuzzy negation, t-norm and s-norm, being the complement, min, max, and complement, algebraic product, algebraic sum, respectively. Reset and set type equations were unified, and the fundamental equations for the hardware circuit were defined in the min max and algebraic system, respectively. The possibility of the hardware implementation of fuzzy flip-flop circuits was also proposed using fuzzy gate circuits; moreover, hardware circuits in voltage mode and discrete mode of both types were presented and tested with illustrations. Finally, a VLSI design of fuzzy register was presented, and optimization of the VLSI was discussed using a VHDL compiler and a logic simulator. Detailed characteristics of the circuit (area information and delay time) were also discussed. Moreover, it is shown that the recursive fuzzy inference can be realized at the hardware level using the fuzzy register.

Section 7.12 of this work has been supported by the Laboratory for International Fuzzy Engineering Research (LIFE).

REFERENCES

1. Yamakawa, T., Miki, T., and Ueno, F. (1985) The design and fabrication of the current mode fuzzy logic semi-custom IC in the standard CMOS IC technology. *Proc. 1985 ISMVL (IEEE)*, pp. 76–82.
2. Yamakawa, T. (1986) High-speed fuzzy controller hardware system. *Proc. 2nd Fuzzy Systems Symposium* (by IFSA Japan Chapter), Tokyo, pp. 122–130.
3. Togai, M., and Watanabe, H. (1985) A VLSI implementation of fuzzy inference engine toward an expert system on a chip. *Proc. 2nd Int. Conf. on AI and Applications (IEEE)*, pp. 192–197.
4. Marinos, P. N. (1969) Fuzzy logic and its application to switching systems. *IEEE Transactions on Computers*, **18**, 343–348.
5. Yamakawa, T., Inoue, T., Ueno, F., and Shirai, Y. (1980) Implementation of fuzzy logic hardware systems—three fundamental arithmetic circuits. *Transactions of The Institute of Electronics and Communication Engineers Japan*, 720–721 (in Japanese).
6. Gray, J. R., and Kitsopoulos, C. (1964) A precision sample and hold circuit with subnanosecond switching. *IEEE Transactions on Circuit Theory*, **11**, 389–396.
7. Hirota K., and Ozawa, K. (1989) The concept of fuzzy flip-flop. *IEEE Transactions on Systems, Man and Cybernetics*, **19**, 980–997.

8. Hirota, K., and Ozawa, K. (1987) Fuzzification of flip-flop based on various fuzzy operations. *Bulletin of the College of Engineering, Hosei University*, No. 23, 70–94.
9. Hirota, K., and Ozawa, K. (1987) Concept of fuzzy flip-flop. *2nd IFSA Congress*, Tokyo, pp. 556–559.
10. Hirota, K., Ozawa, K., Koczy, L. T., and Omori, K. (1988) Discrete mode algebraic fuzzy flip-flop circuit. *Proc. International Workshop on Fuzzy System Applications*, Iizuka, Japan, pp. 39–40.
11. Hirota, K., and Ozawa, K. (1988) Fuzzy flip-flop circuit using digital technique. *Proc. 4th Fuzzy System Symposium*, Tokyo, pp. 115–120.
12. Hirota, K., and Ozawa, K. (1989) Fuzzy flip-flop and fuzzy registers. *Fuzzy Sets and Systems*, **32**, 139–148.

8
Design Automation of Fuzzy Logic Circuits

L. Lemaitre
Integrated Circuit Design Center, Swiss Federal Institute of Lausanne, Switzerland

8.1 INTRODUCTION

The recent explosion of interest in consumer products based on fuzzy logic technology indicates a need for a new generation of microchips. Applications based on fuzzy logic need to implement complex algorithms [3, 12, 17, 18, 19, 34, 35, 42]. Some analogue and digital solutions leading to VLSI implementation of fuzzy units have been proposed during the last eight years. However, it appears to us that no real study resulting in the development of well-suited design techniques of fuzzy units has been published. This issue has been the leading idea of the work presented here.

In Section 8.2 we show that a special fuzzy operator, the so-called bounded difference denoted by $\ominus$, plays a key role in the definition of most of the fuzzy operators. More specifically, we observe that any kind of fuzzy formula can be expressed by means of this operator. We then give some relationships between the bounded difference and the common fuzzy operators.

In Section 8.3, starting from the analogue implementation of the bounded difference, we focus on the analogue current-mode implementation of fuzzy logic circuits. The use of current mirrors as fuzzy logic building blocks perfectly matches this design technique; however, it requires a complete overhaul of classical VLSI circuit design philosophy. That fact stimulated the elaboration of a computer-aided automatic layout system, in which a new strategy for the CMOS implementation of fuzzy logic units is proposed.

Section 8.4 presents the main problems that face such a design system. They relate to the problem of placing a set of basic building blocks on a Manhattan grid and routing a set of connections between the blocks.

8.2 BASIC FUZZY OPERATORS

8.2.1 Terminology and Resolution Principle

This section aims at selecting from among all the fuzzy operators those offering the best reliability to build a design strategy that must be as much as possible comparable to the classical digital design strategies. Finding a solution to this issue will allow us to apply well-known digital automation tricks to the design of fuzzy logic circuits. Therefore, we review the mechanism that permits us to generalize Boolean concepts to fuzzy ones: the so-called Resolution Principle [45]. By applying the Resolution Principle to special families of Boolean sets (the cut families), we remember that most of the common fuzzy operators can be viewed as the fuzzy extension of Boolean operators. We stress, however, that the definition of the fuzzy complement needs to reverse the order of the index of a cut family. The reverse ordering leads to careful examination of the definition of fuzzy inclusion. This examination gives rise in a natural manner to the definition of a new operator, the bounded difference. The bounded difference can serve as the basic building block for the synthesis of a complex fuzzy operator. At the same time this careful examination will clarify the problem of what is for us an integration on silicon of fuzzy logic-based algorithms.

First let us remind ourselves of the basic terminology used in the framework of fuzzy set and fuzzy logic.

A fuzzy subset $\tilde{\mathscr{A}}$ of a given universe $\mathscr{U}$ is a set of all the elements of $\mathscr{U}$ to which is assigned a real number ranging between 0 and 1 [44]. One writes symbolically

$$\tilde{\mathscr{A}} = \{(x, \mu_{\tilde{\mathscr{A}}}(x)) \text{ such that } (x \in \mathscr{U}) \quad \text{and} \quad \mu_{\tilde{\mathscr{A}}}(x) \in [0, 1]\} \tag{8.1}$$

where the variable x describes the universe $\mathscr{U}$ and where the so-called membership function $\mu_{\tilde{\mathscr{A}}}(x)$ takes on x, a real number belonging to the closed interval $[0, 1]$.

A family of crisp sets denoted $\{\mathscr{A}_\alpha\}$, indexed by the real number set $[0, 1]$ and satisfying the abutment property 'if $\alpha \subseteq \beta$ then $\mathscr{A}_\beta \subseteq \mathscr{A}_\alpha$', is a cut family.

It is straightforward to note that to any fuzzy set is attached the corresponding cut family $\{\tilde{\mathscr{A}}_\alpha\}$, where

$$\tilde{\mathscr{A}}_\alpha = \{x \in \mathscr{U}, \text{ such that } \mu_{\tilde{\mathscr{A}}}(x) \geq \alpha\} \tag{8.2}$$

Let now see how the definition of common fuzzy operators can be retrieved by applying the Resolution Principle [49] to the above cut family.

Concerning the Resolution Principle, let $\tilde{\mathscr{A}}$ be a fuzzy set and $\{\tilde{\mathscr{A}}_\alpha\}_{\alpha \in [0,1]}$ its cut family. If the fuzzy sets with constant membership function α are denoted by $\alpha \mathscr{U}$, then $\tilde{\mathscr{A}}$ is equal to the infinite union of the $\alpha \mathscr{U} \cap \tilde{\mathscr{A}}_\alpha$ sets. In other words we have

$$\tilde{\mathscr{A}} = \bigcup_{\alpha \in [0,1]} \{\alpha \mathscr{U} \cap \tilde{\mathscr{A}}_\alpha\} \tag{8.3a}$$

or pointwisely

$$\mu_{\tilde{\mathscr{A}}}(x) = \sup_{\alpha \in [0,1]} \{\alpha \text{ such that } x \in \tilde{\mathscr{A}}_\alpha\} \tag{8.3b}$$

Hence fuzzy sets whose cut families are $\{\tilde{\mathscr{A}}_\alpha \cup \tilde{\mathscr{B}}_\alpha\}$ and $\{\tilde{\mathscr{A}}_\alpha \cap \tilde{\mathscr{B}}_\alpha\}$, respectively, are equal to the fuzzy union $\tilde{\mathscr{A}} \cup \tilde{\mathscr{B}}$ and the fuzzy intersection $\tilde{\mathscr{A}} \cap \tilde{\mathscr{B}}$. In other words the following relations hold:

$$\tilde{\mathscr{A}} \cap \tilde{\mathscr{B}} = \bigcup_{\alpha \in [0,1]} \alpha(\tilde{\mathscr{A}}_\alpha \cap \tilde{\mathscr{B}}_\alpha) \qquad (8.4a)$$

$$\tilde{\mathscr{A}} \cup \tilde{\mathscr{B}} = \bigcup_{\alpha \in [0,1]} \alpha(\tilde{\mathscr{A}}_\alpha \cup \tilde{\mathscr{B}}_\alpha) \qquad (8.4b)$$

Note that in relations (8.4a) and (8.4b) the operators of the right-hand side expressions are Boolean, whereas the nature of the operators on the left-hand side is properly fuzzy.

As previously said, the family $\{\overline{\tilde{\mathscr{A}}_\alpha}\}$ is not a cut family. Actually only $\{\overline{\tilde{\mathscr{A}}_{(1-\alpha)}}\}$ defines a cut family. It is easy to prove that this family of crisp sets satisfies the abutment condition, i.e. 'If $\alpha \leq \beta$ then $\overline{\tilde{\mathscr{A}}_{(1-\beta)}} \subseteq \overline{\tilde{\mathscr{A}}_{(1-\alpha)}}$'. The negation of the fuzzy set is then defined to be equal to

$$\overline{\tilde{\mathscr{A}}} = \bigcup_{\alpha \in [0,1]} \alpha \overline{\tilde{\mathscr{A}}_{(1-\alpha)}} \qquad (8.4c)$$

The operation of reversal ordering performed to define the fuzzy complement plays a major role in the way that the notion of fuzzy inclusion is introduced.

8.2.2 Fuzzy Inclusion as the Natural Extension of Boolean Inclusion

The concept of fuzzy inclusion combined to the concept of fuzzy relation is the fundamental tool of fuzzy reasoning [45, 46]. Fuzzy inclusion can be viewed as an extension of the ordinary Boolean inclusion. However, we need to pay attention when generalizing the Boolean inclusion operator to fuzzy sets. In Boolean set theory there are several equivalent ways to express that the ordinary set $\mathscr{A}$ is included in the ordinary set $\mathscr{B}$, or in short: $\mathscr{A} \subseteq \mathscr{B}$. If $\mathscr{A}$ and $\mathscr{B}$ are subsets of the universe $\mathscr{U}$ the most usual ways are:

$$\mathscr{A} \subseteq \mathscr{B} \text{ if and only if } \mathscr{A} \cap \overline{\mathscr{B}} = \varnothing \qquad (8.5a)$$

or

$$\mathscr{A} \subseteq \mathscr{B} \text{ if and only if } \mathscr{A} \cap \mathscr{B} = \mathscr{A} \qquad (8.5b)$$

The crisp set $\mathscr{A} \cap \overline{\mathscr{B}}$ contains the elements of $\mathscr{A}$ that are not in $\mathscr{B}$; that is the reason it is often referred to as the difference set of $\mathscr{A}$ and $\mathscr{B}$. We write formally

$$\mathscr{A} \ominus \mathscr{B} = \mathscr{A} \cap \overline{\mathscr{B}} \qquad (8.6)$$

It may then be natural generalize the inclusion to fuzzy sets by saying that 'the fuzzy set $\tilde{\mathscr{A}}$ is included in the fuzzy set $\tilde{\mathscr{B}}$ if and only if all their cuts satisfy one the four above conditions'. Formally we write

$$\tilde{\mathscr{A}} \subseteq \tilde{\mathscr{B}} \text{ if and only if } \tilde{\mathscr{A}}_\alpha \cap \overline{\tilde{\mathscr{B}}_\alpha} = \varnothing, \text{ for all } \alpha \in [0,1] \qquad (8.7a)$$

or

$$\tilde{\mathscr{A}} \subseteq \tilde{\mathscr{B}} \text{ if and only if } \tilde{\mathscr{A}}_\alpha \cap \tilde{\mathscr{B}}_\alpha = \tilde{\mathscr{A}}_\alpha, \text{ for all } \alpha \in [0,1] \quad (8.7b)$$

Let us see if we can use either (8.7a) or (8.7b) to define the notion of fuzzy inclusion by means of the resolution principle. We noted in the last section that the Boolean family $\{\overline{\tilde{\mathscr{B}}_\alpha}\}$ is not a cut family. So, only the condition given by the relation (8.7b) can serve to extend consistently the concept of Boolean inclusion to fuzzy sets.

In other words, by use of the resolution principle we can say that the fuzzy set $\tilde{\mathscr{A}}$ is included in the fuzzy set $\tilde{\mathscr{B}}$ if and only if the following relation is satisfied:

$$\tilde{\mathscr{A}} \cap \tilde{\mathscr{B}} = \tilde{\mathscr{A}} \quad (8.8)$$

In terms of the membership function this is equivalent to saying that for all $x \in \mathscr{U}$

$$\mu_{\tilde{\mathscr{A}}}(x) \wedge \mu_{\tilde{\mathscr{B}}}(x) = \mu_{\tilde{\mathscr{A}}}(x) \quad (8.9)$$

or

$$\mu_{\tilde{\mathscr{A}}}(x) \leq \mu_{\tilde{\mathscr{B}}}(x), \text{ for all } x \in \mathscr{U} \quad (8.10)$$

Concerning the bounded difference, let us introduce the new fuzzy set denoted by $\tilde{\mathscr{A}} \ominus \tilde{\mathscr{B}}$ and defined by the membership function

$$\mu_{\tilde{\mathscr{A}} \ominus \tilde{\mathscr{B}}}(x) = \max(\mu_{\tilde{\mathscr{A}}}(x) - \mu_{\tilde{\mathscr{B}}}(x), 0), \text{ for all } x \in \mathscr{U} \quad (8.11a)$$

Hence, the condition '$\tilde{\mathscr{B}}$ contains $\tilde{\mathscr{A}}$' is satisfied if and only if $\tilde{\mathscr{A}} \ominus \tilde{\mathscr{B}}$ is the null set of $\mathscr{U}$, i.e.

$$\tilde{\mathscr{A}} \ominus \tilde{\mathscr{B}} = \emptyset_\mathscr{U} \quad (8.11b)$$

The fuzzy set $\tilde{\mathscr{A}} \ominus \tilde{\mathscr{B}}$ is often called the bounded difference of $\tilde{\mathscr{A}}$ and $\tilde{\mathscr{B}}$ [20, 43]. We have

$$x \ominus y = \max(x - y, 0) \text{ and } \mu_{\tilde{\mathscr{A}} \ominus \tilde{\mathscr{B}}}(x) = \mu_{\tilde{\mathscr{A}}}(x) \ominus \mu_{\tilde{\mathscr{B}}}(x) \quad (8.12)$$

Note that in the first equation of (8.12) the operator $\ominus$ is a real dyadic function defined onto the set of all real numbers $\mathscr{R}$, since in the second equation the same symbol refers to an operation between fuzzy sets. For the sake of simplicity (and if the following convention does not lead to misunderstanding) we will often use the same name to refer to a real dyadic operator or to refer to the fuzzy set that results from the application of the dyadic operator.

It is common to introduce the two following fuzzy sets that describe in a simpler manner the condition of inclusion of a given fuzzy set in another one.

The bounded sum and the bounded product of two fuzzy sets $\tilde{\mathscr{A}}$ and $\tilde{\mathscr{B}}$ are, respectively, the fuzzy sets $\tilde{\mathscr{A}} \oplus \tilde{\mathscr{B}}$ and $\tilde{\mathscr{A}} \odot \tilde{\mathscr{B}}$, membership functions are provided by the two following mathematical functions:

$$\mu_{\tilde{\mathscr{A}} \oplus \tilde{\mathscr{B}}}(x) = \min(\mu_{\tilde{\mathscr{A}}}(x) + \mu_{\tilde{\mathscr{B}}}(x), 1) \quad (8.13a)$$

$$\mu_{\tilde{\mathscr{A}} \odot \tilde{\mathscr{B}}}(x) = \max(\mu_{\tilde{\mathscr{A}}}(x) + \mu_{\tilde{\mathscr{B}}}(x) - 1, 0) \quad (8.13b)$$

BASIC FUZZY OPERATORS 241

Operators $\oplus$ and $\odot$ are not distributive to each other, but they repetitively satisfy the property of the s-norm and t-norm [20, 49]. We will see later that this property of non-distributivity makes difficult the VLSI implementation of fuzzy units. The condition of inclusion of fuzzy set $\tilde{\mathcal{A}}$ in fuzzy set $\tilde{\mathcal{B}}$ may be rewritten as

$$\tilde{\mathcal{A}} \subseteq \tilde{\mathcal{B}} \text{ if and only if } \tilde{\mathcal{A}} \odot \overline{\tilde{\mathcal{B}}} = \emptyset_{\mathcal{U}} \qquad (8.14a)$$

$$\tilde{\mathcal{A}} \subseteq \tilde{\mathcal{B}} \text{ if and only if } \overline{\tilde{\mathcal{A}}} \oplus \tilde{\mathcal{B}} = \mathcal{U} \qquad (8.14b)$$

We have moreover the counterpart of the relation (8.6) concerning the definition of set difference, i.e.

$$\tilde{\mathcal{A}} \ominus \tilde{\mathcal{B}} = \tilde{\mathcal{A}} \odot \overline{\tilde{\mathcal{B}}} \qquad (8.15)$$

Note the analogy between (8.5a) and (8.14a), or (8.6) and (8.15). Analogical reasoning between the Boolean structure and the fuzzy structure leads to the first important result: generalizing to fuzzy sets Boolean expressions containing the complement operator needed to work in the $\oplus - \ominus$ system instead of the $\vee - \wedge$ system (the symbols $\vee$ and $\wedge$ denote the maximum and minimum operators.)

This result forces us to overhaul the traditional view point of fuzzy logic circuit design (mostly based on the integration of the max and min operators.) We deduce, moreover, the following rule to generalize Boolean formulae to fuzzy sets: Boolean expressions $\tilde{x} \cup \tilde{y}$ and $\tilde{x} \cap \tilde{y}$ must be replaced by $\tilde{x} \oplus \tilde{y}$ and $\tilde{x} \odot \tilde{y}$.

The bounded difference plays an important role when modelling the fuzzy inclusion. A more surprising result is given by the fact that basic fuzzy operations may be formulated as the compound of this operator.

If $\tilde{\mathcal{A}}$ and $\tilde{\mathcal{B}}$ are two given fuzzy sets the following relations are verified.

$$\overline{\tilde{\mathcal{A}}} = \mathcal{U} \ominus \tilde{\mathcal{A}} \qquad (8.16)$$

$$\tilde{\mathcal{A}} \cap \tilde{\mathcal{B}} = \tilde{\mathcal{B}} \ominus (\tilde{\mathcal{B}} \ominus \tilde{\mathcal{A}}) \qquad (8.17)$$

$$\tilde{\mathcal{A}} \cup \tilde{\mathcal{B}} = (\mathcal{U} \ominus \tilde{\mathcal{A}}) \ominus (\tilde{\mathcal{B}} \ominus \tilde{\mathcal{A}}) \qquad (8.18)$$

$$\tilde{\mathcal{A}} \odot \tilde{\mathcal{B}} = \tilde{\mathcal{B}} \ominus (\mathcal{U} \ominus \tilde{\mathcal{A}}) \qquad (8.19)$$

$$\tilde{\mathcal{A}} \oplus \tilde{\mathcal{B}} = \overline{\overline{\tilde{\mathcal{B}}} \ominus \tilde{\mathcal{A}}} \qquad (8.20)$$

We deduce the second importat result: The bounded difference can generate any of the basic operations proposed by Zadeh in his seminal paper of fuzzy set theory.

This result is the cornerstone of the automatic design method of fuzzy units that will be introduced in Section 8.4. The bounded difference drastically lowers the difficulty for implementing the fuzzy logic architecture and it serves as the basic building block of our fuzzy logic environment.

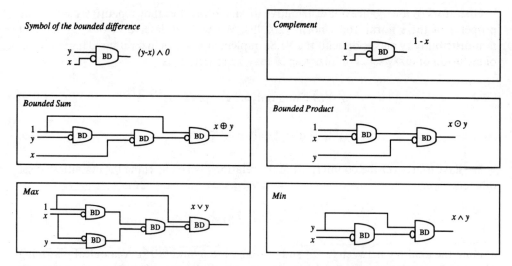

Figure 8.1 Symbolic implementation of the fuzzy operators represented by means of graphs

8.2.3 Symbolic Implementation of Fuzzy Operators

From the above subsection, if we are able to implement the bounded difference then we will theoretically be able to implement any other fuzzy operators. Figure 8.1 illustrates the symbolic representation of the main fuzzy operators by means of graphs whose vertices are the bounded difference. Note the interesting similarity between the structure of the graphs of Figure 8.1 and those that define Boolean operators.

Other basic fuzzy operators can be employed as basic building blocks. Figure 8.2 gives the symbolic implementation of the max and the min in the $+ -$ system, according to (8.21). Following relationships are straightforwardly derived from the set of relations (8.16) to (8.20).

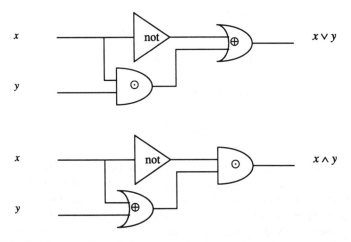

Figure 8.2 Implementation of the max and min operators by means of $\oplus$ and $\odot$

Figure 8.2 gives the symbolic implementations of the max and the min in the $+ -$. system, according to (8.21)

$$\tilde{\mathscr{A}} \cup \tilde{\mathscr{B}} = \tilde{\mathscr{A}} \oplus (\overline{\tilde{\mathscr{A}}} \odot \tilde{\mathscr{B}}) \quad \text{and} \quad \tilde{\mathscr{A}} \cap \tilde{\mathscr{B}} = \tilde{\mathscr{A}} \odot (\overline{\tilde{\mathscr{A}}} \oplus \tilde{\mathscr{B}}) \tag{8.21}$$

Note, however, that most of the Boolean elimination laws do not hold for $\oplus$ and $\odot$ (due to the property of non-distributivity of these operators). More specifically we have in general

$$\tilde{\mathscr{A}} \oplus (\overline{\tilde{\mathscr{A}}} \odot \tilde{\mathscr{B}}) \neq \tilde{\mathscr{A}} \oplus \tilde{\mathscr{B}} \quad \text{and} \quad \tilde{\mathscr{A}} \odot (\overline{\tilde{\mathscr{A}}} \oplus \tilde{\mathscr{B}}) \neq \tilde{\mathscr{A}} \odot \tilde{\mathscr{B}} \tag{8.22}$$

That is why the manipulation of fuzzy expressions is quite tedious compared with the that of Boolean expressions. The lack of simple elimination laws available in the framework of fuzzy sets makes difficult the synthesis of fuzzy expressions, either by means of analogue or digital circuits.

8.3 CMOS IMPLEMENTATION

In Section 8.2 we gave details of a method to synthesize symbolically fuzzy logic based functions. This discussion emphasized the reliability of our mathematical framework to represent in a simple manner some building block-based fuzzy formulae.

We present a CMOS implementation of the symbolic representation of any fuzzy formula. For the sake of simplicity the implementation works in current mode and uses only current mirror devices. We stress that other solutions (digital or analogue-based) could be advantageously explored in the light of the symbolic description shown in Section 8.2. Under these restrictions the sum operator reduces in a connection of wires and the implementation of most usual fuzzy logic operators leads to 'convivial' electrical structures. These structures are convenient to set up a design automation strategy and they are easy to test.

8.3.1 CMOS Implementation of Fuzzy Operators

Since the first application of fuzzy logic theory to the area of control, engineers and scientists have proposed numerous distinct approaches to implement fuzzy units in a VLSI circuit. Let us review the most important of them.

The digital approach offers several immediate benefits, including low-cost processes, fast prototyping and relatively easy design automation. Moreover, automatic regeneration of logic levels in the electronic blocks provide very accurate and reliable data signal processing. Most of the consumer goods companies are involved in the development of digital-based fuzzy logic system implementation [10]. Unfortunately, however, digital implementations of common fuzzy operations rapidly lead to complicated and enormous VLSI circuits; see [13, 36, 41,8, 28] where a full CMOS micro-controller is presented, and [8] where a digital fuzzy accelerator is proposed. For example, the synthesis of a simple MAX function in [41] results in a CMOS unit of about 100 transistors. Furthermore, an A–D converter and a D–A converter must be included to communicate with the real world [28].

With analogue circuitry, nonlinear functions like max and min are easier to synthesize. Both current and voltage are convenient conveyers of data information.

In the voltage mode circuits the max and min operators are easy to build in. However, implementing the elementary sum function necessitates the use of complex circuits (for example, transconductance circuits to implement the sum operator.) This severe limitation explains the weak interest shown by researchers in exploring this technique.

Solutions that combine the current mode and the voltage mode were presented in [6, 14, 15, 16, 33, 29]. In [29] a CMOS transconductance amplifier and bipolar current gain cell structure were investigated. In [6] a voltage-input current-output tunable membership function circuit is proposed. However, these solutions lead to fuzzy units that are difficult to test, and they offer no simple perspective for design automation.

A current mirror-based CMOS design of fuzzy units is revealed, however, to be a reliable means to construct fuzzy logic architecture. The method has already been discussed in [1, 2, 5, 6, 21, 22, 23, 26, 30, 31, 37, 38, 43, 47, 48]. Original structures of fuzzy building blocks were proposed [2] (fuzzy memory element), [26, 30] (current-mode defuzzifier), [37] (max–min circuits). A monolithic CMOS analogue function synthesizer based on a current-mode algorithm applied in fuzzy membership function synthesis was presented in [23]. Unfortunately, due to a lack of a simple relationship between transistor-level circuits and symbolic representation of fuzzy formulae, the authors were not at the stage to perform large-scale implementation of fuzzy algorithms.

Improving on the ideas shown in [43, 48], we present here a rigorous mathematical description of current mirror based fuzzy functions. This description leads to a simple design automation strategy shown in Section 8.3.

The current-based technology for the silicon implementation of multi-valued or fuzzy logic circuits is gaining in strength [9]. The trend towards the current-mode solution can be noticed through the continuous increase in the number of papers that are based on this technique. At the most famous conference [27] dealing with multi-valued logic, a special volume has been devoted to the synthesis of multi-valued circuits. Papers proposing a current-mode solutions are in the majority. Current-mode techniques have found a lot of applications in areas that need massive parallel computation (notably in the areas of neural networks and genetic algorithms.) For a good compromise between complexity and precision, current-mode circuits as seen to be the appropriate candidates (compared with voltage- mode techniques) for applications requiring a high density of interconnection. These reasons encourage us to employ current mirrors as basic building blocks in this section.

8.3.2 Current Mirror-based Approach

In Section 8.2.3 we stressed that any fuzzy formula, as considered here, can be expressed by means of a bounded difference. However, the CMOS implementation of this operator is not straightforward. For this reason we introduce three primitive operators that simplify drastically the implementation process: the sum operator +

$$\text{SUM}(x, y) = x + y \qquad (23a)$$

The positive-cut operator N

$$N(x) = \begin{cases} x, & \text{if } x \text{ is positive} \\ 0, & \text{if } x \text{ is negative} \end{cases} \quad (8.23b)$$

The negative-cut operator P

$$P(x) = \begin{cases} x, & \text{if } x \text{ is negative} \\ 0, & \text{if } x \text{ is positive} \end{cases} \quad (8.23c)$$

Any fuzzy logic formula can be formulated as a combination of the monodic primitive operators P and N by means of the dyadic operator $+$. Indeed we can easily show that the bounded difference satisfies:

$$x \ominus y = P[N(x) + P(x) + y] \quad (8.24a)$$

Or if we assume that we know the sign of the input x

$$x \ominus y = P[N(x) + y], \quad \text{for } x \geq 0 \quad (8.25a)$$

$$x \ominus y = P[P(x) + y], \quad \text{for } x \leq 0 \quad (8.25b)$$

Thus

$$\bar{x} = P[N(1) + x] \quad (8.26a)$$

By assuming that $x \geq 0$ and by use of the set of relations (8.16) to (8.20), we write

$$x \oplus y = \overline{P[N(\bar{x}) + y]} \quad (8.26b)$$

$$x \odot y = P[N(x) + \bar{y}] \quad (8.26c)$$

$$x \vee y = P[y + N(x)] + y \quad (8.26d)$$

$$x \wedge y = P[N(x) + P[N(x) + y]] \quad (8.26e)$$

The corresponding electrical implementation of the operators P and N and $+$ reduces in CMOS current mirrors and wire connection, as shown in Figure 8.3. Before illustrating the method for electrically implementing one of the fuzzy operators provided by (8.26a) to (8.26e) let us define a 'consistent' relationship between the sign of the current flow and its associate label.

The variable x labels a current: its absolute value is equal to the magnitude of the current and its sign is the direction in which the current is flowing. The opposite variable $-x = N(x) + P(x)$ labels the current that flows in the opposite direction to that of the one above. Hence, the following convention will preserve the requirement of consistency of the signs when connecting two blocks together.

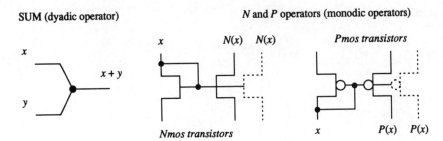

Figure 8.3 Electrical implementation of the basic building blocks

The sign of an input variable of a block is positive when the associate current flows inside the block.

For the sake of symmetry, the sign of an output variable of a block is positive when the associate current flows outside the block.

The above convention simplifies the writing of the equation of a fuzzy block while preserving a certain consistency when connecting two blocks together.

The primitive operators P and N satisfy some properties that may be useful for simplifying expressions such as (8.26b)–(8.26e). By assuming that the input x is positive, let us focus on (8.26e):

$$x \wedge y = P[N(x) + P[N(x) + y]]$$

Since $P(X) = N(X) - X$, where $X = N(x) + y$:

$$x \wedge y = P[N(x) - \{N(x) + y + N[N(x) + y]\}]$$

or

$$x \wedge y = P[-\{y + N[N(x) + y]\}]$$

Since the term $y + N[N(x) + y]$ is always positive

$$x \wedge y = N[y + N(x)] + y \qquad (8.27)$$

Finally, (8.26e) reduces to a new relation (8.27) whose formulation is more simple since it contains less primitive operators.

An appropriate use of the properties of the primitive operators leads to a simplification of the silicon implementation. However the manipulation of these rules is far from straightforward, we give only an overview of the principal ones used by the silicon compiler presented in Section 8.4.

Next we present a case-study that illustrates the method used for synthesizing the schematic circuit of a fuzzy operator from its mathematical expression.

8.3.3 Case-Study: Implementation of the Min Unit

Figure 8.4 gives a hardware-based approach of the implementation of the bounded difference and the min. The degrees of membership are physically implemented by means of the current flow.

In order to verify the proposed approach to fuzzy logic function synthesis, we have developed a fuzzy logic simulation environment utilizing the HP software package VEE-Test [40] and a 'home-made' design automation system. The next section presents the main features of the environment tool. The results of synthesis and simulation of the membership function circuit using the simulation environment are exposed. Finally, we provide a discussion on the utility of our method to generate automatically fuzzy logic blocks.

8.4 FUZZY DEVELOPMENT SYSTEM

As was shown in the preceding section, any class of circuits of performing fuzzy logic operations can be synthesized by means of simple current mirrors. Hand-crafted fuzzy logic cells for most of the principle operators have been designed and published [1, 23, 26,

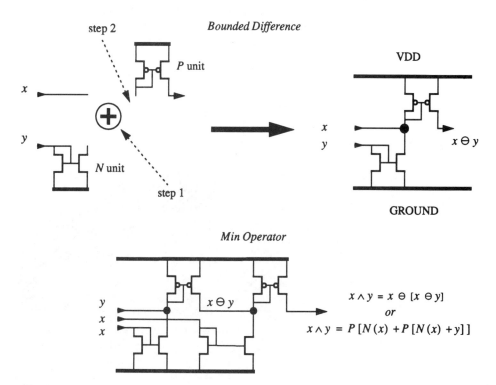

Figure 8.4 Hardware implementation of the bounded difference and the min by means of CMOS transistors working in current mode

47, 48]. However, no automatation has so far been done in this research area. Any fuzzy unit is constructed by means of two basic elements; it results that it may be possible to create a design automation framework to generate the layout of the unit.

Designing fuzzy VLSI circuits rapidly leads to complex structures with a high degree of interconnection. The time required by the design stage becomes preponderant, and some mistake (mainly due to boring repetitive work) can occur, reducing to nothing an important investment of time and money. A computer-aided design tool permits a drastic reduction in the possibility that such a mistake occurs. It speeds up at the same time the turnaround time of the fabrication of a chip. Based on the symbolic description presented in Section 8.2.3 a fuzzy logic development environment has been elaborated.

To develop the fuzzy logic development environment, we work out the framework idea with the concept of a bottom-up system, which means that we treat basic current mirrors as cell compilers to build more complicated fuzzy structures. As a result, cells representing these elementary fuzzy functions may then be assembled into sophisticated fuzzy units.

8.4.1 Basic Framework of the Fuzzy Logic Development Environment

The fuzzy logic development environment, as depicted in Figure 8.5, integrates basically the two following modules: a graphical simulation interface, and a design automation system. The graphical interface facilitates the conceptualization of fuzzy structures by assembling graphically the elementary fuzzy functions N, P and $+$ introduced in

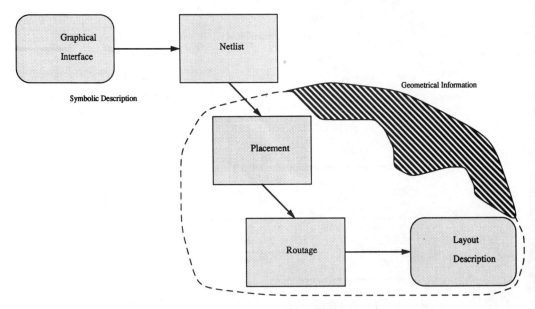

Figure 8.5 Framework of the fuzzy logic development environment

Section 8.3.1. The design automation system adds geometrical information to the output of the graphical interface and synthesizes the corresponding circuit layout.

In the forgoing subsections we briefly present the VEE-TEST graphical interface [40]. We also overview the main required features of the design automation system.

8.4.2 Graphical Simulation Interface

The user-friendly interface of the VEE test tool makes possible the design and simulation of complex hierarchical structures. A basic library consisting of the three primitive operators N, P and $+$ has been created. New fuzzy function can be synthesized and simulated. The VEE test offers the facility to group objects together logically and physically; that facilitates the elaboration of a library of elementary function-level units.

The panel of the simulation interface of the triangular membership function is illustrated in Figure 8.6. We may prove that a possible mathematical expression of the triangular membership function is such that

$$mf(x, i) = P\{P[N(x) + i] + P[N(i) + x] + N(1)\} \tag{8.28}$$

where x and i are the input variable and the central value of the fuzzy function (i.e. for which $mf(x, i) = 1$). The logical-level simulation of the triangular membership function has been performed, and the result is depicted in Figure 8.7.

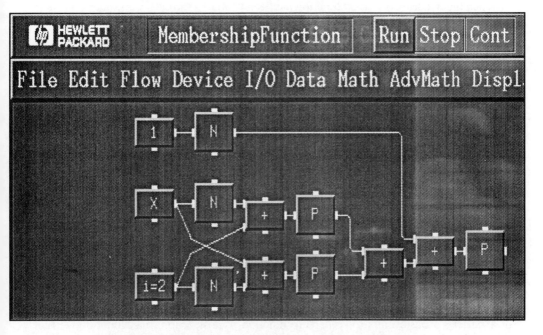

Figure 8.6 Structure of the triangular membership function

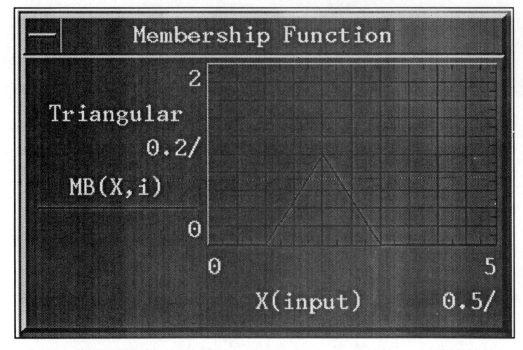

Figure 8.7 Logic-level simulation of the structure depicted in Figure 6

When the structure of a fuzzy function has been validated the design automation system will then be in charge of its implementation on silicon.

8.4.3 Design Automation System

The design automation system basically adds geometrical information to the output that has been synthesized by means of the graphical simulation interface. On the first hand the design automation system has to translate in a SPICE-compatible netlist the formula yielded by the graphical interface. The netlist must contain all the necessary information about the inter-relationships between the elementary modules.

On the other hand the design automation consists in solving the two basic problems of placement and routage. The two problems are closely interdependent. Given certain geometrical constraints (such as global or local symmetry, minimal silicon area, minimization of capacity area, critical path length), the design automation must place and interconnect the elementary modules in a way that is as well optimized as possible. The ideal solution to the two problems consists in developing a self-contained algorithm that will route and place simultaneously and globally the elementary modules. For the sake of simplicity it is, however, useful to treat the two problems as two independent issues, as depicted in Figure 8.8. Indeed the architecture of fuzzy circuits is mostly composed of a battery o basic units (membership function, min function, max function,...) that are working in parallel. At the price of an acceptable loss in area optimization, the

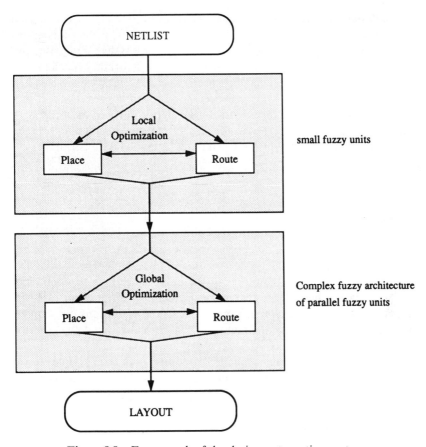

Figure 8.8 Framework of the design automation system

method shown in Figure 8.8 perfectly fits the layout design of current mirror-based fuzzy circuits.

The basic principles considered in the design automation system can be summarized as follows.

Long and short range symmetry (place): the short range symmetry refers to the symmetry at the elementary fuzzy function level. The generic fuzzy function cell is optimized with respect to a short range symmetry. The fuzzy unit, built of several fuzzy functions, should also reflect the long range symmetry.

Symmetrization of a global wiring (place): the requirement for the long term symmetry also involves the symmetrization of wiring, including the wiring within the cells representing elementary fuzzy functions.

Merging of interconnections in the channel (route): the intelligent generator at the level of a unit is capable of merging the redundant interconnections within the channel. This involves modification of the generators at the level of an elementary functions.

Swapping or flipping of basic current mirror cells in order to obtain a global symmetry (route).

Some of above mentioned characteristics may already be found in the existing analogue CAD systems [4], but others (restructing the unit by the hierarchical modifications down to the level of basic cells) are unique for this domain. The system comprises two basic stages: generators for current mirrors, which are used as elementary blocks to build the library of basic fuzzy functions (min, max, $\odot$, $\oplus$,...), the fuzzy unit synthesizer, which takes the description of the fuzzy function and decomposes–synthesizes it according to the structures of available fuzzy functions. After the decomposition and the synthesization of the unit structure, the information about the structure of the designed unit serves as the input set for the unit generator which generates a layout and performs all layout modifications described above.

8.4.4 Netlist

The first step performed by the design automation system is the translation in a SPICE-compatible netlist of the formula yielded by the graphical interface. The netlist must contain all the necessary information about the interconnections between elementary modules. When creating the netlist from a fuzzy formula, a set of operations are performed that prevents the user making electrical mistakes. These operations are similar to the grammatical rules of a natural language. More specifically, we define the following terminology.

The netlist consists of an ordered sequences of terms. Each term belongs to five basic types. Each types is referred to as the vocabulary understood by the design system. Namely, we have

the inputs of the fuzzy unit

the outputs of the fuzzy unit

the N an the P devices (current mirrors with n multiple output transistors are denoted by N_n and P_n)

the interconnections (or + devices)

The combination of terms of the design system can form complex sentences with respect to a certain syntax (i.e. a set of rules.) Syntactical rules must be observed when combining terms together so that electrical requirements are satisfied. For instance, when creating the netlist from a fuzzy formula, a rule prevents the user from creating sentences containing the sequence $N(x)+N(y)$. Indeed implementing electrically the sequence $N(x)+N(y)$ results in a permanent static consumption of current, a resistor being connected between the power supply and the ground.

The semantic of the design system consists in associating to each equation its fuzzy meaning. For instance, the meaning of the sentence (8.28) is referred to as the membership function. For instance, as shown in Section 8.3.3, the sentences $x \wedge y = P[N(x)+P[N(x)+y]]$ and $x \wedge y = N[y+N(x)]+y$ have the same meaning but correspond to two different netlists.

Note that the generators produced by the design automation system will use knowledge extracted from a technology file, which contains a quantitative representation of technology-related parameters (i.e. design rules, technological parameters, etc.). This strategy makes the structure of the sentences portability, independent on the technology. This means that the compound of a set of terms permits of building a high-level language library, instead of a complex silicon library.

In the vocabulary of the design system, subsets of sentences have the same meaning. Hence it may be advantageous to replace all the sentences of a given subgroup by one whose synthesis uses the minimum number of transistors. Two obvious examples are that subset $[N_2(x)]$ has the same meaning as the subset $(N(x); N(x))$; and that the subset $[P_2(x)]$ has the same meaning as the subset $(P(x); P(x))$. (A subscript added to devices N and P refers to the number of output terminals of the current mirror. For instance N_2 is a NMOS current mirror with two outputs.)

Another example is provided by the fact that subset $(N_2(x) + P_2(y); P_2(x) + N_2(y))$ has the same meaning as the subset $(N(x) + P(y); P(x) + N(y))$.

The sentences may be represented by means of a binary-oriented graph [32]. A graph permits one to represent fuzzy formulae by means of two classes of objects. An object is either a vertex or an edge. Each vertex stands either for a terminal of N_n, P_n or $+$, or either for an input or an output. Each edge gives an information on the ordering of the words contained in the sentence. The properties of each object can easily be attached to the graph. So the programming of the design automation algorithm in object-oriented language (C++) is greatly facilitated. Figure 8.9 gives the graph associated with formula (8.28).

In the same way, simplification rules described above can be illustrated as in Figure 8.10.

When the verification–simplification steps of a given sentence are performed, a netlist file of the fuzzy unit is generated. The numbering of the inputs, outputs, operators N_n and P_n is performed by the incrementing process. More specifically, when reading the oriented graph of the min function two successive elements of the same type are labelled by two successive numerical values, the label of the first elements of the graph receiving the value 0. The terminals of a given operator are subnumbered in the same fashion. As an example, the netlist sentence $x \wedge y = P[N(x) + P[N(x) + y]]$ of the membership

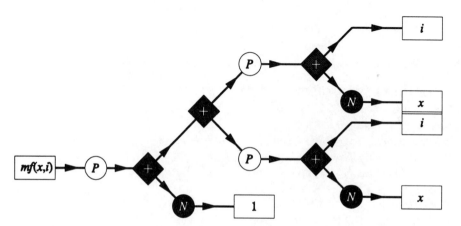

Figure 8.9 Oriented graph of the equation (8.28) representing the membership function

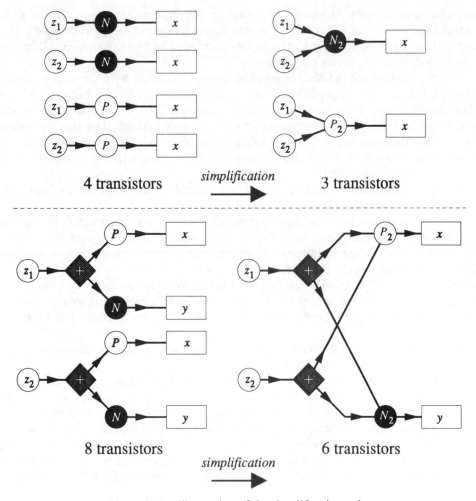

Figure 8.10 Illustration of the simplification rules

function is:

.Global VDD VSS NET0 NET1 NET2 NET3
*subcircuits
.subckt N2 in out1 out2
M1 in in vss vss N w=10u l=10u
M2 out1 in vss vss N w=10u l=10u
M3 out2 in vss vss N w=10u l=10u
.ends N2
.subckt P in out
M1 in in VDD VDD P w=10u l=10u
M2 out in VDD VDD P w=10u l=10u

```
.ends P
*netlist
XN0 NET0 NET1 NET2 N2
XP0 NET1 NET2    P
XP1 NET2 NET3 P
.END
```

Figure 8.11 illustrates the relationship between the layout of the min function, its netlist and its electrical implementation by means of current mirrors. In order to facilitate the placement-route operation, the elements of the graph are constrained to lay on a Manhattan grid.

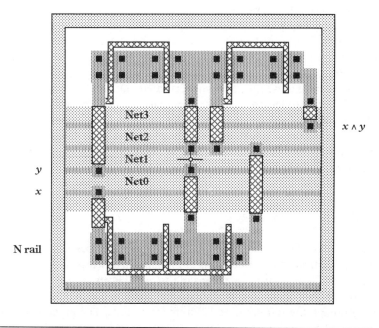

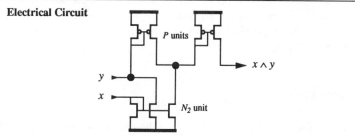

Figure 8.11 Relationship between the layout and the electrical circuit of the min function: upper, layout resulting from the examination of the oriented graph; lower, electrical circuit of the membership function.

8.4.5 Placement

For the sake of convenience the current mirrors are manipulated as monolithic devices, i.e. we assume that they cannot be resized. All the transistors of the same fuzzy architecture will then have identical geometrical characteristics. For example, in a $2\,\mu$m CMOS technology the layout design may consist of P-MOS and N-MOS transistors sized to $W = L = 10\,\mu$m.

Under these assumptions the netlist of a fuzzy formula provides a first solution of the placement of the terminals of the current mirrors, the inputs and the outputs. The layout given in Figure 8.12 illustrates the positioning operation of the N and P devices. The device terminals can be placed according to the following procedure.

The device terminals are positioned on the grid according to the value of their labels. The origin $(0, 0)$ is set at the left-bottom corner of the Manhattan grid. The inputs are placed in the first column $x = 0$ (first column on the right side.) The N and P terminals lie on the lower row and upper row, respectively. These rows are referred to as the N rail and P rail, respectively. The total length of the cell is fixed to the longer abutted row of P or N current mirrors. The total height of the cell is determined by the number of edges contained in the graph of the fuzzy formula. Finally the outputs are laid on at the last column of the grid.

The procedure places close together devices that are linked. As a result it minimizes the total length of wires needed to wire the terminals. In that respect the first placement obtained by the above procedure is partially optimized. In order to ameliorate the symmetry of the structure, a simulated annealing [39] has been developed in [7]. Its goal consists in the optimization of the vertical alignments of the terminals (local symmetry) and the total length of wire, as shown in Figure 8.12.

8.4.6 Route

The routing region is delimited above by the line of P type devices, and below by the line of N type devices. Vertical interconnections are realized by means of the polysilicon layer, and horizontal interconnections by means of the metal layer.

After the placement operation, the horizontal wires are moved until it is no longer possible to approach the N-mirror and P-mirror columns. The moving action is pictorially illustrated in Figure 8.13. It is performed until it is no longer possible to minimize the height of the layout cell (i.e. in the vertical direction.) Figure 8.14 shows the new layout obtained from the random moving actions on the min function. The new layout area is smaller than that of Figure 8.11.

8.4.7 Superphenix

A first draft of the design automation system, called SUPERPHENIX, has been implemented in the C language [7]. It consists of a simulation platform based on the VEE-Test HP graphical tool and a design automation system (or silicon compiler.) The graphical tool provides a logical simulation of fuzzy algorithms and helps in the design of fuzzy system architecture. The silicon compiler generates from the mathematical expression of

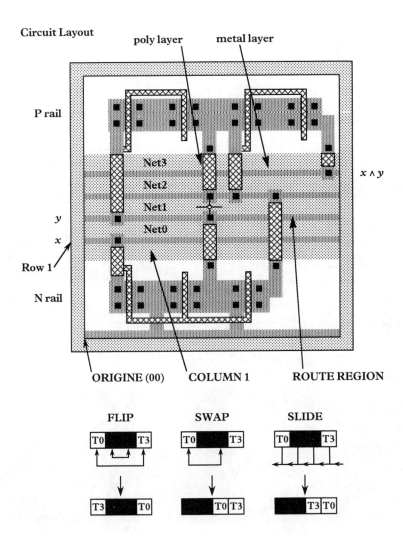

Figure 8.12 Four possible actions optimizing vertical alignment during simulated annealing

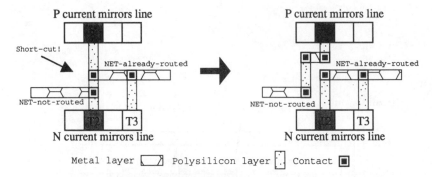

Figure 8.13 Routing on a Manhattan grid of the elementary devices

Figure 8.14 New layout of the min function after some random flip actions

a given fuzzy algorithm its corresponding layout. Its input consists of a mathematical equation that relies on a fuzzy algorithm. The silicon compiler first replaces each term of the equation by its electrical counterpart. Then it tries to reduce and place each electrical blocks by simulated annealing in order to minimize the total length of wires. Finally the Mentor Graphics tools GDT [11] reads the cell generator provided by the silicon compiler and produces the layout of the fuzzy algorithm, as explained in Section 8.3.

8.5 CMOS FUZZY LOGIC-BASED CONTROLLER

Based on the ideas exposed in Section 8.4 a fuzzy logic based controller has been developed. The controller was constructed by means of CMOS current mirror building blocks [25]. The transistors of all the blocks were sized to a width and length equal to $10\,\mu m$. The circuit allows two premises, one conclusion and nine rules; the latter are programmable with current sources. The chip consumes 2 mA at a 5 V power supply for a core area of $0.4\,mm^2$. Furthermore, the performance of the developed fuzzy architecture reached 10 MegaFLIPS (fuzzy inferences per second) for the standard $1.2\,\mu m$ CMOS technology. A real time application that successfully used the chip is presented in [24]. The application deals with the control of a metallic ball by means of an electromagnetic field.

The main features of the chip may be summarized as follows:

Low-power and small-sized integrated circuit: the circuit can be used in a lot of consumer product applications.

Programmable by means of current sources: some applications need to change in real time the values of the fuzzy rules. The circuit proposes such a reliability. We can imagine adapting the rules of the chip by means of a conventional microprocessor addressing an

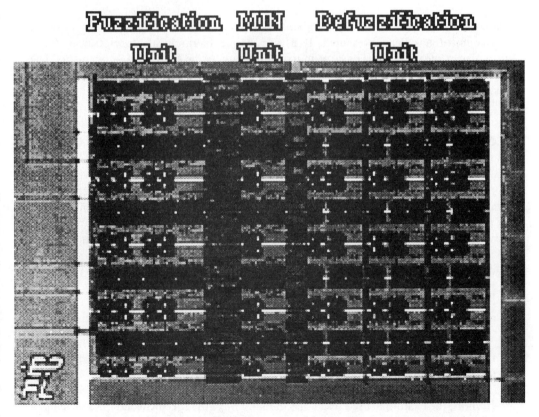

Figure 8.15 Photograph of the CMOS Fuzzy Logic Controller (Die size $500\,\mu m \times 800\,\mu m$)

appropriate number of digital to analogue sources. The program running on the microprocessor might use a neuronal network algorithm or other new similar adaptive algorithms.

Two weeks of development from fuzzy algorithm to layout: the extreme shortness of the phase of chip design results straightforwardly from the strategy adopted and presented in Section 8.4.

Design close to fuzzy algorithm: this feature makes easy the testing of the circuit. Only three days of testing was needed to verify the functionality of the chip. Each functional block is easy to recognize when looking at the core of the chip. That facilitates the comparison between the schematic or fuzzy formula to be implemented and its corresponding layout.

Fuzzy block library independent to the technology: CMOS and bipolar technologies can both be used to synthesise the layout of the chip.

'Sea of gates' or 'gate array' strategy affordable: we can imagine implementing a regular array of basic fuzzy structures and connect them, if necessary and as a last step, by means of the metal layer. A family of fuzzy logic architecture can then be build up whose cost will be very low.

8.6 CONCLUSION

Based on the notion of fuzzy numbers and fuzzy units, a fuzzy logic development environment has been presented. It uses a simple hierarchical description of fuzzy formulae by means of three basic primitive units: namely the P-, N- and $+$ operators. The combination of these operators can lead to complex fuzzy architecture such as that of the circuit briefly described in Section 8.5.

8.6.1 Features of the VLSI Technique Taken for the Integration of Fuzzy Circuits

A technique for the automated design of fuzzy units was shown. Some restrictions that limit the flexibility of the use of the CMOS process were advocated (constant value of the W and L, monolithic structures, layout on a Manhattan grid). We emphasize the numerous advantages resulting from the adoption of these restrictions. The most relevant are:

development of a fuzzy logic function library independent on the technology

test approach similar to the digital ones

a close relationship between the schematic of a fuzzy unit and its corresponding form in terms of basic primitive units

transportability to low-cost solutions such as gate array or sea of gates to facilitate the fast-prototyping of a fuzzy architecture

In short, the technique used resembles in many ways the specific techniques developed for the design of digital circuits.

Note that the rules of the fuzzy logic-based controller presented in Section 8.5 can change in real time (they are programmed by means of external current sources.) We may imagine that the rules would be computed by means of a digital microprocessor connected to a battery of digital to analogue converts (i.e. as set of current mirrors that are switched off or on.) Observation of biological systems recently gave rise to the growth of new adaptive algorithms. The most popular of them are the genetic algorithms and the neural networks. The neural networks aim to reproduce from biological and physiological observations the processing structures of the information in the human brain. Fuzzy logic works by experience and past observations. Hence, it would be interesting to combine the adaptive algorithms to fuzzy ones to create new types of architecture for control [12].

8.6.2 Improvement of the Structure of the Fuzzy Logic Development Environment

A fuzzy logic-based controller was presented. Its synthesis used the fuzzy logic development environment. The chip works like an independent black box. It is, however, easy to imagine a great improvement of the features of the circuit by implementing in the design automation tools the possibility to produce pure digital circuits. The implementation would be consistent with writing fuzzy formulae (additional new primitive units would have to be added in the vocabulary understood by the development environment.) An exchange or memorization of data would be taken in charge by the 'digital core' (no loss of information is permissible), however the tasks that necessitate a low degree of precision would be performed by the 'analogue core' (almost all the devices that interface with the human being are in this category of tasks.)

Now new microstructures can be implemented on silicon (due to the improvement in microtechniques in the last decade). If these structures can easily be characterized they can enrich the vocabulary understood by the home-made development environment. For instance, we can imagine not only summing the N- and P- operators but also new devices such as micromotors, thermal sensors, pressure sensors or phototransistors. An enrichment of the syntax and the semantic of the design automation system would result. The fuzzy logic development environment would then address the realization of low-cost microsystems for smart electronic market (low cost, low performance electronic finger, electronic eye, electronic ear for the robotics, etc.) These low-cost microsystems would be like small portable microlaboratories. Then the technique taken would play an important role in the market of electronic vision or electronic touch sense, the challenge being to bring together as much as possible the sensors and the data processing unit.

Another aspect presented by our approach is the verification of the consistency of fuzzy electrical circuits. We noted that the development environment defines a semantic or meaning of fuzzy formulae. Conversely an electrical circuit may be represented by a fuzzy formulae. It may then be interesting to verify the consistency of an electrical circuit by projecting it into the formal fuzzy logic space. In the case that the semantic of the electrical circuit is determined, the procedure of verification of its syntax can be performed. The output of the verification process would be expressed in terms of linguistic expressions such as 'your circuit is fairly good', 'you are to have critical problems with the temperature dependence', 'if is not possible to connect circuit A and circuit B together ', and so on.

ACKNOWLEDGMENTS

The author would like to expresss his grateful thanks to Philippe Duc and Didier Nicoulaz for their invaluable support in developing a first draft of the automated layout design system.

REFERENCES

1. Abd-El-Barr, M., and Mahroos, M. (1992) On the synthesis of MVL functions for current-mode CMOS circuits implementation. *Proceedings of 22nd International Symposium on Multiple-Valued Logic*, Sendai, Japan, pp. 221–228.
2. Balteanu, F., Opris, I., and Kovacs, G. (1993) Current-mode fuzzy memory-element. *Electronics Letters*, **29**, 236–237.
3. Berenji, H. R. (1992) Fuzzy logic controllers. In *An Introduction to Fuzzy Logic Applications in Intelligent Systems*, Yager R. and Zadeh L. (eds) (Kluwer Academic Publishers).
4. Bowman, R. (1990) Analog integrated circuit design conceptualization. In *Introduction to Analog VLSI Design Automation*, M. Ismail and J. Franca (eds) (Kluwer Academic Publishers: Boston).
5. Chang, Y.-H., and Butler, J. (1991) The design of current mode CMOS multiple-valued circuits. *Proceedings of 21st International Symposium on Multiple-Valued Logic*, Victoria, pp. 130–138.
6. Chen, J. J., Chen, C. C., and Tsao, H. W. (1992) Tunable membership function circuit for fuzzy control systems using CMOS technology. *Electronics Letters*, **28**, 2101–2103.
7. Duc, P., and Nicoulaz, D. (1993) Superphenix: un environment de dévelopment pour contrôleurs flous. Internal Report (in French), Laboratoire d'électronique générale, EPFL.
8. Eichfeld, H., Künemund, T., and Klimke, M. (1993) An 8b fuzzy coprocessor for fuzzy control. *Proceedings of the International Solid State Circuits*, San Francisco.
9. Epstein, G. (1993) *Multi-valued Logic Design: an Introduction* (IOP Publishing, London).
10. *Fuzzy Logic'93, Conference Proceedings*, (1993) (Pennwell).
11. *GDT Designer Manuals* (1990) Mentor Graphics Corporations, San Jose, CA.
12. Harris, C. J., Moore, C. G., and Brown, M. (1993) *Intelligent Control: Aspects of Fuzzy Logic and Neural Nets* (Singapore, World Scientific).
13. Ikeda, H., Kisu, N., Hiramoto, Y., and Nakamura, S. (1992) A fuzzy inference coprocessor using a flexible active-rule-driven architecture. *IEEE Int. Conf. Fuzzy Systems FUZZ-IEEE*, pp. 537–544.
14. Inoue, T., Ueno, F., Motomura, T., Matsuo, R., and Setoguchi, O. (1991) Analysis and design of analog CMOS building blocks for integrated fuzzy inference circuits. *Proceedings of IEEE International Symp. on Circuits and Systems*, pp. 2024–2027.
15. Inoue, T., Ueno, F., Motomura, T., Setoguchi, O., and Matsuo, R. (1991) New high-speed analog MAX and MIN circuits using OTA-based bounded-difference operations. *Electronics Letters*, **27**, 1034–1035.
16. Inoue, T., Motomura, T., Matsuo, R., and Ueno, F. (1991) New OTA-based analog circuits for fuzzy membership functions and MAX MIN operations. *IEICE Transactions*, **74**.
17. *The First International Workshop on Industrial Applications of Fuzzy Control and Intelligent Systems*, Working notes, College Station, Texas, 1991.
18. Jamshidi, M., Vadiee, N., and Ross, T. J. (eds) (1993) *Fuzzy Logic and Control: Software and Hardware Applications* (Englewood Cliffs, Prentice Hall).

19 Kandel, A., and Langholz, G. (eds) (1993) Fuzzy control systems. *Colloquium on Two Decades of Fuzzy Control*, The Institution of Electrical Engineers, London, 1993.
20 Kaufmann, A., and Gupta, M. M. (1985) *Introduction to Fuzzy Arithmetic: Theory and Application* (New York, Van Nostrand Company).
21 Kawahito, S., Mizuno, K., and Nakamura, T. (1991) Multiple-valued current-mode arithmetic circuits based on redundant positive-digit number representation. *Proceedings of 21st International Symposium on Multiple-Valued Logic*, Victoria, Canada, pp. 330–339.
22 Kelber, J., Triebel, S., and Scarbata, G. (1993) Modulgeneratoren fuer Fuzzy-Controller (in German). In *Fuzzy Logic*, B. Reusch (ed.) (Springer-Verlag), pp. 32–41.
23 Kettner, T., Heite, C., and Schumacher, K. (1993) Analog CMOS realization of fuzzy-logic membership functions. *IEEE Journal of Solid-State Circuits*, **28**, 857–861.
24 Lemaitre, L., Patyra, M., and Mlynek, D. (1993) Integrated CMOS fuzzy logic functions: a current mirror based approach. *CICC'93*, San Diego, pp. 25.8.1–25.8.3.
25 Lemaitre, L. (1994) Theoretical aspects of the VLSI implementation of fuzzy algorithms. PhD thesis 1226, Swiss Federal Institute of Lausanne, Switzerland.
26 Liu, S. I., Hwang, Y. S., and Tsay, J. H. (1993) CMOS-based fuzzy membership function and max/min circuits. *Electronics Letters*, **29**, 116–118.
27 *Proceedings of the International Symposium on Multi-Valued Logic* (1994) Boston.
28 Nakamura, K., Sakeshita, N., Nitta, Y., Shimanura, K., Ohono, T., Egushi, K., and Tokuda, T. (1993) A 12b resolution 200kFLIPS inference processor. *Proceedings of the International Solid State Circuits Conference*, San Francisco.
29 Opris, I. E., and Balteanu, F. (1993) BiCMOS tunable membership function circuits: *International Journal of Electronics*, **75**, 689–696.
30 Pélayo, F. J., Rojas, I., Ortega, J., and Prieto, A. (1993) Current-mode analog defuzzifier. *Electronics Letters*, **29**, 743–744.
31 Sasaki, M., Ishikawa, N., Ueno, F., and Inoue, T. (1992) Current-mode analog fuzzy hardware with voltage input interface and normalization locked loop. *IEEE International Conference on Fuzzy Systems, FUZZ-IEEE*.
32 Schmidt, G., and Ströhlein, T. (1993) *Relations and Graphs* (Springer-Verlag, Berlin).
33 Shetty-Wagoner, M., Rattan, K. S., and Siferd, R. (1993) Membership function generator circuit for a fuzzy logic controller. *IEEE National Aerospace and Electronics Conference*, Part 1 (of 2).
34 Sugeno, M. (ed) (1992) Industrial applications of fuzzy control. (Amsterdam, North-Holland).
35 Terano, T., Asai, K., and Sugeno, M. (1991) *Fuzzy Systems Theory and its Applications* (Academic Press).
36 Togai, M., and Watanabe, H. (1986) A VLSI implementation of fuzzy inference engine: Toward an expert system on a chip. *Information Sciences*, **38**, 147–163.
37 Tsukano, K., Inoue, T., Ueno, F., and Fumio (1993) Design of current-mode analog circuits for fuzzy inference hardware systems, Part 2. *Proceedings IEEE International Symposium on Circuits and Systems*, pp. 1385–1388.
38 Ueno, F., Inoue, T., Shirai, Y., and Sasaki, M. (1987) A maximum and minimum circuit with multiple inputs in current mode. *Transactions of IEICE*, **70**, 392–395.
39 van Laarhoven, P. J. M. and Aarts, E. H. L. (1987) *Simulated Annealing: Theory and Applications* (D. Reidel).
40 VEE-Test Reference Manuals (Draft), Hewlett-Packard, 1991.
41 Watanabe, H., Dettloff, W., and Yount, K. (1990) A VLSI fuzzy logic controller with reconfigurable, cascadable architecture. *IEEE JSSC*, **25**, 376–382.
42 Yager, R. R., and Zadeh, L. A. (1992) *An Introduction to Fuzzy Logic Applications in Intelligent Systems* (Kluwer Academic Publishers).

43 Yamakawa, T., and Miki, T. (1986) The current-mode fuzzy-logic integrated circuit. *IEEE Transactions on Computers*, 161–167.
44 Zadeh, L. A. (1965) Fuzzy sets, *Information and Control*, **8**, 338–353.
45 Zadeh, L. A. (1973) Outline of a new approach to the analysis of complex systems and decision process. *IEEE Transactions on Systems, Man and Cybernetics*, **3**, 28–44.
46 Zadeh, L. A. (1975) Fuzzy sets and approximate reasoning. *Synthsis*, **30**, 407–428.
47 Zhijian, L., and Hong, J. (1990) CMOS fuzzy logic circuits in current-mode toward large scale integration. *Proceedings of the Internationl Conference on Fuzzy Logic and Neural Networks*, Iizuka, Japan, pp. 155–158.
48 Zhijian, L., and Hong, J. (1990) A CMOS current-mode, high speed fuzzy logic microprocessor for a real-time expert system. *Proc. 20th Int. Symp. on MVL*, Charlotte, USA.
49 Zimmermann, H. J. (1991) *Fuzzy Set Theory—and its Applications* (Kluwer Academic Publishers).

HYBRID SYSTEMS
AND APPLICATIONS

9
Neuro-fuzzy Systems: Hybrid Configurations

H.-N. L. Teodorescu
Technical University of Iasi, Iasi, Romania

T. Yamakawa
Kyushu Institute of Technology, Iizuka, Fukuoka, Japan

9.1 PRELIMINARIES

The linguistic systems and the fuzzy systems are known to be easy to design and develop. The complexity of their description is combinatorial. A finite number of combinations is needed to completely describe a linguistic system and the corresponding fuzzy system. On the other hand, the neural networks have a proven ability in learning; moreover good learning algorithms are available for them. Therefore, it is practically interesting to combine fuzzy systems and neural networks.

There is no formal and unique definition of neuro-fuzzy systems. By a neuro-fuzzy system one understands a system which involves in some way both fuzzy systems and neural networks, or features of both combined in a single system.

We consider in this chapter several types of neuro-fuzzy configurations and their operation. Sometimes neuro-fuzzy systems are also named hybrid systems. Here, by a neuro-fuzzy system we shall understand systems having either a neural network structure but using fuzzy neurons instead of a classic neuron, or systems composed of both fuzzy systems and neurons, or neural networks, each of these subsystems preserving the classic characteristics of its class in the hybrid configuration. A departure is the single fuzzy neuron structure that can perform nontrivial functions alone [16, 17].

Three categories of simpler systems are mainly used in building the neuro-fuzzy systems discussed in this chapter: neural networks, fuzzy systems and linear recurrent systems. The linear recurrent systems are interpreted either as a particular case of neuron (linear neuron), or a particular case of neural network (namely, as a couple of linear neurons). By the combination of systems from these categories one builds various neuro-fuzzy structures [7, 9, 10].

This chapter presents a systematic analysis of mixed fuzzy system—neural network configurations, including recurrent neuro-fuzzy systems. In the first sections we summarize the main concepts related to the above-mentioned three categories of systems.

Because a neural network and a fuzzy system essentially perform the same function, namely they are nonlinear vectorial systems, they can stand one for the other; moreover they can stand for a classic nonlinear system. This equivalence, between different classes of nonlinear systems, including classic systems, neural networks, fuzzy systems and neuro-fuzzy systems raises the question of why and when to use a system from one class instead of a system from another class. One answer is related to modelling issues: one should use the type of system that best fits the 'meaning' involved in the modelled process. This is why interpretations related to modelling are emphasized. Only a few examples of simple interpretations are provided. The reader will easily find other interpretations, according to his interests.

The technical reasons to use neuro-fuzzy systems instead of classic systems, or instead of fuzzy systems or neural networks can be:

Ease of development, because the fuzzy systems offer an intuitive way to asses nonlinear functions, whereas neural networks allow an automatic way to asses nonlinear functions;

Ease of adaptation, because of the above reasons.

These arguments are mainly related to the use of some knowledge about the problem in hand, knowledge provided by human experts and implemented at the level of the fuzzy system.

On the other hand, there are several methodological reasons to use neuro-fuzzy systems in modelling. These reasons are that:

Fuzzy systems allow modelling of the imprecise but rational behaviour of human experts [3]

Neural networks can be trained to model empirical behaviour of human experts, or to model complex nonlinear systems.

Let us explain the above two points. Human decision-making processes are believed to be rational, i.e. guided by some rules, although numerically imprecise. Such a behaviour is suitably modelled by knowledge-based systems involving fuzzy data and fuzzy inference, i.e. by fuzzy logic systems. On the other hand, there are many situations when humans, mainly in some community-related behaviour, are using empirical knowledge, got by experience. There are no rules or inference procedures in these cases. Such a behaviour is best modelled by neural networks whose training process mimics an empirical accumulation of experience. By mixing the two categories of models, namely fuzzy systems and neural networks, one can hope to obtain complex models that better mimic the human-related processes.

In this way neuro-fuzzy systems allow the modelling of complex processes involving both human experts and non-expert humans. It is probably the only available tool to create sound models that are still related to the modelled process itself.

There is an extensive literature on neuro-fuzzy systems. The basic results included in this chapter are presented in the papers by these authors, as listed in the bibliography. A review of papers dealing with learning a fuzzy system (FS) by neural networks (NN) can be found in [3, 5, 6].

9.2 MAIN CLASSES OF FUZZY SYSTEMS

There are six main classes of neuro-fuzzy systems reported in the literature (Figure 9.1).

Systems in class A have the same structure (topology) as the classic neural networks, but they employ fuzzy neurons and possibly fuzzy weights instead of numerical neurons.

Class B includes adaptive fuzzy systems that are trained by a neural network. Either the rules or the membership functions can be modified during the adaptation process by the neural network.

Class C includes classic neural neurons, but their learning is performed by a fuzzy method; that is, by computing the adaptation values of the weights by a fuzzy system based function. This class is the counterpart of class B.

Class D includes systems that are composed by independent neural networks and fuzzy systems.

Systems in class E have the configuration of a neural network, but they use (more or less complex) fuzzy systems standing for neurons. Of course, class A is a subclass of class E, but we mention it for historical reasons: it was the first introduced.

Class F is a mixing of classic systems and systems from one of the above classes, used to perform an adaptation of the classic system.

Only systems from the classes A, D and E are discussed here. The neuro-fuzzy systems in this chapter are simple networks of subsystems constituted by fuzzy systems, or by neural networks, or by combinations of both. It is not necessary, as in classic neural networks, that all subsystems be identical. Many neuro-fuzzy systems in this chapter are recurrent systems.

Before analyzing the different possible configurations of neuro-fuzzy systems, remember that fuzzy systems can be used either with defuzzifier (when they perform as classic systems) or without defuzzifier. Moreover, two categories of fuzzy systems have to be

---

Classes of neuro-fuzzy systems

A Neural network structure + fuzzy logic neurons (e.g., MIN and MAX neurons)

B Fuzzy system + training (adaptation) by neural network

C Neural network + fuzzy learning method

D Structure including interconnected fuzzy system + (general) neural network

E Neural network topology (parallelism) + fuzzy logic systems (+ mixed nonlinear systems)

F Neural network topology + fuzzy algebraic systems

Figure 9.1 Classes of neuro-fuzzy systems

considered: fuzzy logic systems and fuzzy algebraic systems. These types of fuzzy systems will give rise to different neuro-fuzzy combinations. However, only fuzzy logic systems are considered here.

To be used in combination with a neural network (i.e. a system with vectorial input, not accepting continuous membership functions as inputs), one has to use either fuzzy systems with sampled output membership functions, or a fuzzy system with defuzzified output.

In what follows, we remember some concepts related to fuzzy systems and neural networks.

9.3 FUZZY SYSTEMS

9.3.1 Linguistic Systems and Fuzzy Systems

The linguistic systems use linguistic variables with linguistic labels as values (linguistic values), for instance 'small', 'medium', 'big'. The linguistic systems are described by rules, for instance

⟨ If variable 'input' is 'big', then variable 'output' is 'very small' ⟩

Linguistic systems are finite systems, in the sense that there are only a finite number of linguistic values, for each of the input, output and state variables. Moreover, the operation of these systems is characterized by a finite set of tuples, standing for the input values, state values and the corresponding output values. The operation can read

$$(a_{1i_1}, a_{2i_2}, \ldots, a_{pi_p}, s_{1j_1}, s_{2j_2}, \ldots, s_{qj_q}; b_{1k_1}, b_{2k_2}, \ldots, b_{nj_n})$$

where the labels a stand for the linguistic values of the p input variables, the labels s stand for the values of the states, and the labels b stand for the output.

The system reads, in an equivalent manner, as a rule-based system (knowledge-based system). Denoting by $x_1, x_2, \ldots, x_p$ the input variables, by $\sigma_1, \sigma_2, \ldots, \sigma_q$ the state variables, and by $y_1, y_2, \ldots, y_n$ the output variables, the description by rules reads

IF (input) x_1 is a_{1i_1}, and x_2 is $a_{2i_2}, \ldots$, and x_p is a_{pi_p},

and (state) σ_1 is s_{1j_1}, and σ_2 is $s_{2j_2}, \ldots$, and σ_q is s_{qj_q},

THEN (output) y_1 is b_{1k_1}, and y_2 is $b_{2k_2}, \ldots$, and y_n is b_{nj_n})

For instance, for a linguistic system with only one input variable x, no state variable and only one output variable y, the system is described by the couples (a_i, b_{ki}), or, more explicitly: $(a_1, b_{k1}), (a_2, b_{k2}), \ldots, (a_p, b_{kp})$, where b_{ki} stands for the label of y corresponding to the input label a_i.

Because of the finite character of the variable spaces, the operation of the linguistic systems can be easily computed, as a combinatorial problem.

Fuzzy systems are obtained from linguistic systems by adding membership functions to a linguistic system, in the sense that to each linguistic value one attaches a membership function. The rules are applied according to the fuzy logic. The output of a fuzzy system can be 'defuzzified'. This means that the membership function is converted to a numerical value that 'best characterizes' the membership function. This operation is very similar to the assignment of the average to a probability distribution.

9.3.2 Fuzzy Systems and Memory Processes

Before developing various classes of neuro-fuzzy systems, we briefly discuss the use of fuzzy systems with memory at the input.

Complex human-related processes involve human 'experts', i.e. humans using some inference, rule-based procedure, beyond the use of empirical knowledge. Remember that fuzzy systems are recognized as being a good model for human made inferences. This is due to the fact that human inference is an 'approximate reasoning', best modelled by fuzzy logic. Therefore, a process involving human inference based on rules will be best modelled by fuzzy systems.

Current and previous values of the input variables are used in decision and control. The corresponding configuration of a model involving memory of the input values and fuzzy systems is represented in Figure 9.2.

The rules corresponding to this elementary configuration are

If x (present moment) is A_{1i} and x (previous moment) was A_{2i} and ... and input (n moments before) was A_{ni}
THEN output is B_i

The output membership function is represented by its samples $(H_{1i}, H_{2i}, \ldots, H_{pi})$.

The above rules describe the fuzzy logic system in its classic case (Mamdani type). Similar descriptions are obtained for other types of fuzzy systems. The fuzzy system provided with a memory at the input will be used in subsequent sections as a basic subsystem in neuro-fuzzy systems.

Extensions to recurrence, i.e. memory of the output values, that influence the present behaviour (previous behaviour and performance influence next decisions) are easy to derive.

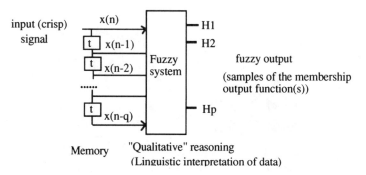

Figure 9.2 A fuzzy system with multiple inputs and memory to store samples of an external process (signal). The samples are the input data. The fuzzy system can be seen as a qualitative reasoning system

9.3.3 Classic Neurons

An artificial neuron is a simple system with many inputs and one or more outputs (see Figure 9.3). Such a system mimics a natural neuron that has many synapse (about 1000 to 10 000), through which it interconnects to other neurons. Some of the synapse of the natural neurons act as input gates for the information, and the others act as output gates. Every synapse is characterized by the efficiency of realizing the interconnection. In natural neural networks, the interconnections are realized by electric currents. In an artificial neuron, the synapses are represented by weights, standing for multiplicative factors.

Let us fix our attention on a neuron with one output only. (This case is easy to generalize to more outputs.) Let us denote by $x_1, x_2, \ldots, x_n$ the inputs, by y the output, and by w_i the weights. Then the equation describing the elementary neuron is

$$y = f\left(\sum_{i=1}^{n} w_i x_i\right) = f(w_1 x_1 + w_2 x_2 + \cdots + w_n x_n)$$

where $f(.)$ stands for the function of the neuron (characteristic function). In this equation one gives no preference to any synapses: their effect is identical if they have equal weights, as their effect sums to the effect of the other synapse. The expression of the output is symmetrical with respect to the (weighted) inputs $w_1 x_1, w_2 x_2, \ldots, w_n x_n$. Hence, one can denote shortly the total input by

$$x = w_1 x_1 + w_2 x_2 + \cdots + w_n x_n$$

We shall call the value of x 'activation'.

Because the neuron sums the inputs and transforms the resulting value according to the nonlinear function f, its implementation is split into two blocks, one being an adder and the other a nonlinear function in one variable only. From now on, by 'neuron' we shall understand this type of artificial neuron, when not otherwise specified.

A neuron can be seen as a multi-input one-output nonlinear system (see Figure 9.4), described by an equation above, symmetric in the weighted inputs.

There is no formal restriction on the function f. In most technical applications the function $f(.)$ is a sigmoidal (S-shaped) function, to model a 'saturation effect' (see Figure 9.5).

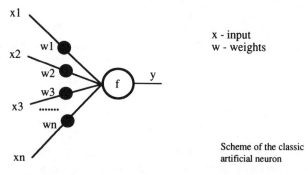

Figure 9.3 Scheme of the classic artificial neuron

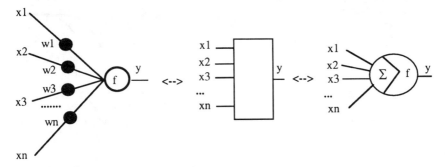

Figure 9.4 An artificial neuron is a classic vectorial nonlinear system characterized by the summation of the weighted input values

There are various classic functions that have an S-shape. The sigmoidal functions most used in neural networks are:

$$f(x) = \frac{2}{\pi} \arctan\left(\frac{x}{a}\right)$$

and the 'logistic', or 'natural growth' function:

$$f(x) = \frac{1}{1+e^{-x}} - \frac{1}{2}, \quad -\infty < x < \infty; \; -1/2 < f(x) < 1/2$$

In the general form, i.e. when not centred on the zero value and not limited to values in a unitary range, the logistic function reads

$$f(x) = \alpha + \frac{\beta}{\gamma + \xi e^{-\zeta x}}$$

Frequently used are the coefficients values $\alpha = 0$ and $\beta = \gamma = \xi = \zeta = 1$. With these coefficients, the function has values in the interval $[0, 1]$. By changing the constant ζ in the exponential, one can control the slope of the sigmoid.

Such a characteristic function can be interpreted as follows: the system is more sensitive to small absolute values of the input variables and it adapts to be less sensitive to changes when these occur for high values of the input variables.

If one introduces one extra input, with positive values only, as in the following equations:

$$f(x) = \frac{2}{\pi} \arctan\left(\frac{x-\theta}{a}\right), \quad \theta \geq 0$$

$$f(x) = \alpha + \frac{\beta}{\gamma + \xi e^{-(\zeta x - \theta)}}$$

then θ can be interpreted as a threshold. This interpretation becomes obvious when the output of the neuron has positive values only (see Figure 9.6).

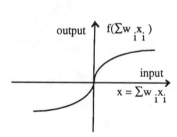
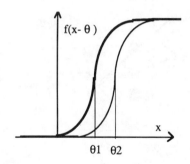

Figure 9.5 Characteristic function of a neuron: sigmoidal function

Figure 9.6 The sigmoidal function with threshold and the effect of threshold value

From the approximation theory point of view, the neurons with sigmoid functions, and correspondingly the neural networks using such neurons, are suitable for global approximation of given nonlinear systems. When a local approximation is important, functions that tend to zero when $x \to -\infty$, or when $x \to \infty$, are used. In this case, the functions of the neurons are 'located' in the space and each neuron is characterized by the coefficients of the corresponding function, showing the location. For instance, if Gaussian functions are used as characteristic functions of the neurons, then the neuron number k in a network is characterized by the couple of coefficients (a_k, b_k), and its characteristic function is

$$y = f(x) = \exp\left[\frac{-(x-a_k)^2}{b_k}\right]$$

This is a formal interpretation of the neuron, as this model has no relation with the natural neuron behaviour.

9.3.4 Linear Combiners as Neurons

The most elementary neuron is a 'linear' neuron, whose characteristic function $f(.)$ is the identity function: $f(x) = x$ for all x. The scheme of such a neuron is presented in Figure 9.7. In this figure and in subsequent figures we shall use the convention of not representing the weights. The equation corresponding to the linear neuron is

$$y = \sum_i w_i x_i = w_1 x_1 + w_2 x_2 + \cdots + w_n x_n$$

The systems described by this equation are sometimes named linear combiners, because they combine in a linear way the values of the inputs. Such systems are well known in the linear system theory.

Above, only 'static' aspects were considered, that is, processes where time is not a parameter. We now discuss a case of behaviour when time is a parameter, and memory effects are important. Suppose that only one input variable is to be considered. Then it is possible that previous values of the variable are also important in the problem in hand. In

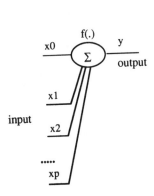

Figure 9.7 The linear neuron

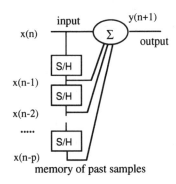

Figure 9.8 Linear neuron with memory. S/H blocks stand for memory cells and introduce a unitary delay

this situation, a 'memory' to store the previous values is needed. Such a memory acts as follows: sample (determine) the value, and hold (store, memorize) it as the last determined value; after a given lapse of time, sample again and store the new value in the 'case' (memory location) of the most recent value, pushing the previous values downward, in the 'previous' cases. Such a memory can be represented as a raw of 'Sample & Hold (S/H)' cells. Let us suppose a simple linear function to represent the process

$$y(n+1) = \sum_{k=0}^{p} a_k x_{n-k}$$

Using the representation of this function as a linear neuron, and combining the neuron with the memory, one obtains the system depicted in Figure 9.8.

9.3.5 Elementary recurrent systems

A recurrent process is a discrete dynamic process: time plays an important part, and the system variables are dependent on time. However, the characteristic of such processes is evidenced by the autonomous representation. The process $y = h(.,.,...)$ is characterized by the fact that the values of some of the input variables are values of the output at previous moments of time. For instance, the system in Figure 9.9a is a recurrent system. Some of the inputs are dependent on (i.e. are functions of) previous values of the output.

There are many processes whose evolution is currently simulated by models represented by equations in the form

$$y_n = \sum_{k=0}^{p} a_k x_{n-k} + \sum_{i=1}^{q} b_i y_{n-1}$$

These models suppose memory at both input and at output. In particular, when the process is dependent only on previous values of the output, the equation is

$$y_n = \sum_{i=1}^{q} b_i y_{n-1}$$

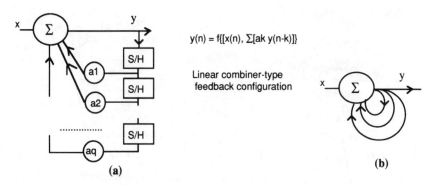

Figure 9.9 (*a*) A recurrent linear system, and (*b*) its equivalence with a recurrent neuron

For instance, in a management situation, the previous performance is perhaps as important as external influences, so the decision takes into account past performances ('outputs') too. Such a process can be represented as in Figure 9.9*a*. This can be seen as a recurrent linear neuron (Figure 9.9*b*).

More complex neurons, with nonlinear functions, or with fuzzy weights, or with fuzzy functions (see following sections) can also be used in such recurrent schemes.

9.4 FUZZY NEURONS

9.4.1 Elementary Fuzzy Neurons

The simplest neuro-fuzzy system is the so-called 'fuzzy neuron'. This is a neuron performing an operation associated with a fuzzy logic function, for example with one of the logic connectives, or to negation.

Above, classical characteristic functions $f(.)$ of the neurons were considered. However, any other type of function can be used for $f(.)$. For instance, the criterion modelled by the neuron can be 'winner takes all'; that is, a MAX function:

$$y = f(x_1, x_2, \ldots, x_n) = \mathrm{MAX}(x_1, x_2, \ldots, x_n)$$

Or one can use the complementary MIN function, for instance inputs are interpreted as 'losses' in a modelling problem.

Such neurons are named fuzzy neurons because they implement the fuzzy logic connectives' functions MAX and MIN. Other interpretations of the fuzzy logic connectives, i.e. other analytic expressions standing for the logic connectives, can also be used. A common notation for such fuzzy neurons is shown in Figure 9.10.

The operation of the MAX fuzzy neurons is similar to that of the classic sigmoidal neurons (as in Figure 9.5): the former extracts the closest to 1 (the most significative) value from the input values, whereas the last assigns a close-to-one value if at least one input is high enough. Hence, the neurons have similar behaviour for input vectors including a high value component. On the other hand, the MIN fuzzy neuron behaves similarly to a classic sigmoidal neuron when the input values are dominated by a low value.

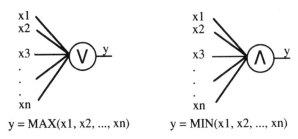

Figure 9.10 Fuzzy neurons standing for logic connectives

The MAX neuron is the equivalent of the linear neuron in the sense that it performs the 'logic sum' (represented by the MAX operation) over the input components. The MIN neuron has the same complexity, but performs the complementary operation.

A more complete fuzzy neuron is obtained by the true fuzzy counterpart of the non-linear neuron, i.e. by providing a nonlinear function block after the logic sum block. Of course, the nonlinear function block can be realized as a one-input one-output fuzzy system.

9.4.2 Neurons with Fuzzy Weights

Another way to fuzzify the neuronal model is to use fuzzy weights instead of numerical (crisp) weights. The fuzzy weights are interpreted as membership functions.

In this model of a fuzzy neuron the linear synaptic connections are replaced with a nonlinearity characterized by a membership function labeled as 'tightly connected', 'loosely connected', etc. The excitatory connections and inhibitory connections are represented by fuzzy logic intersections and fuzzy logic complements followed by fuzzy logic intersections, respectively [14, 15, 17].

9.4.3 Inclusion of Fuzzy Weights into the Conventional Neuron Model

Consider again an ordinary model of the neuron as shown in Figure 9.3, where $x_1, x_2, \ldots, x_n$ and $w_1, w_2, \ldots, W_n$ are crisp numbers (singletons) standing for the inputs and the weighting factors, respectively. Denote by θ and y the threshold level and the output of the neuron, respectively. Positive weighting factors represent excitatory connections and negative weighting factors represent inhibitory connections. The weighted sum of inputs produces a nonlinear output over the threshold level, according to the sigmoidal function. This model has only one parameter, the corresponding weight w_i, characterizing each synaptic junction, and has only one nonlinear (sigmoid) function in the output stage. This small freedom forces us to construct complicated networks to achieve sophisticated tasks, such as pattern recognition.

In order to generate a more powerful neuron model, the classic neuron is modified by assigning a membership function to each synapse, and allowing the inputs and the outputs to be membership functions. In hardware implementations, the membership functions are represented by samples. The operation is as follows.

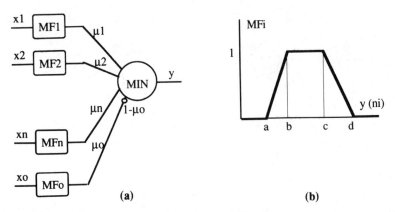

Figure 9.11 (a) Fuzzy neuron. (b) Typical membership function of the weights

The scalar input signal x_i to the ith synapse in a classic neuron is replaced in this type of fuzzy neuron with an n_i element fuzzy vector input signal x_i. This is represented by a train of n_i numbers ranging from 0 to 1. For instance, these numbers can be (0.1, 0.3, 0.5, 0.6, 0.8, 0.5, 0.3, 0.2, 0.1, 0, 0), as shown in Figure 9.11a, where n_i represents the number of elements in each synaptic input.

The deterministic synaptic weight w_i in the classic neuron is replaced by a weight represented by a fuzzy number (fuzzy weight). This is characterized by a membership function MF_i (practically, by the samples of the corresponding membership function, i.e. by a fuzzy vector of n_i elements). The typical shape of membership function is the trapezoidal one, characterized by four parameters a, b, c and d, as shown in Figure 9.11b.

The excitatory connections perform the MIN (minimum) operation, and the inhibitory connection performs the fuzzy logic complement $(1 - \mu_i)$ followed by the MIN operation. Weighting (algebraic product of input x_i and w_i) at the synapse level is performed as a fuzzy inner product of the fuzzy input vector x_i and the fuzzy weight MF_i.

No threshold level is assigned in this fuzzy neuron.

The weighting at a synapse of a fuzzy neuron, described above, produces a scalar grade μ_i through the following calculation:

$$\begin{aligned}
\mu_i &= MF_i \otimes x_i \\
&= (m_{i1}, m_{i2}, \ldots, m_{ik}, \ldots, m_{in_i}) \otimes (x_{i1}, x_{i2}, \ldots, x_{ik}, \ldots, x_{in_i}) \\
&= (m_{i1} \wedge x_{i1}) \vee (m_{i2} \wedge x_{i2}) \vee \cdots \vee (m_{ik} \wedge x_{ik}) \vee \cdots \vee (m_{in_i} \wedge x_{in_i}) \\
&= \bigvee_{k=1}^{n_i} (m_{ik} \wedge x_{ik})
\end{aligned}$$

where m_{ik} and x_{ik} represent the kth fuzzy singletons, standing for MF_i and x_i, respectively. The symbols $\wedge$ and $\vee$ stand for the MIN and MAX operations, respectively, and $\otimes$ is the MIN–MAX composition [25, 26]. In Figure 9.12 the application and physical meaning of the above equation is illustrated. Figure 9.12a illustrates the soft matching between two membership functions x_i and MF_i and μ_i stands for a degree of soft matching or similarity

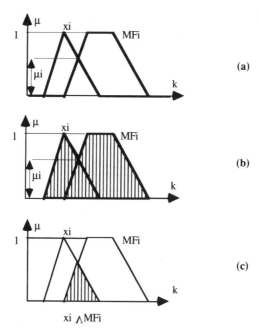

Figure 9.12 (a) A degree of soft matching between two membership functions x_i and MF_i. (b) A sampled membership function of x_i (fuzzy input vector) and of MF_i. (c) A grade of soft matching calculated by minimization and maximization (m_{ik} and x_{ik} represent grades of the kth fuzzy singletons in fuzzy vectors x_i and MF_i, respectively)

measure. Figure 9.12b shows how the membership functions are sampled, to yield a set of singletons m_{ik} for MF_i and x_{ik} for x_i. Each fuzzy singleton (i.e. a singleton with a truth value between 0 and 1) is delivered on one of the signal lines of the data bus of the synapse. Taking the minimum value of the corresponding singletons m_{ik} and x_{ik} is followed by taking the maximum value for over all k, to calculate a scalar grade μ_i, as shown in Figure 9.12d.

In the excitatory synaptic junctions the scalar grade μ_i is delivered directly to the MIN block which aggregates all the synapse outputs. In the inhibitory synapse, a scalar grade is obtained through the fuzzy logic complement $(1 - \mu_i)$. The output of the fuzzy neuron y_j is thus a scalar grade ranging from 0 to 1.

An example of practical operation of this fuzzy neuron is as follows. Fuzzy weights are used as a memory, to store representations of patterns. At one input port of a fuzzy neuron, a fuzzy vector independent of other input fuzzy vectors, e.g. pattern, voice, etc., is compared with the corresponding fuzzy weights in the same synapses to obtain the 'grade' of soft matching. The grade appears at the synapse output as a value μ_i ranging from 0 to 1. When the synapse is inhibitory, the effective synapse output is represented by $\mu'_i = 1 - \mu_i$. The minimum value of all the synapse outputs appears at the output of the fuzzy neuron and represents the possibility of simultaneous soft matching between a set of fuzzy vectors and a set of fuzzy weights. If the grade of soft matching is high, the effective output of the inhibitory synapse is low and suppresses other synapse outputs to produce a low output of the fuzzy neuron.

This fuzzy neuron model has much more flexibility than a conventional neuron model and thus may facilitate pattern recognition with much simpler structure of network (possibly with one neuron). This type of fuzzy neuron was integrated in a monolithic form on a silicon chip to guarantee a compact hardware system.

Now we give an example of an application: pattern recognition. The use of very precise information in pattern recognition is generally not needed; moreover, processing of high precision data reduces the computation and pattern recognition speed. On the other hand, some features of the shapes (patterns) can be more informative than knowing the exact shape.

We consider below handwritten numerals and letters. Features can be extracted from the symbols by employing cross-detecting lines as shown in Figure 9.13. A mesh of orthogonal 'crosslines' are used to extract the important information. To every line in the net, a string of photodetectors is used to determine the handwritten line position. For letters and characters, there are three types of regions on the cross-detecting line: crossing area, forbidden area and don't care area. A crossing area is the portion of the cross-detecting line where the symbol should cross. The forbidden area is the portion of the cross-detecting line where the symbol should not cross. The don't care area is the portion of the cross-detecting line where neither the crossing nor the uncrossing is essential, or where the handwritten line uncertainly crosses.

The boundaries of these three regions are fuzzy, so they should be defined by membership functions as shown in Figures 9.14 and 9.15. If there are two or more crossing areas on one cross-detecting line like in Figure 9.13, a corresponding number of membership functions have to be assigned on that cross-detecting line. In the example corresponding to Figure 9.14, cross-points between a cross-detecting line and a symbol, give three

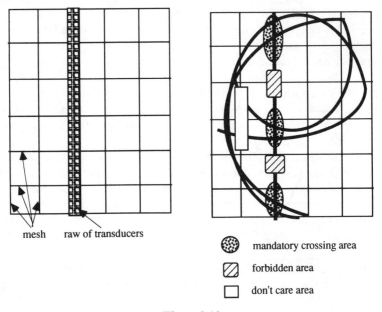

Figure 9.13

FUZZY NEURONS 281

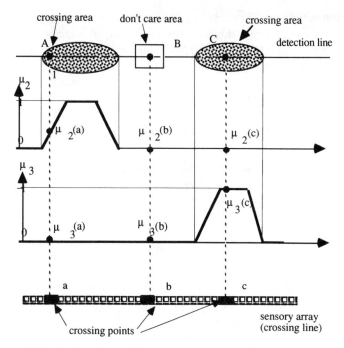

Figure 9.14 Assignment of the membership function of a crossing area

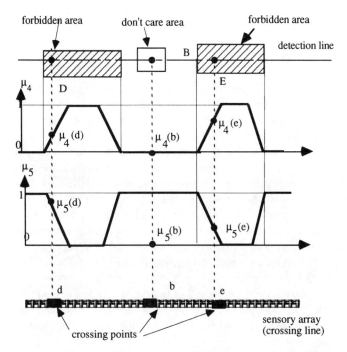

Figure 9.15 Assigment of the membership function of the forbidden area

grades of membership for each membership function, for instance $\mu_2(a)$, $\mu_2(b)$ and $\mu_2(c)$. The minimum value of these three grades does not give the grade at which these cross-points satisfy the feature of the symbol of interest. The cross-point in the don't care area gives zero-grade, though point b is not significant. Only points a and c are significant in this case. The membership functions μ_2 and μ_3 give the grades at which a and c satisfy crossing areas A and B, respectively. Consequently

$$\{\mu_2(a) \vee \mu_2(b) \vee \mu_2(c)\} \wedge \{\mu_3(a) \vee \mu_3(b) \vee \mu_3(c)\}$$

gives the grade of possibility of e concerning the selected cross-detecting line.

Figure 9.15 shows a membership function of two forbiden areas D and E and one don't care area B. The cross-points between the cross-detecting line and the symbol, as determined by the sensor array, are d, b and e. The grade of rejection is obtained by

$$\mu_4(d) \vee \mu_4(b) \vee \mu_4(e)$$

Therefore, the grade of possibility of the symbol is represented by the complement of rejection:

$$\overline{\mu_4(d) \vee \mu_4(b) \vee \mu_4(e)}$$

By De Morgan's law, this can be reduced to

$$\overline{\mu_4(d) \vee \mu_4(b) \vee \mu_4(e)} = \overline{\mu_4(d)} \wedge \overline{\mu_4(b)} \wedge \overline{\mu_4(e)} = \mu_5(d) \wedge \mu_5(b) \wedge \mu_5(e)$$

where μ_5 is a fuzzy logic complement (NOT) of a membership function μ_4 representing forbidden areas D and E.

Thus, the two grades of possibility, one from a crossing area and the other from a forbidden area, are obtained for each cross-detecting line. The final decision of the symbol (pattern) recognition is made by taking the minimum value of these grades on all the cross-detecting lines.

There exists a powerful design algorithm for the design of the membership functions of the weights, in pattern recognition applications [19]. The algorithm utilizes sample data of various hand-written symbols. It categorizes and classifies them to obtain the membership functions for crossing and forbidden areas.

The fuzzy neurons are utilized to recognize a hand-written symbol by assigning one or more fuzzy neurons to each pattern (symbol). Simple symbols, for instance 1, 8 or 0, need only one neuron, whereas two fuzzy neurons are needed to recognize 2, 3, 5, 6, 7, 9 and some of the letters. When two neurons are used, their outputs are logically AND-ed (minimized) to obtain the possibility of each character. So the latter case guarantees the higher score rather than the former case. In the case of 4 and of some letters, three or four neurons are needed. The outputs of three fuzzy neurons are OR-ed (and outputs of four fuzzy neurons are also OR-ed), and the outputs of the two MAX blocks are AND-ed to obtain the symbol possibility. The possibilities range from 0 to 1 for each symbol. The result looks like, for instance: the possibilities are 4 at 0.8, 9 at 0.6, 7 at 0.2, and so on.

9.5 NEURAL NETWORKS

In preceding sections we have discussed only issues related to the meaning of a single neuron. Of course, a single neuron can help in modelling only very elementary processes. Complex processes involve many subprocesses in networks of subsystems.

Many identical neurons are gathered in a neural network by interconnecting them according to a given graph (network topology).

The characteristic of a typical network of artificial neurons is its high degree of interconnectivity. This is not only a good model for natural neural networks (parts of a natural nervous system), but for many other complex systems too.

In the most simple case the neurons act independently, but the inputs are common (see Figure 9.16). The input vector in Figure 9.16 has n components $(x_1, x_2, \ldots, x_n)$, and the output vector has m components $(y_1, y_2, \ldots, y_m)$. Such a 'single layer' neural network is obviously a vectorial nonlinear system. It is not necessary that the input vector and the output vector have the same dimension.

The equations of a single-layer network of linear neurons are

$$y_k = \sum_{i=1}^{m} w_{ki} x_i, \quad k = 1, 2, \ldots, m$$

where w_{ik} is the weight from the input i to the neuron k, and y_k is the output of the neuron with number k. For a neural network with nonlinear, but identical neurons, the equations are

$$y_k = f\left(\sum_{i=1}^{m} w_{ki} x_i\right), \quad k = 1, 2, \ldots, m$$

Again, the function of the neurons is not restricted to any specific class of functions.

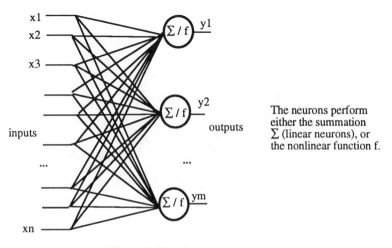

The neurons perform either the summation Σ (linear neurons), or the nonlinear function f.

Figure 9.16 One-layer network

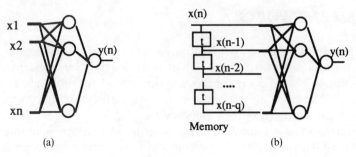

Figure 9.17

If the output of a single-layer neural network has to be used as a basis for aggregating the results, for instance to make a decision, then the first layer of neurons is followed by a second layer possibly consisting of one neuron only (see Figure 9.17a). The output of the network is

$$y^{(2)} = f_2\left(\sum_{k=1}^{m} w_k^{(2)} y_k^{(1)}\right)$$

$$y_k^{(1)} = f_1\left(\sum_{i=1}^{n} w_{ki}^{(1)} x_i\right), \quad k = 1, \ldots, m$$

where $f_1(.)$ is the characteristic function of the neurons on the first (input) layer, and f_2) stands for the characteristic function of the output (aggregating) neuron. The weights corresponding to the input layer are $w_{ki}^{(1)}$, and the weights of the output neuron are $w_k^{(2)}$.

In Figure 9.17b the same neural network is provided with a memory, to store the samples of the input signal.

A neural network can have any number of layers. A multilayer neural network is shown in Figure 9.18. It is not necessary that all layers have the same number of neurons. The equations of the network are

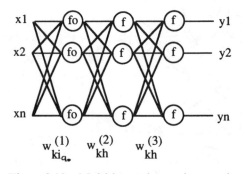

Figure 9.18 Multi-layered neural network

$$y_k^{(1)} = f_0\left(\sum_{i=1}^{n} w_{ki}x_i\right), \quad k = 1,\ldots,n$$

$$y_h^{(j)} = f\left(\sum_{k=1}^{n} w_{kh}^{(j)} y_k^{(j-1)}\right), \quad j = 2, 3$$

where all neurons are supposed to have the same characteristic function f, except the neurons on the first layer, with the characteristic function f_0. (In general, f_0 is a linear function.) To simplify the notation, all the layers are supposed to have the same number n of neurons.

Note a feature of the models based on neurons and on neural networks: they are not "rational", but analytical, in the sense that in their description one does not explain by some logic rules the way one determines the output values, as for fuzzy systems. Moreover, there is no way to say how to (easily) determine the values of the weights. The weights in a neural network are determined by successive adaptation steps. So a neuron and a neural network can stand for a representation of the empirical formula, and of empirically determined (learned, trained) controls and decision-making processes. They cannot stand for rational rule-based processes. This is a methodological advantage, because it allows the modelling of such empirical processes, based on empirical learning and adaptation of the behaviour. Indeed, the learning is done according to heuristic rules (by search of the best solution), not by solving equations. Moreover, they allow modelling of the way one 'learns', i.e. how one accumulates the experience that helps to establish the empirical knowledge. Parts of the human problem-solving processes closely resemble the empirical learning in a neural network.

9.6 DISCRETE SYSTEMS AND GENERALIZATION TO NEURO-FUZZY SYSTEMS

In this section we emphasize the connection between the discrete-time neuro-fuzzy systems obtained by combining fuzzy systems and neural networks, and some classes of usual discrete-type systems. This helps in systemizing the approach of neuro-fuzzy systems; moreover it shows the connections with the classic theory.

Let us consider some of the main classes of usual (crisp) discrete systems, to establish their possible generalizations.

The linear systems (linear combiners), are described by

$$y_n = \sum_{k=0}^{p} a_k x_{n-k} + \sum_{i=1}^{q} b_i y_{n-i}$$

where x stands for the input signal, y for the output signal, n is the current moment of time, and a_k and b_i are the system parameters (weights).

Remember that a linear moving average (MA) combiner is described by the general equation $y_n = \sum_k a_k x_{n-k}$ is equivalent to a 'linear neuron'. An auto-regressive (AR) combiner is described by the general equation $y_n = (\beta x_n +)\Sigma_i b_k y_{n-i}$ and is equivalent to a linear recurrent neuron. Finally, an auto-regressive moving average (ARMA) combiner is equivalent to a system of two linear neurons in a recurrent connection: $y_n = \Sigma_k a_k x_{n-k} + \Sigma_i b_i y_{n-i}$. The generalization of these structures leads to neurons that exhibit characteristic functions that are nonlinear, as discussed below.

The nonlinear combiners are obtained from the linear ones by introducing nonlinear functions instead of the linear functions. They are described by equations such as

$$y_n = \sum_{k=0}^{p} f_k(a_k x_{n-k}) + \sum_{i=1}^{q} g_i(b_i y_{n-i})$$

If all functions f are identical ($f_i = f_k$, for all i, k), moreover, if the functions g are identical, then the nonlinear combiner equation is

$$y_n = \sum_{k=0}^{p} f(a_k x_{n-k}) + \sum_{i=1}^{q} g(b_i y_{n-i})$$

For instance, $f(.)$ and $g(.)$ can be power functions or polynomial functions. By interchanging the places of Σ and f, one has the equation

$$y_n = f\left(\sum_{k=0}^{p} a_k x_{n-k}\right) + g\left(\sum_{i=1}^{q} b_i y_{n-i}\right)$$

Note that this is the equation of a two-neuron (recurrent) network, with the function f, respectively g, and with delayed inputs. In particular

$$y_n = x_n + g\left(\sum_{i=1}^{q} b_i y_{n-1}\right)$$

This equation can stand for the mixed linear system (neuron configuration, depicted in Figure 9.19a) or a fuzzy system (linear neuron configuration, as in Figure 9.19b). In the last case, the function $f(.)$ is realized by the fuzzy system with defuzzifier.

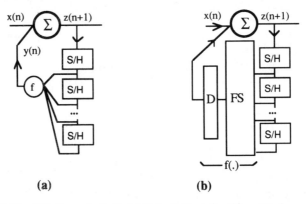

Figure 9.19 SH, Sample & Hold (delay); D, defuzzifier; FS, fuzzy system

In this way, by interpreting the nonlinear functions as standing either for fuzzy systems with defuzzifier, or for neural networks, one can generate various neuro-fuzzy configurations.

A more general type of nonlinear discrete systems has the general equation

$$y_n = g(\ldots, x_{n-k}, \ldots, x_n; \ldots y_{n-i}; \ldots, y_{n+1})$$

where $g(.,.,\ldots)$ is any function (usually from C^∞). For instance

$$y_n = g[x_n, f(y_{n-1}, \ldots, y_{n-k}, \ldots y_{n-q})]$$

describes a system that can stand for a fuzzy system with two inputs, one being x_n and the other being $f(y_{n-1}, \ldots, y_{n-k}, \ldots, y_{n-q})$. Also, the function $f(.,.)$ can stand for the characteristic function of a neuron. Moreover, the function g can stand for a neuron, replacing in Figure 9.19 the adder. So such systems can be suitably modelled by of couple of systems, one being a neuron and the other a fuzzy system with defuzzifier, or both being neurons, or both being fuzzy systems with defuzzifiers.

Taking into account this discussion, different classes of neuro-fuzzy systems can be developed and used as models of complex phenomena described by the above recurrent formula.

9.7 INVARIANT NEURO-FUZZY SYSTEMS

9.7.1 Main Configurations

In this section we present some of the main possibilities of interconnecting neuralnetworks and fuzzy systems, with a focus on loop-type (recurrent) connections. From the architectural point of view, some of the main variants of combinations of fuzzy systems with neural networks are sketched in Figure 9.20. Other derived configurations are easy to imagine.

All fuzzy systems in Figure 9.20 have sampled output membership functions.

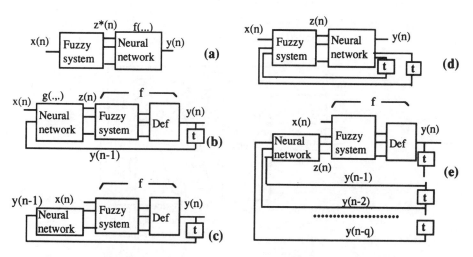

Figure 9.20 Basic configurations for fuzzy system—neural network connections. Def stands for defuzzifier; t stands for unitary delay (sample and hold operation); n stands for the current time moment

The equations of the systems sketched in Figure 9.20, and a brief discussion of the implementation of these equations, are given below. Note that different values of the delays (i.e. values of delays equal to $2, 3, \ldots$, introduced between, or by the neural network, by the fuzzy system, by the defuzzifier or by the connecting paths) can be considered in the equations. This is dependent on the memory process implemented by the system.

Case a This is a series fuzzy system, neural network architecture, represented in a general form. The equation is

$$y_n = f(z_n^1, z_n^2, \ldots, z_n^r)$$

where $z_n^1, z_n^2, \ldots, z_n^r$ represent the first, second, $\ldots$, rth values of the sampled output membership function of the fuzzy system at time moment n. In this case, the fuzzy system is not provided with a defuzzifier. In this example, for simplicity, the fuzzy system is considered to have only one output variable. Obviously, z_n^j are functions of the input x_n. The function performed by the neural network, supposed to have only one neuron on the output layer, is denoted by $f(.)$. The fuzzy system has no defuzzifier.

One simple (but not the only) 'interpretation' of this structure is as follows. One has to build a nonlinear system, and some raw basic knowledge exists about the desired behaviour. This knowledge is incorporated into the fuzzy system. However, a tuning of the obtained characteristic function is necessary (may be for the purpose of adaptability or robustness). The tuning is performed by the neural network that is trained to perform the best defuzzifying–tuning mixed operation.

A simple extension of the above case is obtained by memorizing samples of the input signal and using a multiple-input fuzzy system. One obtains the natural architecture mixing a fuzzy system and neurons in a two systems in series configuration, as sketched in Figure 9.21. This is a simple case using delayed samples of the incoming signal.

Case b This type of system (Figure 9.20b) has a loop configuration, with the input signal entering the neural network. The output of the fuzzy signal is defuzzified. The

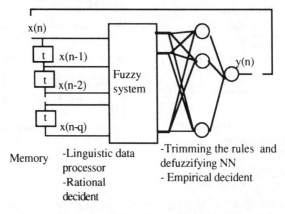

Figure 9.21 Basic series configuration. Note some interpretation of the roles of the blocks. The possibility of making the system recurrent by adding a feedback is also shown

equations of the system are

$$y_n = f(z_n^1, z_n^2, \ldots, z_n^r) \quad \text{with } z_n^j = g_j(x_n, y_{n-1})$$

so

$$y_n = f(g_1(x_n, y_{n-1}), g_2(x_n, y_{n-1}), \ldots)$$

where f is the function of the fuzzy system (defuzzifier included), and g_i are the functions performed by the neural network, corresponding to each of its outputs. Note that the neural network has only two neurons on the input layer. The number of neurons on the output layer should be equal to the number of inputs of the fuzzy system.

Variants of this configuration, with neural networks with more than two inputs, the inputs to the neural network being $x_n, y_n, y_{n-1}, \ldots$, are easy to derive.

Case c These systems (Figure 9.20c) also have a loop configuration, with the input signal entering the fuzzy system. The output of the fuzzy signal is defuzzified. The equation is

$$y_n = f[x_n, g_1(y_{n-1}), g_2(y_{n-1}), \ldots]$$

where f is the function of the fuzzy system (defuzzifier included), and g_i are the functions performed by the neural network, corresponding to each output. Note that the neural network has only one neuron on the input layer.

There are numerous variants of this configuration, using neural networks and fuzzy systems with more than one input. The reader will find them easy.

Case d In this loop configuration (Figure 9.20d), the fuzzy system output is not defuzzified. The equations are

$$y_n = g_1(z_n^1, z_n^2, \ldots, z_n^r)$$

$$z_n^j = f_j[x_n, g_1(z_{n-1}^1, z_{n-1}^2, \ldots, z_{n-1}^r), g_1(z_{n-1}^1, z_{n-1}^2, \ldots, z_{n-1}^r), \ldots]$$

where g_1 is the function performed by the neural network at output 1, etc.

In all the above cases, mainly when loops are used, one can generalize the structure by introducing more delays (not necessary unitary delays). For instance

Case e This case (Figure 9.20e) generalizes Case c, by introducing delays. The equations can be easily derived from the above ones.

As the systems in the classes *b–e* are recurrent systems, and because of the nonlinear (and possibly nonmonotonic) character of neuro-fuzzy systems, these systems will often exhibit chaotic behaviour

Different types of neuro-fuzzy loops in the above categories were introduced in [8, 9, 10] and in other subsequent papers as systems generating specific chaotic behaviours.

In the above descriptions, classic fuzzy systems (fuzzy input–fuzzy output, i.e. Mamdani-type) were used. Similar configurations can be easily derived by using fuzzy input–numerical output (Sugeno-type) fuzzy systems. Details are skipped.

9.7.2 Series Neuro-fuzzy Systems and their Interpretation

In this section we shall briefly discuss some series-connection (nonrecurrent) neuro-fuzzy systems, introduced in [8, 9, 10]. Of course, by adding to these systems a feedback loop, one obtains recurrent systems. Moreover, such systems can stand as 'elementary' subsystems in a network of systems.

In applications, it is crucial to give an appropriate interpretation to the elements used in the system; either it is a model, or not. Moreover, it is essential to establish the modelling capabilities of the elementary subsystems used in the model.

The interpretation of some series configurations will be discussed below.

9.7.2.1 Fuzzy system tuned and defuzzified by a neural network

This scheme was basically introduced in Figure 9.20a and Figure 9.21. Here we provide another interpretation, disregarding delayed inputs, in connection with schemes in Figure 9.22.

Consider that some knowledge was incorporated in the fuzzy system. However, this basic knowledge does not allow the fuzzy system to fit well enough the application in hand, and a tuning of the fuzzy system has to be done. There are several methods to perform this tuning, for instance by changing the rules, by changing the membership functions' shapes, by changing the interpretation (expression) of the logic connectives, by changing the belief degrees in the rules, or by changing the defuzzification expression

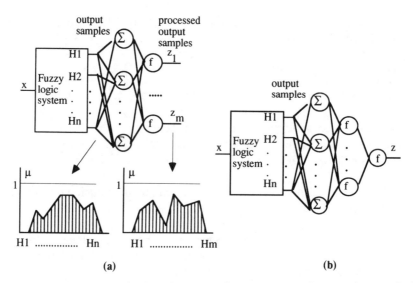

Figure 9.22 Direct series connections between a fuzzy system and a neural network. (*a*) The neural multiple-output network transforms the output membership function of the fuzzy system. (*b*) The one output neural network transforms the output of the fuzzy systems, moreover the last layer neuron performs the defuzzification. $H_1, \ldots, H_n$ denote the samples of the membership function

(when defuzzification is used). The scheme in Figure 9.22a offers an alternative to these methods, by performing an external tuning using a 'correcting' neural network.

The scheme in Figure 9.22b has the advantage that it allows the implementation by the neural network of two functions: the 'defuzzification' function, in fact, of any mapping that stands for a $R^{[0,1]} \to R$ function, even if it does not obey the classic requirements for the defuzzification; adjustment (tuning) of the function performed by the fuzzy system.

Moreover, it offers the flexibility of determining, by adjusting (learning) the neural network only, the result of the corresponding belief degrees in the rules, and of the defuzzification method.

The derivation of the corresponding equations is left to the reader.

9.7.2.2 Architecture with defuzzification after the neural network

Another hybrid configuration is shown in Figure 9.23. Here, the defuzzification task is performed by a classic defuzzifier block, whereas the neural network performs an adaptation equivalent to changing the rules, the belief degrees in the rules, the meaning of the logic connectives, or all of them.

Moreover, the neural network is able to adapt itself (learn) and to find the best modification of a given fuzzy system in a specified application, as above described. A memory at the input, as sketched in Figure 9.23, allows the use of this system in signal processing. (For simplicity, a fuzzy system with only one output is considered.)

By performing the defuzzification after the neural network, one obtains a truly hybrid system, as the neural network can be interpreted as being a part of the fuzzy system, or vice versa.

The scheme behaves, in a simple interpretation, as a two-state approximator, the fuzzy system performing the heavy part of the task, while the neural network tunes the output membership function. As each value at the output of the neural system is a function of all sample values of the fuzzy system output, this is also equivalent to adding more rules to the fuzzy system, or adding belief degrees (different of 0 and 1) to the rules in the fuzzy system.

As an example, consider that the number of outputs of the neural network is equal to that of the inputs (i.e. to the number of outputs of the fuzzy system). Consider that for a given x only two output membership functions are fired. Let us say that the correspond-

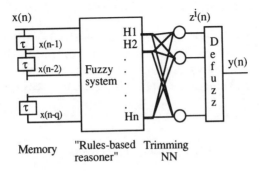

Figure 9.23 Serried fuzzy system, neural network with neural network regarded as part of the fuzzy inference (defuzzifier after the neural network)

ing output samples that are not vanishing are $B_q, \ldots, B_p$. Because of the mixing effect of the neural network, generally all neural network output will be non-vanishing. Consequently, this is equivalent to the situation, when for any x there are enough rules to yield all membership functions. This reads

If x is A_r
Then y is B_1 and y is $B_2 \ldots$ and $\ldots$ and y is B_n

This is possible in a nontrivial way in a typical fuzzy system only by using different degrees of belief in the rules

If x is A_r Then y is B_1 with belief degree β_1
and y is B_2 with belief degree $\beta_2 \ldots$
and $\ldots$
and y is B_n with belief degree β_n

Consequently, we have the following property. A fuzzy system with belief degrees in the rules (weighted rules) is implemented by a simple fuzzy system (with belief degrees in rules equal to 1) followed by a neural network.

The neural network is easily trimmed to determine the belief degrees for a specific application.

The above property also shows the equivalence between the chaotic fuzzy systems with belief degrees in the rules and the chaotic neuro-fuzzy systems involving fuzzy systems with complete belief (belief degree equal to 1) in the rules.

9.8 SEVERAL RECURRENT NEURO-FUZZY SYSTEM CONFIGURATIONS

We consider in this section several types of neuro-fuzzy configurations and their loop interconnections, with emphasis on their use in modelling. These recurrent systems exhibit an interesting dynamic behaviour, including chaos [8–11].

9.8.1 Elementary Loops: Models of Memory Effects (Output Memory)

In Figure 9.24 is sketched a very simple but still nontrivial system mixing fuzzy systems and neurons in a loop. The fuzzy system is supposed to operate in a discrete manner, i.e. in discrete time it generates a new value at the output after a determined laps of time from the moment of getting a new value at the input. Moreover, the fuzzy system has a defuzzified output (Defuz block).

The equations of the loop are

$$z_{n+1} = g(x_n, y_n) \quad \text{and} \quad y_n = f(z_n, z_{n-1}, \ldots, z_{n-q})$$

In the above equation, $f(.)$ is supposed to stand for a neuron, i.e. the inputs are summed and the function $f(z_n, z_{n-1}, \ldots, z_{n-q})$ can be written as a function in $\sum a_k z_{n-k}$.

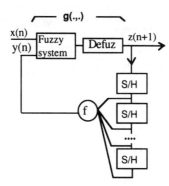

Figure 9.24 An elementary, non-trivial fuzzy-neural association

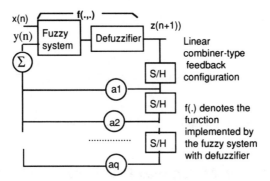

Figure 9.25 Fuzzy logic linear combiner

This system is transformed in that shown in Figure 9.25 by using a linear neuron (simple adder), instead of a nonlinear neuron. In this way, one obtains the 'fuzzy logic linear combiner', sketched in Figure 9.25.

A variant of this configuration is obtained by using an elementary fuzzy neuron (MAX-type), as in Figure 9.26.

Here, the system implements the equation

$$z_{n+1} = f(x_n, y_n) = f[x_n, \text{MAX}[a_1 z_n; a_2 z_{n-1}; a_k z_{n-k}]]$$

where y is the defuzzified output of the system, f represents the nonlinear function of the fuzzy system plus defuzzifier, n is the current moment of time, k is the corresponding delay, and $a_1, \ldots, a_k$ are the multiplication factors.

The interpretation of the system in Figure 9.26, from the modelling point of view, is as follows: the rational decider uses imprecise inference; he is represented by the fuzzy system; he considers the best past performance only (MAX operation) in his decision process; a forgetting process is modelled by the weights decreasing from a_0 to a_q. (Other interpretations of the weights, such as data reliability degree, are also possible.)

All above configurations can model a single decider that considers in the decision process current input (external, stimulus) values as well as a function of previous values.

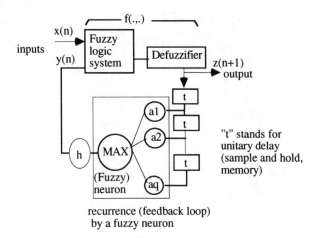

Figure 9.26

This function, which can be linear or nonlinear, stands for the process of aggregation (weighting) of the past performances.

In the next section, examples of implementation of systems with coupling between the x and y variables, or between x and z variables, are shown.

9.8.2 Implementation of Complex Equations and Connections with Chaos in Classic Systems

As is well known, all classic chaotic discrete systems are recurrent systems. If the aim is to create more powerful, more general and flexible chaotic systems by using fuzzy and neuro-fuzzy systems, the equations of classic chaotic systems have to be generalized and implemented in a suitable way. This issue is dealt with below.

In this section, examples of how to implement classic systems of discrete equations by neuro-fuzzy systems are indicated. The examples show the power of neuro-fuzzy systems in modelling complex systems and in other applications.

The advantage of these neuro-fuzzy systems is that one can shape the function $f(.,..)$ standing for the fuzzy system with defuzzifier by changing the rules and the membership functions in a easily understandable way.

The next figure presents a generalized ARMA-type configurations, but with coupled inputs.

The system in Figure 9.27 has the equations

$$x(n) = cy(n-1) + bz(n)$$
$$y(n) = \sum a_k z(n-k)$$
$$z(n+1) = f(x(n), y(n)) = f[cy(n-1) + bz(n)$$
$$\sum a_k z(n-k)] = f[c\sum a_k z(n-k-1) + bz(n)$$

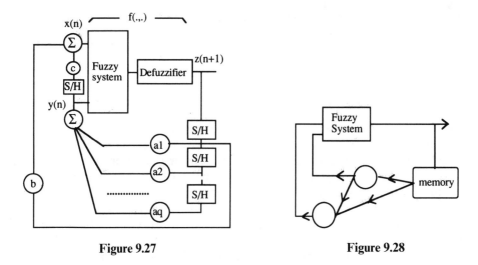

Figure 9.27   Figure 9.28

Obviously, this is a nonlinear recurrent system. For different types of nonlinear functions $f(.)$ implemented by the fuzzy system, and for specific values of the coefficients, the behaviour of the system exhibits chaos.

One can say that the above configuration includes a fuzzy system and two neurons, both attached to the inputs, as in Figure 9.28.

The scheme in Figure 9.29 has the equations

$$x(n) = cy(n-1) + bz(n) + x(n-1)$$
$$y(n) = \sum a_k z(n-k) + y(n-1)$$
$$z(n+1) = f(x(n), y(n)) + z_n$$

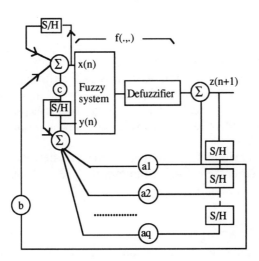

Figure 9.29 A neuro-fuzzy system that stands for a generalization of classic chaotic processes

The system can implement the coupled discrete equations

$$z(n+1) - z(n) = f(x(n), y(n))$$
$$x(n) - x(n-1) = cy(n-1) + bz(n)$$
$$y(n) - y(n-1) = \sum a_k z(n-k)$$

that generalize almost all classic systems of chaotic systems.

The scheme in Figure 9.29 can be regarded as a mixed system, consisting of a fuzzy system and three neurons, one for each input and each having a memory attached to store the history of the process (i.e. the previous values of the variables). One can see the neuron attached to the y input and to the output as forming a single layer, the output layer being the neuron attached to input x.

It is easy to see that, in the way shown above, one can develop neuro-fuzzy models for any classic system described by linear or nonlinear systems. This is also true for the development of neuro-fuzzy models of chaotic processes described by discrete equations [8, 9, 10].

9.9 FINAL REMARKS

Neuro-fuzzy systems including either fuzzy neurons in a neural network structure, or neural networks and fuzzy systems as distinct subsystems in a hybrid structure can be useful in various applications such as

> Modelling of complex processes from various fields, when the modelled process includes human decision-making based on both rational knowledge (rules) and empirical-driven, experience-based knowledge
> Control of complex processes
> Pattern recognition
> Development of systems that mimic classic systems described by discrete coupled equations
> Chaos generation

Due to the complexity of the systems involved, the study of these systems is often performed by computer simulation. Also, it is difficult to determine analytically the stability of the complex neuro-fuzzy systems.

The networks of fuzzy systems, or neuro-fuzzy systems as presented above, possibly also including 'independent' neural networks, can stand as models for complex systems, with many subsystems.

In general, the behaviour of fuzzy and neuro-fuzzy models of chaotic classic systems is very similarly to that of their classic counterpart.

REFERENCES

1 Ahn, T., Oh, S., and Woo, K. (1993) Automatic generation of fuzzy rules using the fuzzy-neural networks. *Proc. Fifth IFSA Congress*, pp. 1181–1185.

2 Buckley, J., and Hayashi, Y. (1993) Hybrid neural nets can be fuzzy controllers and fuzzy expert systems. *Fuzzy Sets and Systems*, **60**, 135–142.
3 Gupta, M. M., and Rao, D. H. (1994) On the principles of fuzzy neural networks. *Fuzzy Sets and Systems*, **61**, 1–18.
4 Katayama, R. *et al.* (1993) Dimension analysis of chaotic time series using self generating neuro-fuzzy model *Proc. Fifth IFSA World Congress*, pp. 857–860.
5 Lee, M. H., Lee, S. Y., and Park, C. H. (1993) Neuro-fuzzy identifiers and controllers for fuzzy systems. *Proc. Fifth IFSA Congress*, pp. 177–180.
6 Lee, M. H., Lee, S. Y., and Park, C. H. (1994) Neuro-fuzzy identifiers and controllers for fuzzy systems. *Journal of Intelligent and Fuzzy Systems*, **2**, 1–14.
7 Teodorescu, H. N. (1992) Generalized fuzzy neural networks. In H. N. Teodorescu, T. Yamakawa, A. Rascanu (eds) *Fuzzy Systems and Applications* (Iasi Polytechnic Press, Iasi).
8 Teodorescu, H. N. (1993) *Chaos in networks of fuzzy systems*. Tutorial, EUFIT'93, Aachen, Germany.
9 Teodorescu, H. N. (1994a) Architecture for neuro-fuzzy discrete systems (filters). *Proc. Int. Conf. IIZUKA'94*.
10 Teodorescu, H. N. (1994b) Non-linear systems, fuzzy systems, and neural networks. *Proc. Int. Conf. IIZUKA'94*.
11 Teodorescu, H. N. *et al.* (1994a) Analysis of a chaotic trade model and improved chaotic fuzzy trade model. *Proc. Int. Conf. IIZUKA'94*.
12 Teodorescu, H.-N. L. *et al.* (1994b) Interpretation of neuro-fuzzy systems in models in management and creativity, chaos generation. *Proc. First SIGEF Conference*, Reus, Spain.
13 Tokunaga, M. *et al.* (1992) Learning mechanism and an application of FFS-network reasoning system. *Fuzzy Systems and A.I.*, **1**, 33–39.
14 Yamakawa, T. (1989a) Japanese Patent Application, A Fuzzy Neuron, TOKUGANHEI No. 1-133690 (TOKUKAIHEI No. 2-310782), May 1989.
15 Yamakawa, T., and Tomoda, S. (1989b) A fuzzy neuron and its application to pattern recognition. *Proc. Third IFSA Congress*, Seattle, pp. 30–38.
16 Yamakawa, T. (1990) Pattern recognition hardware system employing a fuzzy neuron. *Proc. International Conference on Fuzzy Logic and Neural Networks*, Iizuka, Japan, pp. 943–948.
17 Yamakawa, T. (1991a) A fuzzy neuron and its application to a hand-written character recognition system. *Proc. IEEE International Symposium on Circuits and Systems*, Singapore, pp. 1369–1372.
18 Yamakawa, T. *et al.* (1991b) Japanese Patent Applications, TOKUGANHEI No. 3-101392, No. 3-101393, No. 3-102757, No. 3-102758, May 1991.
19 Yamakawa, T. (1992a) Japanese Patent Application, 'A Nonlinear-Synapse Neuron,' TOKUGANHEI No. 4-132897, May 1992.
20 Yamakawa, T. (1992b) A fuzzy logic controller. *Journal of Biotechnology*, **24**, 1–32.
21 Yamakawa, T. (1992c) A fusion of fuzzy logic and neuroscience—a fuzzy neuron and its application to a pattern recognition. *Proceedings of the 36th Annual Conference of the Institute of the Systems, Control and Information Engineers*, ISCIE, pp. 45–48.
22 Yamakawa, T., and Furukawa, M. (1992) A design algorithm of membership functions for a fuzzy neuron using example-based learning. *Proc. IEEE International Conference on Fuzzy Systems*, San Diego, pp. 75–82.
23 Yamakawa, T. *et al.* (1992a) Identification of nonlinear dynamical systems by a neo fuzzy neuron and prediction of their behaviour. *Proc. 8th Fuzzy System Symposium*, Hiroshima, pp. 249–252.

24 Yamakawa, T. *et al.* (1992b) A neo fuzzy neuron and its applications to system identification and prediction of the system behavior. *Proceedings of the 2nd International Conference on Fuzzy Logic and Neural Networks*, Iizuka, Japan, pp. 477–483.
25 Zadeh, L. A. (1973) Outline of a new approach to the analysis of complex systems and decision process. *IEEE Trans. Systems, Man and Cybernetics*, **3**, 28–44.
26 Zimmermann, H.-J. (1985) *Fuzzy Set Theory—and Its Applications* (Kluwer-Nijhoff Publishing, Boston-Dordrecht-Lancaster).

10
A Fuzzy Logic Approach to Handwriting Recognition

D. J. Ostrowski and **P. Y. K. Cheung**
Imperial College, University of London, UK

10.1 INTRODUCTION

This paper describes an approach to handwriting recognition using fuzzy logic, and is concerned with cursive script recognition that puts no constraint on the writer. Various other techniques enforce some limitations in order to simplify the recognition task. Commercial experience now shows that although this may be acceptable for notepad type applications, it is essential that the writer should use his or her most natural writing style for any more than brief (about 10 words) notes. The basic styles are illustrated in Figure 10.1. In other recognizers the writer may be constrained to input letters in boxes, making recognition easier, but input slow and tedious. Or writing may be partially or wholly unconstrained, making input fast and natural, but making recognition considerably more difficult.

The fundamental problem of cursive script is segmentation and the other variations that occur in normal handwriting, which include size and shape, stroke number and stroke order and the presence or absence of retraces. (This topic is thoroughly treated by Ward and Kuklinski, 1988.) The nature of cursive script makes reliable separation of individual letters very difficult, and hence the use of prototypical models tends to be unsatisfactory. The problem of cursive script recognition is considerably compounded by variability of handwriting, both between subjects and between separate examples from the same subject. The approach described here is to recognize letters in terms of features of individual letters, and to accept that separation of features too is unreliable. The proposed solution uses a fuzzy rule based approach which is able to accept the variability of data and its unreliable extraction from the original image, and derives its knowledge of characters and the ability to differentiate them entirely from training.

The first section is a general discussion about the human reading process and describes a few important aspects of this. The second, briefly describes some of the current methods of machine recognition. The third describes a fuzzy logic approach to handwriting

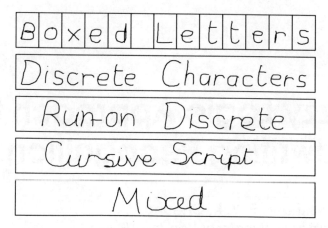

Figure 10.1. Handwriting styles

recognition, in particular why it is essential to emulate the human reader as closely as possible and how this may be done; the fourth describes the approach to training a rulebase, and the final section explains some useful metrics for measuring and monitoring the progress of training the fuzzy rulebase.

10.2 HUMAN READING

Methods used by human readers are reviewed here, so that those aspects similar to a machine model can be identified. Two fundamental points are made by Blesser *et al.* (1974). The first is that if a machine recognition tool is to work well outside its training set, it must use specific knowledge about human character recognition. The second point, concerning the recognition of ambiguous character shapes, is that these must be separated by recognition of their essential features (Figure 10.2), not by comparison with archetypal prototypes. Though there has been much research into reading, it has been unable to uncover the actual characteristics of individual letters that trigger recognition and differentiate one from another. There is, however, a general consensus about letter recognition being feature-based and the processing mechanisms used. Some of these are naturally expressed in the design of a fuzzy recognizer and the similarities are described.

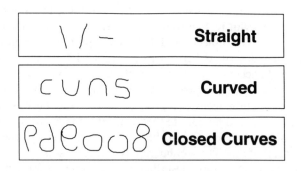

Figure 10.2 Character features

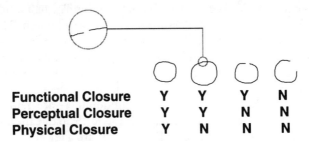

Figure 10.3 Imperfect feature connection

Cursive script recognition cannot proceed without an attempted segmentation of the characters within a word. It is likely that several possible segmentations are attempted (Well and Pollatsek, 1981), based on a potential character on either side. This proceeds from left to right until several characters or syllables are recognized with reasonable confidence. Recognition of these 'potential characters' must be character feature based (e.g. closed loops 'o', ascenders 'b', descenders 'j', etc., double arches such as 'm' must have more than one segmentation attempted as 'm' may be followed by another 'm' or 'n'). So feature recognition must first be used, followed by good character or good syllable recognition, then segmentation. This feature-based recognition looks for the degree of matching of the features of an individual letter and accepts the recognition above a threshold. The feature set is probably not minimal, allowing some redundancy and to incorporate style variations.

Handwritten characters do not usually conform to their definition. An 'o' should be a closed circle, many handwritten 'o's will be nearly closed. This may fall in to one of three cases (Figure 10.3), physical closure where the character fulfils the definition, perceptual closure where it does not, but cannot easily be seen not to by a reader, and functional closure where the character clearly falls short of its definition but in some other way is accepted as fulfilling the necessary conditions to be correctly identified. The latter two cases must be incorporated in fuzzy rule definitions of characters.

Handwriting includes ligatures (Figure 10.4) that physically join the letters. These may be an elongation of the last feature of the first character to link it to the second, or

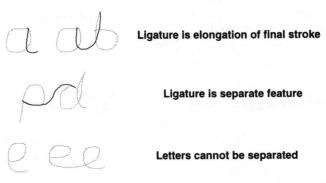

Figure 10.4 Letter joining ligatures

a particular non-feature which is drawn as a joining stroke. The first kind of join must be incorporated in the fuzzy rule for all possible types of character join. This is a relatively small set, small letter to small letter, tall to small, etc., and a few special cases (e.g. 'ee'). The latter type of join that uses a special stroke not normally part of the character set, must be defined as a separate character, so that the rulebase will not insist on its inclusion in one or other letter. Again there is a relatively small set of these.

On top of the normal feature-based letter recognition there are many techniques that help us recognize words and cope with variations in style. Probably the most important of these are 'sight words' (Otto and Stallard, 1976). These are a few hundred words that we simply recognize as objects, because they tend to be used so often. Sight words tend to be short (1–6 characters) and we do not verify that the letter order or content is correct. Context alone assures us that we have correctly recognized the word. After sight words, the next most important technique is syllables (Sakiey *et al.*, 1980). Syllables are common letter constructions with one (or may be two) specific sounds. There recognition is very like sight words, and are recognized as images, and individual letters are not matched. This process is strongly linked to syllable frequency. These two techniques are used by all experienced readers and linked with context where necessary to narrow the list of permissible words. Sight words and syllables can be considered as an essential supporter of feature-based letter recognition (see Table 10.1).

Table 10.1

The 20 most common sight words...									
the	a	of	to	and	in	is	I	be	you
it	that	n	he	for	was	an	as	re	with
The 20 most common syllables...									
ing	er	ly	ed	es	re	ion	in	con	ter
ex	al	e	com	di	en	an	ty	y	ti

It is particularly clear from the above descriptions of human reading mechanisms that the sequence of processes and what they accomplish is generally agreed upon. However, there is a lack of detail about what is actually tested for, and in particular, concerning the futures of individual letters there is little or no consensus. Within this framework, though, it is possible to describe a machine model similar to the human reader. It is essential to consider syllables and whole words, to attempt several segmentations, to have a very flexible definition (not a single model) of letters that is able to handle perceptual and functional closure, and to have rules that differentiate probable confusions and to adapt itself quickly to an individual writer.

10.3 HANDWRITING RECOGNITION: CURRENT APPROACHES

Most modern approaches to recognition use several stages of preprocessing, smoothing (removing unintended wobbles), filtering (point reduction), wild point correction, dehooking (at stroke ends), dot reduction and stroke connection (to correct unintended pen lifts). Nearly all methods require much data reduction and that the information be reduced to

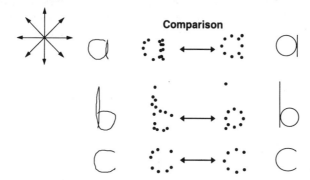

Figure 10.5 Chain codes and recognition

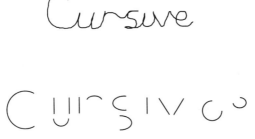

Figure 10.6 Analysis by synthesis

a known set of variables. The first stages are often followed by deskewing (slant correction), baseline drift correction, size normalization and stroke length normalization. Almost all techniques use many of these stages. Most 'clean-up' algorithms are very simple; a good summary can be found in Nair and Leedham (1991).

Many recognition techniques make use of chain codes. Principle among these is curve matching and for the roman alphabet elastic matching (Figure 10.5). This is essentially a comparison with a prototype letter of chain code points and measuring the amount of distortion necessary to fit the example to the prototype. The opposite approach is analysis by synthesis, where the example is reconstructed from a basic library of features (Figure 10.6). Most of these methods use a prototype model of characters (or features for analysis by synthesis) and all require some kind of segmentation of the script, and that the example can always be related to a prototypical model (see also Tappert, Suen and Wakahara, 1990).

The methods described in this section show the differences between the approach used by the human reader and that employed in most machine recognition tools. One of the most obvious differences is the many stages of pre-processing used by most machine recognition methods. These impose artificial constraints and structures on script to force compatibility with the machine algorithm. Similarly the value of slope and slant correction is questioned, as these processes are not carried out by the human reader. A real recognition tool will encounter examples not covered by its prototypes (or training sets). Only if the recognition method emulates human reading is there the possibility of it

304 FUZZY LOGIC APPROACH TO HANDWRITING RECOGNITION

reaching the same conclusion as a human reader in difficult cases. Only one method has been encountered that makes use of some of the characteristics of a human reader. This is the Senior (unpublished) method which uses an off-line neural net approach that maps the image of cursive letters to character codes.

10.4 A FUZZY PROCESSOR FOR HANDWRITING RECOGNITION

A fuzzy processor that attempts to emulate some aspects of the known human reading process will now be described in more detail. First the preprocessing of the points is described, then the feature measures are explained. This is followed by an explanation of the structure of the fuzzy rules. In the following section, the training method that is essential to finding the key measures of features is explained.

10.4.1 Data Extraction and Preprocessing

It is argued that the preprocessing seen with other recognition methods has no parallel in human reading and is best avoided. The preprocessing used in this case is intended only for

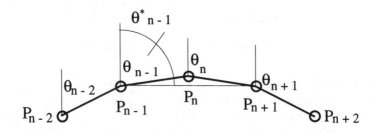

Consider a stroke of points $p_0 \ldots p_z$.

For points $p_1 \ldots p_{z-1}$, find all $d\theta$'s below angular limit ϕ,
where $\quad d\theta_n = |\theta_n - \theta_{n-1}|$
and $\quad d\theta^*_{n-1} = |\theta^*_{n-1} - \theta_{n-2}|$
and $\quad d\theta^*_{n+1} = |\theta_{n+1} - \theta^*_{n-1}|$
$\qquad$ (* indicates new value with θ_n removed)

For smallest $d\theta$:
If $(1 > n > z - 1)$ & $(d\theta_n < \phi)$ & $(d\theta^*_{n-1} < \phi)$ & $(d\theta^*_{n+1} < \phi)$
$\qquad$ remove p_n and modify $d\theta_{n-1}$ and $d\theta_{n+1}$
or if $(n = 1)$ & $(d\theta_n < \phi)$ & $(d\theta^*_{n+1} < \phi)$
$\qquad$ remove p_n and modify $d\theta_{n+1}$
or if $(n = z - 1)$ & $(d\theta_n < \phi)$ & $(d\theta^*_{n-1} < \phi)$
$\qquad$ remove p_n and modify $d\theta_{n-1}$
Find next smallest $d\theta$ and repeat.

Figure 10.7 Point reduction by polygon fitting

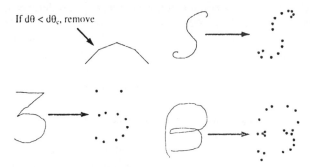

Figure 10.8 Polygon fitting

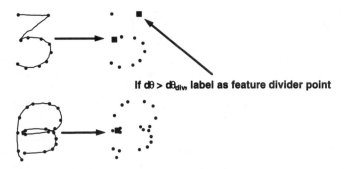

Figure 10.9 Labelling points

data reduction. The preprocessing steps used are polygon fitting (Figure 10.8) and feature divider point identification (Figure 10.9). In polygon fitting, the change in direction between adjacent points is calculated and points are removed when below a threshold. Points on either side are recalculated and this is performed iteratively until removal of further points would push its neighbours above the threshold (Figure 10.7). Note that the number and position of points in the simplified image is variable.

Feature divider points are usually sharp changes in direction. In more rounded handwriting styles these may be represented only as sharper curves. Points in regions of high curvature are labelled as feature segmentation points (some may also be character divider points). It is intrinsic to this recognition method that occasional missed or misplaced segmentation points will be acceptable, as training will have included typical examples of the shortcomings of assigning segmentation points.

10.4.2 The Feature Measures

Feature measures are intended to describe the characteristics of features in a way intuitively understandable to a reader. This will help with feedback to the user when recognition fails, but more importantly it maintains the emulation by the rulebase of

306 FUZZY LOGIC APPROACH TO HANDWRITING RECOGNITION

recognition of similar characteristics as human readers. The difference between the set of feature measures taken from the processed data for letter identification, and the fuzzy sets of feature measures in the rulebase should be explained. On raw data all possible measures are taken on every feature, regardless of whether it is appropriate or not, but as some measures are appropriate to some letters and not to others (e.g. slope is good for 'I' but not for 'o'), these useless measures (for reasons discussed later) are removed from the rulebase.

After a character has been broken down into its component strokes, the characteristics of these features must be measured. There are three broad classes of features: straight, open curves, or closed curves and circles, and different measures are appropriate for each category. A large set of measures can be assigned to any feature. Some rule antecedents will only require one or two measures, others may require many, or a different one or two from the selection. The correct measures that are able to differentiate one character from another are initially unknown. A set of measures were devised and implemented (Figure 10.10). These included the average slope of straight lines, orientation of curves, average $d\theta$, average absolute $d\theta$, length of lines compared to distance between end points and slope of endpoints. For closed curves these included measures of crossings in lines; these are distance to first cross and last cross pair, next pair and a measure of perceptual closures where one end of the line must be extended to produce a crossing. These measures can all be ratioed to make them dimensionless and constrained within the range zero to one. This is a rather simple and straightforward set intended to start off training.

Rules are composed from several of these antecedent measures and linked to a consequent that is usually a single letter. There is a wide range of antecedents, some unique to identifying a single letter, or distinguishing one particular pair of letters. During the initial

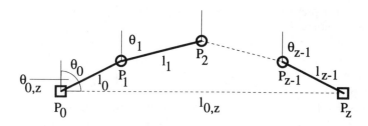

For each stroke of points $p_0 \ldots p_z$, where p_0 & p_z are feature divider points,

$$\text{Average Length} = l_{0,z} \Big/ \sum_{n=0}^{n=z-1} l_n \quad \text{Average Absolute } d\theta = \sum_{n=1}^{n=z-1} |\delta\theta_n|.l_n \Big/ \sum_{n=1}^{n=z-1} l_n$$

$$\text{Average Slope} = \sum_{n=0}^{n=z-1} \theta_n.l_n \Big/ \sum_{n=0}^{n=z-1} l_n \quad \text{Average } d\theta = \sum_{n=1}^{n=z-1} \delta\theta_n.l_n \Big/ \sum_{n=1}^{n=z-1} l_n$$

$$\text{Orientation} = \sum_{n=0}^{n<z/2} \theta_n.l_n + \sum_{n>z/2}^{n=z-1} (\pi + \theta_n)l_n \Big/ \sum_{n=0}^{n=z-1} l_n \quad \text{Slope} = \theta_{0,z}$$

Figure 10.10 Feature measures

phase of training the rulebase, all antecedent measures are connected to all the rule consequents, but some measures that are inappropriate to a letter actually interfere with recognition, due to the random distribution and finite sample size.

10.4.3 The Fuzzy Recognition Process

The recognition process starts on a pen up and time out (which allows for penlifts to complete some letters). The data is reduced to the representative points, some of which are labelled as feature dividers. The measures for the first few features are taken and passed to the rulebase; enough features are passed for the largest syllable or sight word in the rulebase. The highest scorer is accepted and the end of the last feature is marked as the letter divider. The actual process of fuzzy inference is simple: by comparing the current feature measures (which are fuzzy singletons) with the rulebase (which is composed of fuzzy sets), this becomes a simple look-up (equivalent to a min operation). This is followed by finding the min of all measures for each letter. Then, to find a ranking order of letters, the max of all these mins is taken (Figure 10.11).

It is necessary to introduce some redundancy into the recognition process as the basic process will, in practice, always be less than perfect. The first way is to provide some extra techniques when the normal recognition process fails (low scores on recognition), the most likely cause of failure being poor segmentation. Sight words and syllables provide this first technique (working alongside the letter recognition). More redundancy is intrinsic to the

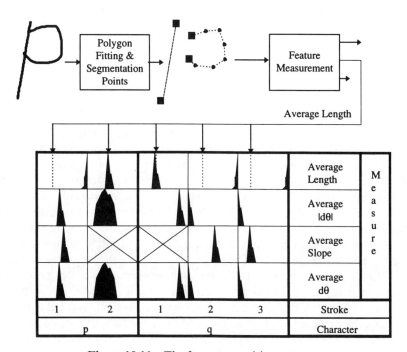

Figure 10.11 The fuzzy recognition process

training process. This takes the measurement set from whole letters (ignoring feature segmentation points) as well as the component features. Training also takes place on many examples of each letter, so typical examples of missed segmentation points (caused by the limitations of the labelling) are naturally included in the rulebase.

10.5 TRAINING

Training is derived from a database of handwriting samples. The current training has been carried out with a small sample, but a large sample will become available through the UNIPEN collaborative data collection/recognition evaluation project. The training process can be described as an engine rather than an interactive process. Using the basic measures described, the values for every character in the database are taken and the frequency of occurrence of each value for each target letter is recorded. The frequency of occurrence is used as the basis for the fuzzy sets (e.g. for letter 'A' and a measure 1, the frequency of occurrence (n) of each value in the range 0 to x $N_{A1} = (n_{A1(0)} \ldots n_{A1(x)})$). The peak number of occurrences of any value for a measure is scaled to 1, and all the other values in this measure are scaled relative to it (to give $A1 = (\mu_{A1(0)} \ldots \mu_{A1(x)})$). This is followed by assessment of the quality of the rulebase with the techniques described in the next section. The metrics described can identify shortcomings in the training, and where modification is necessary. After modification, the training process is re-run on the raw data until the target level of recognition is achieved. Concerning the actual process of data collection, initial samples are of discrete letters and each sample contains the entire character set, but in random order. The subject is prompted to draw the example letter, then explicitly discards or adds that example to the database. Later training will prompt for specific pairs of letters (or triplets), but only suitable letters need be prompted for (e.g. Aa, Ab, aa, ab, Ba, ba, bb).

10.5.1 Fuzziness and Statistics

A careful examination will show that the training and recognition process described may be understood both statistically and as fuzzy inference, but it will be shown that fuzzy inference, better describes the recognition process. Clearly a statistical method of building fuzzy sets from training examples is being used. The frequency of occurrence of each value of each measure is being logged, but when these sets are normalized they represent a degree of acceptability of the conditions necessary for that measure of that character. As training is carried out all examples fulfil (excluding erroneous input) the conditions necessary for the membership of the antecedent set for that measure of that letter, but mutually exclusive consequents do not necessarily demand 100% fulfilment of antecedents from samples for recognition. (The sample, if it is truly the letter it is recognized as, completely fulfils the definition of that letter, but when compared to the training sets (which must be accepted as less than perfect) it need not.) So how can a sample have incomplete membership of its class when its class is mutually exclusive? Let us consider the case of a badly drawn character that might an 'e' or an 'l'. Although the consequent must be one or the other (and presumably the writer intended one or the other), it must be accepted that the example somewhat fulfils the conditions of both letters. So it is best described by its degree of membership of each consequent. This in particular is where a statistical approach is unsuitable. An *a priori* likelihood of each letter may be found, but, in an *a posteriori* case, the two target letters are mutually exclusive and the example must be committed to one or the other. Training examples represent the *a posteriori* statistics of the frequency of each value of each measure for each letter. However, when evaluating unknown examples they are best considered as belonging to that consequent set to a degree dependent upon the matching with the training examples. The difference here is between stating a likelihood of one or the other letter and actually accepting that the example is, to a degree, both letters.

10.6 RULEBASE QUALITY

Measuring the quality of the rulebase and the progress of training can be characterized by three factors: discriminability of consequents, usefulness of measures and completeness of training. The following section forms an original scheme for developing rulebases (such as this one) that have mutually exclusive consequents. As has been explained earlier, this fuzzy rulebase for handwriting recognition belongs to that class of fuzzy applications that have crisp and mutually exclusive consequents. The methods described here are only relevant to fuzzy rulebases of this type, but provide a useful tool for developing a fuzzy rulebase from training.

One of the classic problems of rulebases derived from training (and many other training situations) is in detecting how well the training is progressing. (e.g. How close is training to completion? Are the measurements of the data appropriate and sufficient? How much redundancy does the rulebase contain?) The three factors listed here provide a method within which progress may be measured.

10.6.1 Discriminability

For every possible pair of letters, and all the measures they have in common, the overlap in each antecedent set may be found. The smallest (min) of these values found is the degree of equivalence $\mu_{doe(n,m)}$ of that letter pair. This is used to build an equivalence matrix (Figure 10.12). The overall degree of equivalence $\mu_{doe(av)}$ is found based on letter frequency, and the discriminance μ_{dod} is the complement of equivalence. The root of this rulebase development technique is that letters are the mutually exclusive consequents of the handwritten text. (This is what is meant by mutually exclusive consequents, although the input may be found to be a member of more than one consequent; the consequents themselves have no overlap.) Ideally, for unambiguous handwritten input, all possible pairs of antecedent sets (for any two consequents) will have a zero degree of intersection. Of course a real rulebase will fall short of this ideal, and discriminability is a measure of how close to the ideal the actual rulebase is. Using the method described below, the degree of equivalence of every letter pair may be found, as well as an overall value weighted for normal letter frequency. Each cell in the matrix is the equivalence of two characters and during the training process attention must be paid to values in the matrix that remain persistently high.

> ### Measuring Discriminance
>
> For letters A and B and for a measure 1 where antecedent sets $A1 = (\mu_{A1(0)} \ldots \mu_{A1(x)})$ and $B1 = (\mu_{B1(0)} \ldots \mu_{B1(x)})$, the degree of equivalence is
>
> $$\mu_{doe(A1,B1)} = (\sum \min(\mu_{A1(n)}, \mu_{B1(n)})^2 / \sum \mu_{A1(n)} \cdot \sum \mu_{B1(n)}$$
>
> The degree of equivalence $\mu_{doe(A, B)}$ of this letter pair is then the min of all the pairs of measures they have in common:
>
> $$\mu_{doe(A, B)} = \min(\mu_{doe(A1,B1)} \ldots \mu_{doe(An,Bn)})$$
>
> The overall degree of equivalence is the average of the μ_{doe} for every possible letter pair, weighted by letter frequency f_A, and x the number of letters in the character set:
>
> $$\mu_{doe(av)} = \sum (\mu_{doe(A, B)} \cdot f_A \cdot f_B + \ldots \mu_{doe(Y, Z)} \cdot f_Y \cdot f_Z) / ((x^2 - x)/2)$$
>
> For any pair of letters (or on average) discriminance $\mu_{dod} = 1 - \mu_{doe}$.

10.6.2 Usefulness of Measures and Rulebase Reduction

Usefulness of measures is described here because of its importance in the cycles of refining the rulebase. The usefulness of a character measure for a particular character is tested here with the chi square statistic, as a character measure may contain one or more clusters or be random. The chi square statistic is used to measure its closeness to a random distribution, and may be used to minimize the rulebase by eliminating unnecessary antecedent sets. Where values for chi square close to zero are found, that measure should be removed.

> ### Measuring Chi Square
>
> Chi square χ^2 is a test of how well the data fits a hypothesis of the expected distribution. If it is hypothesized that a measure is random, then values for chi square close to zero indicate that this is true and that the measure should be removed from the rulebase.
>
> If the number of samples of each value for a letter A and a measure 1 is $N_{A1} = (n_{A1(0)} \ldots n_{A1(x)})$ and it is hypothesized that an average number of samples falls in each band, then
>
> $$\chi^2 = \sum \frac{(n_{A1(n)} - \overline{n_{A1(n)}})^2}{n_{A1(n)}}$$

10.6.3 Completeness

It is possible that with the measures used and for the training carried out that the overall discriminance is sufficiently high. However, all drawing methods and complete variation

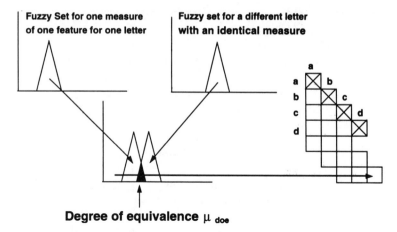

Figure 10.12 Equivalence matrix

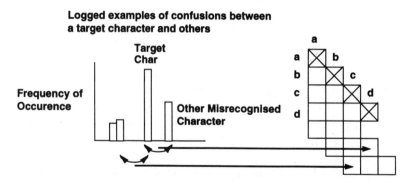

Figure 10.13 Misrecognition matrix

in antecedents may not have been trained on. Only by logging the results from evaluation can a misrecognition matrix (Figure 10.13) be generated that indicates undertraining and where further training must be added for completeness.

Although the completeness matrix appears superficially similar to the discriminance matrix, it is derived quite differently. Discriminance is derived from training samples (where the sample is identified for the recognizer by the writer), whereas completeness is derived from test samples (which are identified by the rulebase and then verified by the writer). (In practice the computer prompts for a particular letter; in training it accepts the example as correct blindly and in testing attempts to identify it as if it was unknown input and keeps a record of the differences between expected and identified input.) Misrecognition will of course occur more frequently than the discriminance of the rulebase would suggest, the amount of difference being related to the amount (or lack of) training.

Measuring Completeness

A misrecognition matrix may be generated from actual observed behaviour by logging each misrecognition as it occurs. For letters A and B, where n_{AA} is the number of samples correctly identified as A and n_{AB} is the number of samples incorrectly identified as B, the degree of misrecognition is $\mu_{dom(A,B)}$. The average must be taken of B misidentified as A and of A as B.

$$\mu_{dom(A,B)} = ((n_{AB}/n_{AA}) + (n_{BA}/n_{BB}))/2$$

Degree of misrecognition then may be defined as $\mu_{dom(av)}$ where f_A is the frequency of letter A and x is the number of characters in the set:

$$\mu_{dom(av)} = \sum(\mu_{dom(A,B)} \cdot f_A \cdot f_B + \ldots \mu_{dom(Y,Z)} \cdot f_Y, f_Z)/((x^2 - x)/2)$$

Just as the discriminance for every possible letter pair is the complement of equivalence, so completeness is the complement of degree of misrecognition:

$$\mu_{doc} = 1 - \mu_{dom}$$

10.6.4 Overall Quality and Self-tuning

The maximum possible value for completeness is equal to the maximum for discriminability. This indicates that the rulebase is fully trained and that no further improvement is possible with the current measures. So the quality of training μ_{doq} is equal to the ratio of completeness μ_{doc} to discriminability μ_{dod}. This measure is useful in allowing self tuning.

Once evaluation indicates that the rulebase has approached the ideal of completeness (i.e. the level incompleteness approaches that for the overall discriminability that is acceptable) the rulebase may be left to self-tune to the writer. Self-tuning can only modify antecedent sets, not add new ones or remove existing ones. The antecedents are slowly modified as the samples from a particular user replace the general training set, so focusing the rulebase onto that specific writer.

10.7 RESULTS

At this stage only preliminary results can be quoted. These are from a small number of training samples and relate to the basic recognition technique used with discrete letters only. The UNIPEN project is still in the data-collection phase, where participants are invited to make their contributions, in the specified format, to a write only database. At the end of this phase a labelled proportion will be released as training data, whereas the remainder is released as unlabelled test data.

Using the initial data collected, and only employing the basic recognition technique described, recognition of discrete letters was about 80% of input. The initial tool developed allowed interactive training and testing; although this is good to ensure the quality of training input, it causes the user to adapt to the tools. So the result quoted can only be thought of as a guide to show the basic validity of the technique used. Similarly, results quoted for other recognizers are suspect unless they use unknown independent test data.

10.8 CONCLUSIONS

Preliminary results show strong evidence that a fuzzy rule based handwriting recognizer is capable of emulating a human reader. As training data become available from UNIPEN this technique will be completely evaluated on a wide ranging sample of styles and detailed results will be presented. This will be compared with other commercial tools, which are successful using letter prototypes and matching algorithms. They tend to achieve this somewhat inelegantly, either by trading of the results from several techniques or making use of whole-word recognizers with lexicons, causing them to have intrinsic limitations.

A very important mechanism, and as yet not seen in any handwriting recognition tools, is feeding back information to the user about why a particular sample failed to be recognized. The psychological aspects of this have been explored by Frankish, Morgan and Noyes (1994). As the recognition method described here uses a fuzzy comparison of feature measures that are similar to human perception of the characteristics of features, this is a feasible and possibly very useful further development that will be included in later work.

ACKNOWLEDGMENTS

This work was supported by a Science and Engineering Research Council (UK) grant; thanks also to LSI Logic Europe plc for advice and support.

REFERENCES

1. Ward, J. R., and Kuklinski, T. (1988) A model for variability effects in handprinting with implications for the design of handwriting character recognition systems. *IEEE Transactions on Systems, Man and Cybernetics*, **18**, 438–451.
2. Blesser, B., Shillman, R., Kuklinski, T., Cox, C., Eden, M., and Ventura, J. (1973) A theoretical approach for character recognition based on phenomenological attributes. *International Journal of Man-Machine Studies*, **6**, 701–714.
3. Well, A. D., and Pollatsek, A. (1981) *Word processing in reading: a commentary on the papers. Visible Language*, **XV**, 287–309.
4. Otto, W., and Stallard, C. (1976) One hundred essential sight words. *Visible Language*, **X**, 247–252.
5. Sakiey, E., Fry, E., Goss, A., and Loigman, B. (1980) A syllable frequency count. *Visible Language*, **XIV**, 137–150.
6. Nair, A., and Leedham, C. G. (1991) Preprocessing of line codes for on-line recognition purposes. *Electronics Letters*, **27**, 1–2.

7 Tappert, C. C., Suen, C. Y., and Wakahara, T. (1990) The State of the Art in On-line Handwriting Recognition. *IEEE Transactions on Pattern Analysis and Machine Intelligence*, **12**, 787–808.
8 Senior, A. W., Off-line Handwriting Recognition: A Review of Experiments, Cambridge University Engineering Department, Cambridge, UK (unpublished).
9 First UNIPEN Benchmark of On-line Handwriting Recognizers Organized by NIST (ftp. cis. upenn. edu).
10 Frankish, C., Morgan, P., and Noyes, J. (1994) Pen Computing: Some Human Factors Issues. *Handwriting and pen-based input. IEE Colloquium.*

Index

automotive engine model 53–54, 59
axiom of extensionality 7

boundary layer (BL) 64, 66–67, 70–71, 73, 77, 91, 103

certainty potential 13
certainty qualification 14–15, 18
chaotic system 294–296
characters handwritten 301
completeness 311–312
control sliding mode (SMC) 64–69, 71, 99–103
control sliding model boundary layer 103–104
correlation coefficient 84, 87, 89

data base 308
defuzzification 7, 46–47, 64, 71, 79, 117, 119, 149, 270, 291
defuzzification Center of Area 149
defuzzification Center of Gravity 149
defuzzification Center of Largest Area 150
defuzzification Height 150
defuzzification Mean of Maxima 150
defuzzifier 244, 286–287, 288–291, 292, 294
denormalization 69–70, 79, 84
design automation system 250–258
design automation system netlist generation 252–255
design automation system placement 255
design automation system routing 255
discriminability 309–312
discriminance 309–312
discrimination power 30
DISO system 151, 155

equivalence 16, 309–310
Euclidean distance 44, 72
extension principle 5

flip flop J-K 197–199
flip-flop MIN-MAX type 210–211, 221–222
fuzzification 64, 78, 90, 144–145
fuzzy attributes 72, 88
fuzzy computer 197
fuzzy control rule 131
fuzzy development system 247–248
fuzzy flip-flop 217–218, 221–222
fuzzy flip-flop algebraic circuit 219–221
fuzzy flip-flop J-K 197, 201–207
fuzzy flip-flop J-K implementation 202, 213
fuzzy function 16, 100
fuzzy hypercube 55
fuzzy implication 178
fuzzy inference 8, 17, 42, 47, 146, 186–187, 197, 268, 308
fuzzy input 90–91, 112, 178, 183–184
fuzzy interval 5–6, 8
fuzzy logic 21, 39, 238, 299
fuzzy logic complement 277, 279
fuzzy logic control 6, 15, 21, 23, 39–43, 48–50, 56, 58–60, 63–65, 70, 73, 84–87, 91, 104, 127, 144, 177
fuzzy logic controller DISO 151, 155–157
fuzzy logic controller MIMO 80, 82–83, 151, 159–163
fuzzy logic controller MISO 80, 151, 157–159
fuzzy logic controller SISO 80–83, 151, 153–155
fuzzy logic controller 43, 45, 50, 52, 54–56, 63–64, 67, 71–74, 77–79, 80, 84–85, 87, 106–113, 117–118, 127, 129, 143–144, 151–153, 163, 171–172
fuzzy logic controller clock frequency 152, 172
fuzzy logic controller CMOS implementation 237, 244–246, 259–260
fuzzy logic controller crisp type 118–127
fuzzy logic controller design rules 68

fuzzy logic controller digital implementation 143
fuzzy logic controller Direct Data Stream
    Architecture 191
fuzzy logic controller hardware accelerator
    191–192
fuzzy logic controller hardware cost 171
fuzzy logic controller hardware implementation
    143, 163–172
fuzzy logic controller hardware implementation
    DISO 166–168, 172
fuzzy logic controller hardware implementation
    MIMO 168–171, 190–191
fuzzy logic controller hardware implementation
    MISO 190–191
fuzzy logic controller hardware implementation
    SISO 164–166, 172–173, 187–190, 192
fuzzy logic controller memory 172
fuzzy logic controller parameter adaptive
    134–136, 138
fuzzy logic controller performance 143, 171,
    192–194
fuzzy logic controller PID type 130–134
fuzzy logic controller processing rate 192, 194
fuzzy logic controller scaling 78, 80
fuzzy logic controller sliding mode 75–77, 91,
    104–105, 107–108
fuzzy logic controller VLSI implementation
    192, 194, 237
fuzzy logic development system 247–250, 261
fuzzy logic processor 304
fuzzy logic system 270
fuzzy Lyapunov function 100
fuzzy neuron 267, 276–280, 282, 293, 296
fuzzy number 278
fuzzy operator addition 11
fuzzy operator aggregation 117
fuzzy operator bounded difference 240
fuzzy operator inclusion 239–241
fuzzy operator intersection 239, 277
fuzzy operator negation 199, 201, 239
fuzzy operator substraction 11
fuzzy operator union 238–239
fuzzy operators CMOS representation
    243–244
fuzzy operators current mirror representation
    244–246
fuzzy operators MIN, MAX circuit 212, 242
fuzzy operators MIN, MAX representation 247
fuzzy operators symbolic representation 242,
    247
fuzzy output 55, 105, 178, 183–184

fuzzy processor 47, 304
fuzzy register 223–224
fuzzy register circuit 226–227, 234
fuzzy relation 9, 186, 239
fuzzy robustness 60
fuzzy sensitivity 52, 60
fuzzy set 3–5, 7–8, 11, 17–18, 22, 28–29, 32, 40,
    56, 63–64, 76, 91–95, 97, 117, 129, 185,
    238–241
fuzzy set convex 97
fuzzy set differentiation 91–95
fuzzy set twofold 29
fuzzy signal 63–64, 99, 289
fuzzy system 40, 63, 267–271, 285–286, 288,
    289–295
fuzzy vector 145
fuzzy weight 269, 277

input scaling 83–86
interpolation linear 9

knowledge base 23–24
knowledge base possibilistic 26

linduistic model 184
linguistic system 267, 270, 324

measure feature 305–306
MIMO system 47, 80, 82, 177, 180–183, 194
minimal specificity principle 18–19, 25
MISO system 80, 177, 178–180, 182–183, 194
modifier function 17
Modus Ponens 3, 20, 146
Modus Ponens generalized 20

necessity 4, 11
necessity degree 23
necessity measure 12, 21–22, 28
necessity theory 11
negligibility 11
neural network 267–268, 273–274, 283–285,
    287–291, 296
neuro-fuzzy system 267–269, 271, 285, 287–290,
    292–294, 296
neuron 267, 269, 272–275, 277, 292, 295
nonlinear system 268
norm triangular 5
normalization 69–70, 72, 78

PD 104, 126–127, 130–135, 138, 173
PI 104

PID 39, 48, 118, 126–127, 130–135, 137–138, 173
possibilistic inference 26
possibilistic knowledge 26
possibilistic logic 21, 23
possibility 4
possibility degree 12, 20–21, 282
possibility distribution 11–12, 14–15, 18–19, 20–23, 40
possibility guaranteed 13
possibility measure 12, 18, 22
possibility qualification 14–15, 19
possibility rules 18, 20
possibility theory 32
preferential entailment 22–23
principle of maximum entropy 18
principle of parsimony 27
principle of resolution 238
principle on minimum specificity 18–19, 20, 25
probability distribution 11, 92
probability theory 18, 20
process nonlinear 42
proximity 4

reasoning abductive 26–27
reasoning approximate 3, 7, 18, 20, 32
reasoning atomated 21
reasoning default 24
reasoning fuzzy 146–148
reasoning hypothetical 24
reasoning interpolative 6–9
reasoning nonmonotonic 23
reasoning qualitative 9–11
recognition character 300, 302, 307
recognition fuzzy 300, 307
recognition handwriting 280, 299, 302–304, 309, 313
recognition machine 300, 303
recognition pattern 277, 280, 296, 308
recurrent system 275, 289
relation duality 13
relation fuzzy 4–7, 16–17, 21, 29, 186
relation fuzzy binary 4
relation negligibility 11

relation possibility 24
relation proximity 9–10
relation similarity 6–7
robot arm 74, 87
robustness 42, 49, 51–53, 60, 288
rule base 33, 39, 46, 56, 89, 302, 309–310, 313
rule base building 182
rule base fuzzy 42, 46, 51, 89, 300
rule fuzzy 3, 15, 18, 39, 50, 300, 304
rule of inference 10, 46, 178
rule of inference compositional 46
rule resolution 22
rules certainty 18, 20
rules gradual 3, 7, 9, 16, 18
rules possibility 18, 20
rules with uncertain conclusion 3

s-norm 200–201
similarity 4
SISO system 64, 81, 83, 177, 178, 180, 194
sliding mode control 64–68, 77, 99
stability 42, 44, 49–52, 53, 60, 107
standard deviation 92
statistic chi-square 310
statistics aposteriori 309
statistics apriori 309
system closed-loop 42, 50
system dynamic nonlinear 40, 53–54
system nonlinear 42, 64–66
system sliding mode 63–66
system variable structure 63

t-norm 200–201
training 308–309, 311

uncertainty 11, 41
uncertainty grading 31
uncertainty measure 21
uncertainty modeling 30

VHDL 222, 224–225, 227, 235
VLSI design 177, 192, 198, 213, 222, 224, 235, 237, 243, 260